Socioecosystems

SCIENCES

Ecosystems and Environment
Field Directors – Françoise Gaill and Dominique Joly

Social Ecosystems, Subject Head – Dominique Joly

Socioecosystems

Indiscipline as a Requirement of the Field

Coordinated by
Patrick Giraudoux

iSTE

WILEY

First published 2022 in Great Britain and the United States by ISTE Ltd and John Wiley & Sons, Inc.

ISTE Ltd
27-37 St George's Road
London SW19 4EU
UK

www.iste.co.uk

John Wiley & Sons, Inc.
111 River Street
Hoboken, NJ 07030
USA

www.wiley.com

Library of Congress Control Number: 2022930810

British Library Cataloguing-in-Publication Data
A CIP record for this book is available from the British Library
ISBN 978-1-78945-052-1

ERC code:
LS8 Ecology, Evolution and Environmental Biology
LS8_1 Ecosystem and community ecology, macroecology
SH2 Institutions, Values, Environment and Space
SH2_6 Sustainability sciences, environment and resources

Contents

Foreword

The Knowledge Community at the Service of Communities

Arnaud Macé
Logiques de l'agir, Université Bourgogne Franche-Comté, Besançon, France

The book you have in your hands invites us to renew our ways of thinking and acting in the world in which we live; perhaps it will even renew the way you see the unity of this world and the way we can build our future within it. You will begin an adventure, that of scientists who have learned to work together despite the diversity of their disciplines and the heterogeneity of their subjects, and discover that by pooling their concerns in this way, they are teaching us to organize collectively in order to take care of this world and ourselves. If I may be permitted a detour, I would say that the approach reflected in this book revives a very old idea, that knowledge is at the service of the public in a very particular way, in the sense that it is the guarantor of the unity of the world in which we find ourselves, so that the diversity of the needs of a community living in a given location can find its place among things. Yes, the ecologists, biologists, epidemiologists, virologists, physicians, socio-anthropologists and a few others whose common endeavors are outlined in this book are following in the footsteps of the ancient Greek sages. We must therefore try to explain why, and to do so, we must start with Homer, as always.

In Canto XVII of the *Odyssey*, a Greek language epic that is often dated to around the 8th century BCE, the hero Odysseus, thought to be dead after an absence of 20 years, finally returns home to find his house occupied by a crowd of men seeking to marry his wife Penelope. He enters his home in the garb of a beggar and thus arouses the indignation of one of the suitors, who thinks that a "wretch" should not be brought into one's home, as opposed to other strangers whom he calls *dêmioergoi* (this is the pronunciation in Ionian; the Greeks of Attica would say *dêmiourgoi*, a term in which we recognize the origin of "demiurge" in English). There follows a list of examples of these strangers whom it is in our interest to welcome because they all provide a valuable service: there is the soothsayer – the one who will tell us whether or not the actions we undertake have the favor of the gods, which, for the ancient Greeks, indicated whether or not they would be successful; there is "the physician for evils", from which he delivers us; there is the "carpenter for wood", from which houses are made; and finally, there is the "singer inspired by the gods, whose song is enchanting", thanks to whom one forgets the ills endured and remembers all that it is important to remember[1]. A *demiourgos* is literally someone who acts, works and strives (this is the meaning of the suffix *ourgos*) for the public: something that is *dêmios* (in classical Greek, we would say *dêmosios*) is what concerns everyone, the whole community, and not the particular interest of a given individual. There are two aspects to be distinguished here in what I would call "epistemic publicity" (Macé 2013, pp. 16–31): on the one hand, it is know-how, a technique, a skill that allows the physician, the soothsayer, the poet or the carpenter to offer his or her services to an audience (as opposed to the amateur, the "Sunday" carpenter, who would do better to keep his or her talents private); on the other hand, the public to which knowledge is offered is very broad, broader than all other communities, for the physician treats all the sick, rich or poor, children or adults, young or old, citizens or strangers, just as the poet sings for all. The "demiurge" is thus someone with knowledge, or know-how, and also someone who offers a universal service.

This last aspect explains why the scholar is often a stranger, an itinerant, belonging to no one and to everyone, even when he or she settles somewhere. Nevertheless, something new happens when the demiurges settle down together in the same place and put themselves at the service of the same community, the same city or, as the Greeks said and perhaps we would say today, of the same territory and its community of inhabitants. The latter then becomes the recipients of the attention of a multiplicity of men and women who are custodians of know-how. The multiplicity of their objects – from the roofs of houses to shared memories, from the health of bodies to the joy of souls – begins to outline the contours of a set of beneficial things that we might call the common good. In ancient Greece, as new knowledge and disciplines developed, from geometry to economy, botany to astronomy, the contours of this common good became increasingly complex, while disagreements among scholars about how to map it grew. The internal quarrels within the city redoubled the conflict: some brandished the expertise of others

1 *Odyssey*, 17, 382–385.

against expertise aligned with their interests. It was in this context that "philosophy" was born, which was initially nothing more than a taste for all this knowledge, as its name indicates ("love of knowledge"), driven by the desire to find a way of arranging it so that a harmony would appear within the heterogeneous multiplicity of their objects and give rise to renewed unity for the city.

Let us return to the present work and its demiurges. There are many of them, all of them itinerant by the very strength of their knowledge, even when they stay in one place. This book follows them to the bedside of different groups around the world: the population of Inongo or Bukama, in the Democratic Republic of the Congo, plagued by cholera epidemics; the herders of the plains and mountains of the Jura and the farmers of Zhang County, north of the Min Mountains in southern Gansu, all of whom are confronted with alveolar echinococcosis; groups of black-and-white snub-nosed monkeys in Tibet and Yunnan, threatened by habitat destruction owing to deforestation; and fish in the Loue River mysteriously dying by the dozen. In this book, there are even scientists who look at other scientists, watch them work and wonder what conditions lead them to be able to collaborate in the same field. The conclusion will tell you all about how their paths have led them to nourish their "indiscipline". Not that they deny their disciplines, quite the contrary: it is understood that no scholar will be useful if they do not know the secrets of their art. The carpenter does not become a soothsayer. The indiscipline is elsewhere – it is that which allows these scholars to accept only one rule, that of following all the threads of the problem that confronts the community to which they have come. Chapter 7 on cholera in the Congo is perhaps the most striking example of this. The young doctor becomes an investigator: he understands that there is no chance of stopping the epidemic without going back to its source in the lakeside camps the sick come from, and that this must be done against the advice of the authorities and the hierarchy. The situation reveals the people: a logistician of the camp also enters into indiscipline by organizing a salutary expedition.

Chapter 2 on the pollution of the Loue River clearly shows the extent to which, without such indiscipline, scientists can fail to tie up the threads of a situation. Like the proliferation of voles in Chapter 1, the episodes of fish mortality in the Loue, observed from 2009 onward, elude single-factor solutions: the problem unfolds in multiple dimensions and each requires new scientists capable of following the different paths – of cyanobacteria, toxic molecules and agricultural nutrients, respectively. The sociological investigation patiently identifies all of the obstacles that prevent the experts from each contributing their piece to the puzzle. Rivalries between teams emerge, fueled by the discordance of administrative bodies and clashing interests. The patchwork of the "ordinary" knowledge of those who know the places, the "scientific" knowledge of the experts who come and go between the river and the laboratory, and the "legal" and "administrative" knowledge of the institutions plays on its disharmony; at times, certain actors manage to articulate them – for example, when the figure of the investigator appears under the guise of ecologist/fisherman/volunteer. The region is struggling with itself around its river, and the people and the fish would need the indiscipline of the scientists, the citizens and even the administrators.

We understand that this indiscipline is in fact nourished by a higher respect, that of the multidimensionality of the situation. We could introduce this with a timeless fable, which we could call the fable of the voles. It all begins with voles: in Chapter 1, we are told how a landscape, the mid-altitude meadows of the Jura mountains, is changing under the impact of intensified production of Comté cheese and land consolidation, and that one day, farmers find themselves confronted with "traveling waves" of voles. It is the resilience of the problem, continuing for more than 40 years, that brings a multiplicity of scientists to the meadows: It is already necessary to multiply the approaches to simply make the problem appear, measure its extent; carry out studies to understand why the first approaches using rodenticides fail and cause cascading harmful effects; and open up other avenues capable of explaining the peaks and declines in vole density – land consolidation, felling of hedges, fluctuation in the predator population, the role of various pathogens, stress, etc. It is clear that this is not a matter of interdisciplinarity, but rather of pooling different types of knowledge: the problem has many facets and very specific knowledge must be brought together to address it. Moreover, the conclusion, looking back at the careers of the researchers involved in this book, insists on the fact that their "indiscipline" is because they have sometimes crossed disciplinary boundaries to acquire an additional discipline, even though they had already mastered one. It is a question of hybridization, of adding up the disciplines: of crossing boundaries, not erasing them. Let us look at the story of the voles again. It appears, after testing all the hypotheses, that there is no other effective action than that which is based on the recognition of the multifactorial dimension of the phenomenon; this also implies a renunciation of curative treatments based on a single variable (chemical action) and the adoption of a preventive perspective which translates the multiple approach into a toolbox bringing together all the relevant levers. Multiple problems require multiple actions. However, multiple actions in turn require the creation of communities of actors. No farmer can solve the problem of voles by relying on himself alone; he has no choice but to become a member of a collective that is itself capable of understanding what knowledge can only communicate by also finding a way to collectively outline a problem. This is the moral of the fable of the voles: there are problems that can only be solved when the scholars, on the one hand, and the inhabitants, on the other hand, agree to be brought together by the multidimensionality of a situation. A community of knowledge must be built – all heterogeneous, all based on their own requirements, but brought together by a common task – and this community must make it clear to the inhabitants of a territory that they can only solve their problem by forming collectives capable of taking on forms of action that respond to the complexity of the situation.

One of the central arcs of the book is probably the moment when two different teams from the University of Franche-Comté realized, in the 1980s, that their objects were aligned. On the one hand, the team of ecologists who were studying the proliferation of voles on the Jura plateaus and who, in the midst of their multifactorial investigations, came across the presence of *Echinococcus multilocularis*, a parasite, in these small animals (Chapter 1); and, on the other hand, the medical team that follows the recrudescence of the disease, alveolar echinococcosis, owing to the presence of this parasite in humans and the mystery of its local appearance in clusters, when the disease is not transmitted from human to human (Chapter 4). Each of these teams had already had to bring together a multiplicity of specialties to investigate such a complex subject, but then this subject took on a scope that led them to join together in an approach that they call "eco-epidemiological" and that truly takes into account the scale of a territory – there is no limit to the number of mathematical, physico-chemical, ecological, medical, and humanities and social sciences knowledge that can be called upon to draw out all of the strands of a problem of this kind. Wherever this method is used in the chapters of the book, all over the world, we find the alliance between multifactorial approach and preventive action, particularly through the identification of "risk landscapes", for example, the one that increases the surface of exposure between rodents and foxes and thus offers the parasite a complete development cycle from one organism to another. The identification of this landscape, once done in the Jura, led to an international collaboration with British, Chinese and Japanese researchers to explore the same avenues in regions of China affected by the same phenomenon, in Xinjiang, Qinghai, Gansu, Sichuan and Ningxia. The result is a panoramic picture of the development of the disease and an exploration of the socio-economic factors that promote it, and, ultimately, a robust toolkit for human communities seeking to prevent the emergence of "risk landscapes" and the onset of cycles that lead to humans.

The exploration of cholera risk landscapes in the Bukama region also leads to the description of preventive actions and a form of toolkit, ranging from awareness-raising to the use of fishing nets, through the distribution of water storage devices. This method is then transposed to other epidemic areas and leads to the construction of a vast prevention program based on a database listing cases and their contexts (Chapter 7). The difficulty of modifying ancestral practices in the fight against cholera in the Congo echoes the investigation in Chapter 5 on the specific resistance of the populations of the Jura to prevention campaigns for alveolar echinococcosis. Socio-anthropologists undertake the collection and analysis of representations: the relationship to health, animals and food, as well as to the environment, disease and health risks. The survey shows that the perception of the risk of contamination is strongly correlated with the way farmers see their environment, "nature", depending on whether they perceive it as intact and beneficial or as corrupted, polluted and a source of new dangers; these representations are linked to know-how and inherited practices, and have an impact on the modification of practices. The perception of risk is also expressed in terms of relationships to proximity, domestic and wild space, and focuses on the intermediate space of the garden. Proximity to the disease, contracted by a close person, is a strong factor in modifying practices. The lessons of anthropology thus add to the precautions to be included in the toolkit: the collective construction of solutions presupposes that the representations through which human collectives approach the situation are taken into account.

A toolkit again gradually emerges in the investigation of Chapter 6 on the black-and-white snub-nosed monkey, in Tibet and Yunnan. The arrangement of knowledge is again daring, by bringing together the ecological study of monkey population movements, the resistance and facilitation of this movement by the environment (the measurement of its "viscosity") with the genetic study of populations over time, the comparison of which is based on the use of the remains of members of the species preserved in museums since the 19th century, with a considerable amount of work carried out to identify ancient capture sites. Once again, the picture that has been drawn up provides a toolkit for the inhabitants and local authorities who wish to prevent the black-and-white snub-nosed monkey from disappearing. Collective action is therefore based on a range of means and actions that could only be identified by bringing heterogeneous knowledge into contact and discovering the concatenation of their objects.

The inclusion of Franche-Comté in the regions where the European PASTURE study was deployed has made it possible to explore the environmental factors of human health in greater depth (Chapter 3). Following cohorts of children across Europe for 20 years, the study established the protective effect of three factors: life on a farm, contact with animals and consumption of raw milk. It leads to the idea of "lifestyle" as a combination of independent factors, the list of which remains open. There is a delicacy to these causalities that only work in correlation with our lifestyles and are only triggered by early exposure, that of mothers during gestation and of young children. The study establishes the role of diversity of exposure, including microbiological, in the development of a balanced immune system. We now know that each of us, individually, are ecosystems unto ourselves, with more microbiotic cells than human cells. The lifestyle described by the PASTURE study is therefore remarkable in that it allows us to make the hypothesis of a correlation between the community of factors gathered in the environment and the balance achieved by the microbiotic community within the individuals who develop there. We understand how the situations we are struggling with involve a plurality of disparate elements, whose variations may be independent but whose chance encounter in the environment is necessary for the balance and continuity of its effects.

In the course of these pages, an equation thus emerges that links the community of things, the community of knowledge and communities of action. The commonality should be understood here not as unification or identification, a way of putting all one's eggs in one basket, but, on the contrary, as the fragile correlation between heterogeneous things, the emergence of a "divisible community" (Macé 2019), in which each thing and each actor must do their part or else the edifice will fall apart. We must first manage to arrange together the heterogeneous objects of specific knowledge in order to understand how our environments are complex architectures where each disparate and independent element has its own role to play. Then, on the basis of this community of things, we must learn to act in a plural way by bringing together communities of actors with disparate points of view, but who will not solve anything without uniting and benefiting from the contribution of each. Let us give the Ancients the last word. As the author of the Hippocratic treatise *On the Nature of Man* (VII, 8) says, a year unfolds with all the qualities, such as hot, cold, dry and wet, but it would be enough for one to disappear for all to disappear: "none would remain the least time without all the others that are in the world" (we translate here the term *kosmos*, which means the arrangement of things). The author concludes that all of these things nourish each other, and are thus bound "by the same necessity". This fragile community, because it depends on the integrity and solidarity of each of its members, is the uncertain world that this book invites us to tame.

January 2022

References

Macé, A. (2013). *Le Savoir public : la vocation politique du savoir en Grèce ancienne.* Presses universitaires de Franche-Comté, Besançon.

Macé, A. (2019). La Communauté divisible : pour une réappropriation contemporaine de la notion ancienne de commun (koinon). In *Vie bonne, vulnérabilité, commun(s) : schèmes anciens et usages contemporains*, Alexandre, S., Gueguen, H., Renaut, O. (eds). Presses universitaires de Paris-Nanterre, Nanterre.

Preface

Patrick GIRAUDOUX[1,2]

[1] *Chrono-environnement UFC-CNRS, Université Bourgogne Franche-Comté, Besançon, France*

[2] *Yunnan University of Finance and Economics, Kunming, China*

The world is going from crisis to crisis: financial crises, economic crises, health crises and, of course, ecological crises, including those caused by climate change and the erosion of biodiversity, the sixth extinction. They are not independent. These crises are perceived on many scales where interests seem irreconcilable. The "end of the month" (the short term) is theatrically opposed to the "end of the world" (the long term), while the local is opposed to the global. These oppositions deny the necessary coexistence of processes that take place at and through each of these scales. They are largely the result of a demographic and social crisis, on a planet with limited resources, which is seeing growing inequalities at all levels where they are measured – rich and poor in each country, rich and poor countries at the global level. This sociodemographic crisis is concomitant with a breakdown of solidarity, where community and individualistic logics are supposed to determine, along with the "invisible hand of the market", "competition", the dogmas of "free trade" and eternal growth of GDP, the best of all possible worlds, in an Anthropocene dynamic described as a "great acceleration" (European Environmental Agency 2019). One or more answers to these crises are desperately sought, without, more often than not, making a distinction between what belongs to the solution and what is the problem. However, our species needs to change its paradigm by ensuring, with increasing urgency, the recovery and maintenance of dignified and secure subsistence conditions for all the elements and processes of the biogeosphere that are necessary for the planet to be habitable for humans (Magny 2019; Latour 2020).

There are regular forums urging scientists to bring their special ability to characterize the ecological and health situation to the political debate. As a result, scientific advice is being multiplied in every respect. Science has a real vocation to produce a discourse free of conflicts of interest, where, in theory at least, only the quality of the data, methodically collected, and their interpretation, rigorously discussed and collectively validated, count. As soon as the issue is environmental, that is, macroscopic while being multi-scale, it comes up against the complexity of the problems to be solved. Their study requires the integration of knowledge divided into different disciplines, many of which have developed on the basis of concepts and methods so different that they can only communicate with each other with great difficulty, or even no longer at all.

Leonardo da Vinci (1452–1519) was the archetypal Renaissance man, some of whom were artists as well as scientists and engineers, capable of combining several orders of thought in a wide range of achievements. Like them, the scientists of the Enlightenment were part of that long line of philosopher-scientists, whose roots go back to antiquity, describing the objects and processes of nature without exclusivity. The major divisions into scientific disciplines, such as those known today, were established only in the 19th century, starting with the work of Auguste Comte (1798–1857) (Legay 2006). At the same time, Louis Pasteur (1822–1895), a chemist with a degree in physics and later a doctorate in science, continued to break free from neighboring disciplinary barriers. This is how he was able to discover molecular dissymmetry, specify the microbiological conditions of alcoholic fermentation and the "diseases" of beer and wine, disprove spontaneous generation and make it possible to fight more effectively against pébrine, a silkworm disease. Moreover, it was by mobilizing a physicist-biologist, Charles Chamberland (1851–1908), and a physician, Émile Roux (1853–1933), that Pasteur was able to discover the process of transmission of anthrax and its vaccine, and then to invent a vaccination against rabies (Gascar 1986). His notes to the Academy of Sciences, without going so far as to include elements of the humanities, were presented under headings as diverse as biology, chemistry and microbiology, and were added to those presented to the Academy of Medicine. He was elected to the French Academy of Sciences (*Académie des sciences*) in 1862 for the field of mineral science and was appointed Permanent Secretary for the physical sciences in 1887. This mobilization of the skills of an eminent scientist, albeit limited to the neighboring fields of biology and chemistry, appears by today's standards to be an admirable transgression of a school model. It cannot, however, illustrate what is known today as multidisciplinarity, insofar as the focal point of the research and the methods used were each time related to a single discipline or to a few closely related disciplines, the contents of which the scientist mastered perfectly. However, his starting point was a science that was less specialized than it is today, where the questioning was determined by an applied issue to solve problems in fields as varied as the mastery of beer production, silk production and disease prevention. In the end, Pasteur was still part of a tradition that was then already thousands of years old, in which, although he did not escape the quarrels of schools and guilds (the Academy of Medicine did not recognize his merits for some time, because he was not a medical doctor), a scientist, in order to solve a problem, was not bothered by disciplinary divisions which were, however, increasingly formalizing his era.

The need for multidisciplinarity shared between specialists then appeared, first in an industry that had experienced the benefits of the Taylorization of tasks. Aeronautics provides a very good example of the outcome of complex but concurrent research, linking engineering sciences and human sciences, both fundamental and applied, ensuring in the end the transport of passengers over very long distances, in conditions of great safety, comfortable for a minority, and of bearable inconvenience for a majority. This confluence of disciplines specialized in research and innovation, led by scientists and engineers specialized in their discipline, is now practically the rule in all industrial or even craft production. The implementation of scientific and technical ways of thinking is then determined by an explicit final object whose characteristics are defined *a priori*, thus outside the disciplinary contents.

The exhortation to multidisciplinarity and multisectoriality has accompanied most calls for research on environmental issues for the past 50 years. This exhortation has been the permanent background music of the career of the authors of this book, without them finding, except by the greatest of chance and the strength of their determination, the human and financial conditions that would allow them to carry it out, at least partially, and, in any case, without experiencing it as anything other than a contingent consequence of the problems to be solved in relation to a clearly defined object of study. It is also remarkable that after such a long period of exhortation, and despite reflection and the production of a substantial bibliography spanning several decades, no work tracing what could be a general theory of operational multidisciplinarity independent of the object of study has emerged, to the point that many of its practitioners currently consider this concept to be inoperative and outdated (Frodeman 2019). Legay (2006), in a collective work entitled *L'Interdisciplinarité dans les sciences de la vie* (Interdisciplinarity in Life Sciences), noted that it was:

> ultimately constructed in the face of a question posed, most often in the course of research. Depending on the case, biologists, for example, will begin the work and will realize along the way that sociology or psychology would provide them with new, perhaps original, views, to the point of modifying their own program[1].

---

1 This example, which calls on the human sciences secondarily to resolve questions that the biologist cannot deal with, without the multidisciplinary approach having been defined

It is therefore the *a priori* invention of objects allowing scientists to cooperate on a subject that is central, the summoning of such and such a discipline to the resolution of the problem being only the consequence. Clearly, despite a plethora of literature on the subject, no design model or engineering has yet emerged for multidisciplinary research. Only the object of study "obliges".

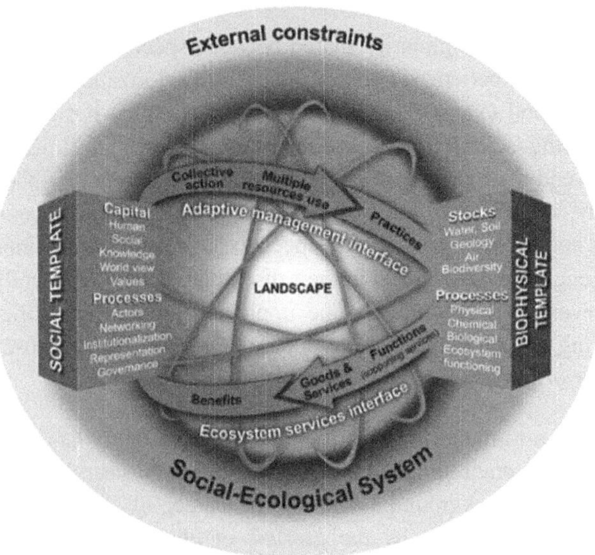

Figure P.1. *Conceptual diagram of a socioecosystem as proposed for the CNRS Workshop Zones, from Bretagnolle* et al. *(2019). For a color version of this figure, see www.iste.co.uk/giraudoux/socioecosystems.zip*

This is evident, for example, in the theorization of regional socioecosystems that has been proposed so far to solve environmental problems. All attempts lack clear operational links explicitly linking several disciplines that can be used as levers. None of them completely shed the ball and chain of the nature/culture dichotomy, as evidenced by the founding conceptual schemes and their most recent derivatives (Collins *et al.* 2011; Bretagnolle *et al.* 2019), an example of which is presented in Figure P.1 with, on one side, the box of the "social template", on the other side, well separated, that of the "biophysical template", and, in between, depending on the version, a more or less furious storm of arrows and feedback, which can hardly

upstream on the basis of a common and shared problem, is problematic insofar as one discipline may then appear to be the applied provider of the other in a restricted field.

account for their real number, the multiple spatial and temporal factors involved, and even less so the interweaving of one into the other.

These integrative conceptual systems have at least the merit of trying to structure all of the disciplines around a common concept on an identified object. Their history is interesting to follow. The Brundtland report (World Commission on Environment and Development 1987) provides one of the first archetypes. It is made up of three intersecting circles representing the economic, social and environmental spheres (to which we sometimes add, suspended above this theoretical tripod, governance) (Figure P.2(b)). Only the overlap is important here, as none of the circles seems to constrain any other: each appears separately dilatable in the non-overlapping part. Humanity (the social and economic) is in no way limited by the non-human (the environmental), the emphasis here is on a balanced governance between three poles. More than 30 years later, a multitude of warnings have been effectively ignored by governments and peoples (Union of Concerned Scientists 1992; Ripple *et al.* 2017). IPBES formalizes the dangerous and unprecedented decline in biodiversity and, following the IPCC on climate, makes it known that one more planetary limit has been exceeded (IPBES 2019). The economic and social spheres are then admitted as constrained by the biogeosphere. The vital necessity of an "ecological transition" is now generally accepted, leaving open the question of "where to land?" (Latour 2017). In the interplay of constraints, the biosphere includes the social, itself including the economic (Giddings *et al.* 2002; Morandín-Ahuerma *et al.* 2019) (Figure P.2(c)). The organization of the real world in the Capitalocene, another proposed name for the Anthropocene (Campagne 2017; Magny 2019), is the antithesis of these models. The hierarchy of circles is simply the reverse: the economic constrains the social, and both weigh irresponsibly and deleteriously on the biogeosphere (Figure P.2(a)). While attention is drawn in all these examples to the need to consider several fields of application simultaneously and in a complementary manner, and implicitly to many disciplinary fields, no science or meta-science formally allows for the organization of convergences.

Reducing the scope of application, for example to the field of health (Patz *et al.* 2004), does not, however, reduce this difficulty, which appears fractal. Figure P.3 shows how the upstream resolution of a health issue cannot be reduced to studying the pathogen-infection-disease causal chain alone (in red in the figure). The context in which pathogen and disease emergence occurs is usually much more complex, being multifactorial and involving a large number of combined distal causes (black). The prevention and sustainable elimination of disease requires understanding and action on key factors in the system. A large number of disciplines should therefore be mobilized in a complementary manner (blue).

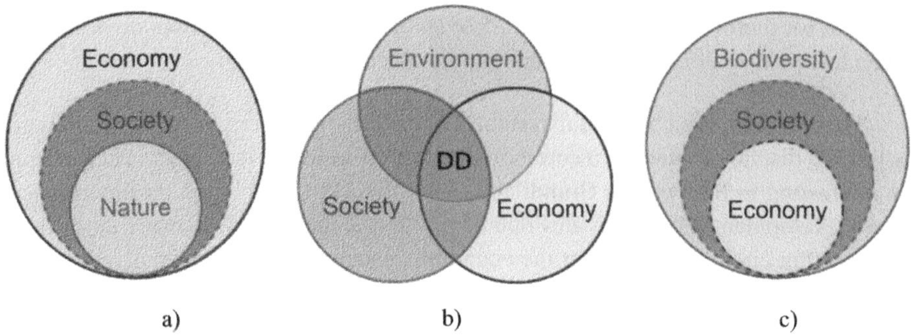

Figure P.2. *Representations of the social, economic and living spheres. a) As currently perceived by world citizens. b) As derived from the Brundtland Report promoting the concept of sustainable development. c) As imposed by the IPBES and IPCC findings and the socio-ecological emergency, according to Morandín-Ahuerma* et al. *(2019), modified. For a color version of this figure, see www.iste.co.uk/giraudoux/socioecosystems.zip*

The scientific and educational challenge is to prepare professionals from each discipline to work in a multidisciplinary context designed to understand and control the processes leading to disease elimination. In addition, it would be necessary to extend this framework to consider the consequences of a change to one of the parameters of the system on other spheres than those of health. For example, to measure the multiple possible collateral consequences of controlling the abundance of a vector, or the release of antibiotics into the environment, on the other balances and "services" of the socioecosystem. This integrative approach, still conducted in a very partial and modest way (Destoumieux-Garzón *et al.* 2018), has underpinned the concepts of *One Health* and *Ecohealth* for the past 20 years (Gibbs 2014). Its most limited meaning, in the concept of *One Health* as currently in use at the WHO, is essentially to cross the contributions of the medical sciences with those of the veterinary sciences, but the real ambition must go much further. *Ecohealth* researchers, that is, researchers in ecology, humanities and health sciences, and the practitioners who apply the results, are committed to promoting human, animal and ecosystem health together, and to conducting research that recognizes the inextricable links between the health of all species and their environments. A core principle is that health and well-being cannot be sustained on a resource-depleted, polluted and socially unstable planet. They implement systemic and integrated approaches to sustainably optimize ecosystem services related to the concept of health (human, animal and ecosystem) and social stability, and seek to foster and promote the interdependence, coexistence and evolution of living things and their environment (Lerner and Berg 2017). Again, no science or meta-science formally organizes disciplinary convergences.

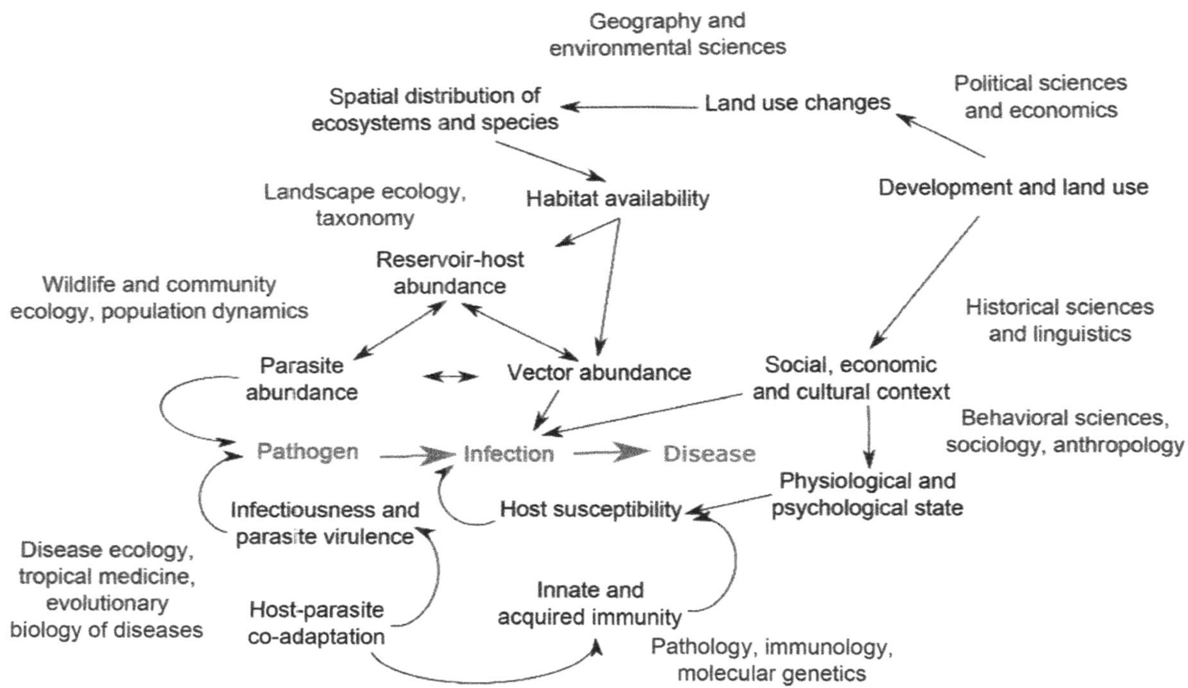

Figure P.3. *Set of causalities and disciplines related to the upstream resolution of a health issue, according to Giraudoux (2019). For a color version of this figure, see www.iste.co.uk/giraudoux/socioecosystems.zip*

Research and disciplinary methods have become so complex in their details that no scientist can master them all at once and they easily cross borders. In the environmental sciences and especially in ecology, the complication of gathering sufficient human resources to carry out the numerous measurements necessary to study regional dynamics is added to this difficulty. This is the case, for example, with measurements, even if only of abundance, made on wild animal populations, where the researcher depends to a large extent on collaboration with technical organizations interested in the results of the research (agricultural, sanitary, forestry, hunting, fish farming, naturalists, etc.) in order to gather the necessary data at the appropriate scales. The multidisciplinary injunction is then coupled with a contingent injunction, that of multisectoriality, which forces the research subject to be placed on the borderline between fundamental and finalized. Participatory sciences can further increase the circle of people who, through their territorial networking, constitute an irreplaceable observation network for measuring and raising public awareness of the state and dynamics of biodiversity and socioecosystems. The disciplines and sectors thus constitute human resources for collectively defining a problem that cannot emerge from the disciplines and sectors considered separately. The formulation and resolution of the problems call for design engineering to coordinate them, but it is often a matter of course and the chance of encounters that organize things. It can also be noted that, very often, it is the emergence of new technologies and methods at a given moment in the researchers' understanding of the problem (e.g. developments in molecular biology, geolocation, computer programming and data analysis, and ultrasound diagnosis) that makes joint advances possible (or easier) and enables a certain type of collaboration. In this apparent complexity, however, the residual common platform of all sciences continues to be summarized in few words and is within everyone's reach: critical doubt and proposals derived from shared evidence, knowledge that is always updated, admitted as provisional, in an open corpus, in perpetual construction.

In the absence of an established theory and science of multidisciplinarity, there are, however, practices, some of which have lasted for decades, around *ad hoc* defined objects, most often motivated by the resolution of a problem that each discipline or sector cannot solve alone. Much of the work presented in this book is inspired by that conducted in the 1960s by French physicians and parasitologists whose training included zoology and botany (ecology was rarely taught as an organized discipline at that time). Following the example of the work of the Pastorians near Chartres, in the "cursed fields" on the transmission of anthrax in the 19th century, these researchers deliberately used systemic approaches to understand the transmission mechanisms of certain zoonotic diseases. Golvan and Rioux (1961, 1963), for example, showed that plague outbreaks in Iranian Kurdistan were caused by a stable and resistant population of gerbils, *Meriones persicus*, living in rocky areas and serving as a reservoir, which transferred the bacillus (*Yersinia pestis*) to an unstable susceptible population of *Meriones vinogradovi* living in agricultural lands. During peak outbreaks of the latter species, the dispersal of a few individuals from farmland to sub-optimal rocky habitats brings them into contact with the fleas of *M. persicus*, present in the burrows, causing a wave of infection and an epizootic in the *M. vinogradovi* population. This epizootic then leads the *M. vinogradovi* population to near extinction. During phases of high density of *M. vinogradovi*, damage to crops is such that people compensate for their losses by digging up caches of grain harvested by the gerbils (several kilos per burrow). This practice, and the careless handling of dead gerbils, then increases the risk of human contamination (Combes 2001). In this case, most of the key issues that challenge multidisciplinary research were already present: the importance of space (transmission is explained by the movement of species between habitats), the importance of time (the dynamics of disease transmission depend on the temporal dynamics of the host population), the importance of species susceptibility (plague is stable because a population is resistant) and the importance of human behavior (which explains human exposure). The term "landscape epidemiology" was coined at the same time by Pavlowski (1964). Later, Darling (1970) emphasized some of the interactions between ecology and medicine. This type of research was continued by Jean-Antoine Rioux (Houin *et al.* 2018), who implemented it in the study of intestinal schistosomiasis in Guadeloupe (Rioux *et al.* 1977), leishmaniases in southern France (Rioux and Golvan 1969) and North Africa (Rioux *et al.* 1997), and the ecology of mosquitoes for their control in southern France (Rioux *et al.* 1967). In this work, he conceptualized how ecological and epidemiological paradigms and methods could be combined with studies on parasite transmission and control (Rioux *et al.* 1981). He coined the term "eco-epidemiology" for systems approaches integrating methods from both ecology and epidemiology. His approach focused on defining spatial, temporal and population risks (related to the age structure of the population, its immune responses, etc.), and stressed the importance of spatial and temporal scales in understanding transmission phenomena (Rioux *et al.* 1990). It was mainly based on ecological strata defined according to composite maps developed from existing thematic maps (topography, soils, vegetation, etc.) (Ozenda 1986) and field measurements (vector population densities, etc.). The link of his work with the Zürich-Montpellier school of phytosociology, which was at its peak, is evident. For example, the use of multivariate exploratory analysis methods, which were popular in phytosociology for analyzing "habitat x species" matrices, was favored for the analysis of insect vector distribution (Rioux *et al.* 1997; Rioux 2001). The multidisciplinarity that resulted from the work as a whole was evident from the quality of the authors associated with the publications (Figure P.4).

Finally, as early as 1958, the creation of joint operational bodies with an applied vocation, such as the *Entente interdépartementale pour la démoustication du littoral méditerranéen* (Interdepartmental Agreement for Mosquito Control on the Mediterranean Coast) in Languedoc-Roussillon, provides an example of successful integration between politicians, scientists and sectoral operators.

***Rev. Epidém. et Santé Publ.* 1977, 25, 483-519**
Écologie de la schistosomiase en Guadeloupe
Analyse du système épidémiologique.
Documents pour un essai de modélisation
J.A. Rioux, H. Decamps, G. Lanotte, **C. Combes, A. Théron,** J.F. Pointier,
S. Seytor, P. Delattre, C. Bourgerol
Authors' disciplines:
Physician & parasitologist Ecologist Immunologist **Parasitologist** Malacologist
Public health Agronomist & ecologist Socio-anthropologist

Figure P.4. *Diversity of specialties involved in the work edited by Jean-Antoine Rioux. For a color version of this figure, see www.iste.co.uk/giraudoux/socioecosystems.zip*

The aim of this book is to illustrate, from a human perspective, how, in a regional socioecosystem, the Jura Mountains, a certain number of answers could be found to questions jointly affecting agriculture, human health and conservation, based on an eco-epidemiological approach and through action-based research. Here, it was the field and a form of pragmatism in the resolution of the questions posed that led disciplined researchers, in the sense of "each having a clear disciplinary root", to indiscipline, that is to extract themselves from the mainstream of their discipline, often operating in spite of it, in order to cultivate the interactions with the disciplines useful for the resolution of the questions. The aim was to gain a better understanding of the functioning of certain processes in the regional socioecosystems of the Jura Mountains, and then to answer similar questions on other continents and socioecosystems. Beyond the case, the aim is to learn from the experience and, if possible, to infer a style and practice *a posteriori*. Each chapter is preceded by a list of guiding questions that motivate it and followed by key points drawn from the experience.

Chapter 1 shows how research in the 1980s on a series of problems linked to the appearance of population outbreaks of grassland voles in the Jura Mountains in the 1970s led to results that brought together the agricultural world, wildlife managers, ecologists, geographers and anthropologists, and ultimately led to a description and understanding of the coherence of a regional socioecosystem and its evolution, which was essentially guided by changes in agricultural practices under socio-economic constraints.

Chapter 2 shows another view of the same socioecosystem. River pollution has led to more frequent episodes of fish mortality. The actors concerned by this situation have mobilized different resources to understand the causes and control them. The anthropologist's work shows the complexity of the phenomenon, which goes beyond the singular resolution capacity of each actor, whether scientific or not.

Chapter 3 shows how health research on children's exposure to the farm environment and their consumption of raw milk, in relation to the predisposition to allergic diseases, leads to general reflections on biodiversity and to ambiguities in the perception of the risk/benefit ratio, including in the medical world. At the heart of these ambiguities, sociological representations, economic concerns and opposing (and often implicit) conceptions of animal welfare and rural development are working hypotheses that have yet to be addressed and that epidemiologists, microbiologists and immunologists alone cannot resolve.

Chapter 4 shows how findings from previous regional studies on landscape ecology and the eco-epidemiology of a rare disease, alveolar echinococcosis, conducted in collaboration with medical doctors and veterinarians, have been confirmed in other ecosystems in China, and how the know-how has then been adapted to other regional realities.

Chapter 5 shows how the representation of risk in this regional socioecosystem is linked to the perception that each person has of nature and their territory, and explores the consequences of this observation on the evolution of practices and the adaptability of the socioecosystem to future changes.

Chapter 6 shows how the use of existing expertise helped to address conservation biology questions relating to a species of primate endemic to Yunnan and Tibet, the black-and-white snub-nosed monkey (*Rhinopithecus bieti*).

Chapter 7, based on the personal itinerary of a doctor "at the end of the trail", shows how an eco-epidemiological approach makes it possible to understand the socio-ecology of cholera transmission in the African Great Lakes region, and on this basis inspires the implementation of the multisectoral strategic plan for the elimination of cholera in the Democratic Republic of the Congo.

Finally, the conclusion presents a socio-anthropological analysis of the scientific position adopted to answer the questions raised in the previous chapters, and thus analyzes the conditions that prevailed in order to address questions relating to a regional socioecosystem through an indisciplinary approach.

January 2022

References

Bretagnolle, V., Benoit, M., Bonnefond, M., Breton, V., Church, J.M., Gaba, S., Gilbert, D., Gillet, F., Glatron, S., Guerbois, C. *et al.* (2019). Action-orientated research and framework: Insights from the French long-term social-ecological research network. *Ecology and Society*, 24(3), 10.

Campagne, A. (2017). *Le Capitalocène. Aux racines historiques du dérèglement climatique*. Éditions Divergences, Paris.

Collins, S.L., Carpenter, S.R., Swinton, S.M., Orenstein, D.E., Childers, D.L., Gragson, T.L., Grimm, N.B., Grove, J.M., Harlan, S.L., Kaye, J.P. *et al.* (2011). An integrated conceptual framework for long-term social–ecological research. *Frontiers in Ecology and the Environment*, 9(6), 351–357.

Combes, C. (2001). *Parasitism – The Ecology & Evolution of Intimate Interactions*. University of Chicago Press, Chicago.

Darling, F.F. (1970). Borderlines of medicine and ecology. *Proceedings of the Royal Society of Medicine*, 63(11 Part 2), 1164–1168.

Destoumieux-Garzón, D., Mavingui, P., Boetsch, G., Boissier, J., Darriet, F., Duboz, P., Fritsch, C., Giraudoux, P., Le Roux, F., Morand, S. *et al.* (2018). The One Health concept: 10 years old and a long road ahead. *Frontiers in Veterinary Science*, 5(14) [Online]. Available at: https://www.frontiersin.org/article/10.3389/fvets.2018.00014 [Accessed January 4, 2020].

European Environmental Agency (2019). The European environment – State and outlook 2020. Knowledge for transition to a sustainable Europe. Report, Publications Office of the European Union, Luxembourg [Online]. Available at: https://www.eea.europa.eu/publications/soer-2020 [Accessed May 28, 2020].

Frodeman, R. (2019). *Pour un savoir soutenable : une théorie de l'interdisciplinarité*. QUAE, Versailles.

Gascar, P. (1986). *Du côté de chez Monsieur Pasteur*. Odile Jacob, Paris.

Gibbs, E.P.J. (2014). The evolution of One Health: A decade of progress and challenges for the future. *Veterinary Record*, 174(4), 85–91.

Giddings, B., Hopwood, B., O'Brien, G. (2002). Environment, economy and society: Fitting them together into sustainable development. *Sustainable Development*, 10(4), 187–196.

Giraudoux, P. (2019). Pour une médecine globale, préventive et écologique…/For a global, preventive and ecological medicine… *Annales africaines de médecine*, 12(4) [Online]. Available at: https://anafrimed.net/editorial-pour-une-medecine-globale-preventive-et-ecologique [Accessed September 17, 2019].

Golvan, Y.-J. and Rioux, J.-A. (1961). Écologie des mérions du Kurdistan iranien. Relation avec l'épidémiologie de la peste rurale. *Annales de parasitologie humaine et comparée*, 36(4), 449–558.

Golvan, Y.-J. and Rioux, J.-A. (1963). La peste, facteur de régulation des populations de mérions au Kurdistan iranien. *Revue d'écologie – Terre et vie*, 1, 3–34.

Houin, R., Léger, N., Dupouy-Camet, J., Bastien, P., Luffau, G. (2018). In memoriam Professor Jean-Antoine Rioux (1925–2017). *Parasite*, 25, 13.

IPBES (2019). Le dangereux déclin de la nature : un taux d'extinction des espèces "sans précédent" et qui s'accélère. Press release, IPBES [Online]. Available at: https://ipbes.net/news/Media-Release-Global-Assessment-Fr [Accessed January 3, 2020].

Latour, B. (2017). *Où atterrir?* La Découverte, Paris [Online]. Available at: https://www.editionsladecouverte.fr/catalogue/index-O_atterrir_9782707197009.html [Accessed December 23, 2019].

Latour, B. (2020). Le surplus de subsistance. *Revue Esprit*, May [Online]. Available at: https://esprit.presse.fr/actualites/bruno-latour/le-surplus-de-subsistance-42765?fbclid=IwAR2NJtrHCEHG9kG9K6YRSiAasSZIkMabITNJuU5EYUeE_dizGPgnLtFmpU4 [Accessed May 27, 2020].

Legay, J.-M. (ed.) (2006). *L'interdisciplinarité dans les sciences de la vie*. Cemagref/CIRAD/Ifremer/INRA.

Lerner, H. and Berg, C. (2017). A comparison of three holistic approaches to health: One Health, EcoHealth, and Planetary Health. *Frontiers in Veterinary Science*, 4 [Online]. Available at: https://www.frontiersin.org/articles/10.3389/fvets.2017.00163/full [Accessed May 24, 2020].

Magny, M. (2019). *Aux racines de l'anthropocène. Une crise écologique reflet d'une crise de l'homme*. Le Bord de l'Eau, Lormont.

Morandín-Ahuerma, I., Contreras-Hernández, A., Ayala-Ortiz, D.A., Pérez-Maqueo, O. (2019). Socio-ecosystemic sustainability. *Sustainability*, 11(12), 3354.

Ozenda, P. (1986). *La cartographie écologique et ses applications*. Masson, Paris.

Passeron, J.-C. (2006). *Le Raisonnement sociologique : un espace non poppérien de l'argumentation*. Albin Michel, Paris.

Patz, J.A., Daszak, P., Tabor, G.M., Aguirre, A.A., Pearl, M., Epstein, J., Wolfe, N.D., Kilpatrick, A.M., Foufopoulos, J., Molyneux, D., Bradley, D.J. (2004). Unhealthy landscapes: Policy recommendations on land use change and infectious disease emergence. *Environ Health Perspectives*, 112(10), 1092–1098.

Pavlowski, E.N. (1964). *Natural Foci of Transmission Diseases in Connexion with the Landscape Epidemiology of Zooanthropozoonses*. Nauka, Moscow/Leningrad.

Petit, S., Fleury, P., Michel, V., Mougenot, C. (2008). Raconter la recherche-intervention. Retour sur trois opérations de gestion de la biodiversité. *Natures sciences sociétés*, 16(4), 326–336.

Ricœur, P. (1985). *Temps et récit. Tome 3 : Le temps raconté*. Le Seuil, Paris.

Rioux, J.-A. (2001). Trente ans de coopération franco-marocaine sur les leishmanioses : dépistage et analyse de foyers, facteurs de risque. Changements climatiques et dynamique noso-géographique. *Association des anciens élèves de l'institut Pasteur*, 168, 90–100.

Rioux, J.-A. and Golvan, Y.J. (1969). *Épidémiologie des leishmanioses dans le sud de la France : enquête écologique*. INSERM, Paris [Online]. Available at: http://www.documentation.ird.fr/hor/fdi:13894 [Accessed May 28, 2020].

Rioux, J.-A., Croset, H., Corre, J.-J., Simoneau, P., Gras, G. (1967). Les bases phytoécologiques de la lutte anticulicidienne – cartographie des biotopes larvaires. Ses applications opérationnelles dans le "Midi" méditerranéen. *Annales de parasitologie humaine et comparée*, 42(6), 665–680.

Rioux, J.-A., Decamps, H., Lanotte, G., Combes, C., Théron, A., Pointier, J.F., Seytor, S., Delattre, P., Bougerol, C. (1977). Écologie de la schistosomiase intestinale en Guadeloupe. Analyse du système épidémiologique. Documents pour un essai de modélisation. *Rev. Epidém. et Santé Publ.*, 25, 483–519.

Rioux, J.-A., Lanotte, G., Cousserans, J., Gabinaud, A., Corre, J.-J. (1981). Écologie, développement et santé publique. Bases écologiques de l'analyse d'un système épidémiologique, foyers d'infection et parasitocoenose. Stratégies d'enquête et d'intervention. In *Écologie et développement des ressources de l'espace*. CNRS, Paris.

Rioux, J.-A., Dereure, J., Perières, J. (1990). Approche écologique du "risque épidémiologique". L'exemple des leishmanioses. *Bulletin d'écologie*, 21(1), 1–9.

Rioux, J.-A., Akalay, O., Perrieres, J., Dereure, J., Mahjour, J., Le Houérou, H.N., Léger, N., Desjeux, P., Gallego, M., Saddiki, A., Barkia, A., Nachi, H. (1997). L'évaluation écoépidémiologique du "risque leishmanien" au Sahara atlantique marocain. Intérêt heuristique de la relation "Phlébotomes-bioclimats". *Ecologia Mediterranea*, 23(3), 73–92.

Ripple, W.J., Wolf, C., Newsome, T.M., Galetti, M., Alamgir, M., Crist, E., Mahmoud, M.I., Laurance, W.F. (2017). World scientists' warning to humanity: A second notice. *BioScience*, bix125–bix125.

Turner, S. (2000). What are disciplines? And how is interdisciplinarity different? In *Practising Interdisciplinarity*, Stehr, N. and Weingart, P. (eds). University of Toronto Press, Toronto.

Union of Concerned Scientists (1992). 1992 world scientists' warning to humanity. *Union of Concerned Scientists* [Online]. Available at: https://www.ucsusa.org/resources/1992-world-scientists-warning-humanity [Accessed December 23, 2019].

Waquet, F. (2015). *L'Ordre matériel du savoir : comment les savants travaillent, XVIe-XXIe siècles*. CNRS, Paris.

Waquet, F. (2019). *Une histoire émotionnelle du savoir : XVIIe-XXIe siècle*. CNRS, Paris.

World Commission on Environment and Development (1987). *Our Common Future*. Oxford University Press, Oxford [Online]. Available at: https://sustainabledevelopment.un.org/content/documents/5987our-common-future.pdf [Accessed January 4, 2020].

1

Agricultural Changes and Population Outbreaks of Grassland Voles

Patrick GIRAUDOUX

[1] *Chrono-environnement UFC-CNRS, Université Bourgogne Franche-Comté, Besançon, France*

[2] *Yunnan University of Finance and Economics, Kunming, China*

GUIDING QUESTIONS. –

How do landscape, land use and their evolution determine, on several scales, the population dynamics of agricultural pests?

How does the predator community respond to changes in the abundance of main prey, and what are the risks to secondary prey?

What are the variations in pathogen community patterns in such systems?

How does long-term observation create the conditions for quasi-experiments?

How can this knowledge be used to implement an integrated control plan for grassland vole populations?

How do these processes fit into a regional "system"?

What new constraints are emerging with global warming?

What are the socio-anthropological and epistemological necessities and limitations to conducting and applying such studies?

1.1. Introduction

Rodent outbreaks have, in historical memory and at more or less regular intervals, massively affected crops and stored goods locally and sometimes even regionally. They have also carried diseases that are transmissible to humans, such as plague, and continue to carry both old and new emerging diseases.

However, they still elude most explanations and no simple remedy has been found to date. Beyond the myths and legends inspired by such proliferations, they have attracted the attention of scholars since the 19th century.

Charles Elton was one of the historical contributors to this research in the early 1920s. In 1942, he described the phenomenon with humor:

> Voles multiply. Destruction reigns. There is dismay, followed by outcry, and demands to Authority. Authority remembers its experts or appoints some: they ought to know. The experts advise a Cure. The Cure can be almost anything: golden mice, holy water from Mecca, a Government Commission, a culture of bacteria, poison, prayers denunciatory or tactful, a new god, a trap, a Pied Piper. The Cures have only one thing in common: with a little patience they always work. They have never been known entirely to fail. Likewise, they have never been known to prevent the next outbreak. For the cycle of abundance and scarcity has a rhythm of its own, and the Cures are applied just when the plague of voles is going to abate through its own loss of momentum. (Elton 1942)

The story told in this chapter spans more than 70 years, including some 40 years of research that is still ongoing. It is therefore part of a worldwide research process, still unresolved, which has led to a colossal body of literature in ecology and led some critics to think that the problem was unsolvable and therefore of no interest. We shall see later that this is not the case. Population fluctuations in rodents are still one of the classic problems of animal ecology (Krebs 2013).

The damage that species in this group cause to agrosystems and stored commodities can equal almost the entirety of some crops (Buckle and Smith 2015). This is the case in the Jura Mountains, where outbreaks of the montane water vole, *Arvicola amphibius*, formerly known as *Arvicola terrestris*, a taxonomically complex grassland species (Chevret *et al.* 2020), can reach a density of several thousand individuals per hectare, such that an entire grassland is devastated (Figure 1.1).

Figure 1.1. *Water vole tumuli in the commune of Les Combes in February 2012, during a population peak (photo: Fredon de Bourgogne Franche-Comté). For a color version of this figure, see www.iste.co.uk/giraudoux/socioecosystems.zip*

When, at the end of the 1970s, researchers from the National Institute for Agricultural Research (*Institut national de recherche agronomique*; now INRAE, *Institut national de recherche pour l'agriculture, l'alimentation et l'environnement*) were called in by the Ministry of Agriculture and requested to find a solution to the new recurrence of regional vole population outbreaks in the Jura Mountains, Charles Elton's description was still relevant. They did not find any result in the scientific literature that would allow them to propose a sustainable control solution based on the knowledge of the causes of the outbreaks and their cyclicity. The agronomic crisis could only be addressed by phytosanitary measures at the level of the individual plot of land, in accordance with the simplistic responses of agronomy at the time. It was based on burying bait containing a rodenticide anticoagulant, often too late in the outbreak cycle to be effective in preserving the crop (Delattre and Giraudoux 2005). The implementation of large-scale collective chemical control (several thousand hectares in the Doubs department alone), supposedly curative, only during outbreak phases, continued for 30 years, until the beginning of the 21st century. The environmental consequences of these treatments were evident from the end of the 1980s until the beginning of the 2000s with the mortality of hundreds of carnivores consuming rodents, by secondary poisoning (buzzards, red kites, foxes, etc.) or directly by consuming bait (wild boar, etc.) (Berny *et al.* 1997). A paroxysmal phase in 1998 constituted a tipping point toward changes in practices and prevention, based on the results of research conducted during an earlier multisectoral regional program from 1992 to 1996, but which had been deliberately neglected by farmers. Naturalist associations and hunting federations have increased the mobilization of public opinion and raised the awareness of the agricultural world to the environmental and socio-economic risks of the "all-chemical" approach, which led to a period of broad consultation between all stakeholders from 2000 to 2006, within the framework of a State-Region plan contract (Delattre and Giraudoux 2009). The research conducted under this multisectoral program and thereafter complemented the main results obtained in the 1990s. They have more strongly integrated economic and socio-anthropological components and have led to the representation of a regional multi-scale system explaining outbreaks, in a context taking into account their multiple causes and consequences, and to a profound revision of control methods, then preventive, now proven effective and sustainable (Giraudoux *et al.* 2017).

The purpose of this chapter is to present this regional system and to discuss the human, structural and epistemological conditions that prevailed in its discovery and implementation, and the various constraints that must be taken into account because of its multifunctionality. We will specify the research obstacles that need to be overcome in order to understand the processes involved and explore the consequences. We will also see how this system is linked to other components of the regional socioecosystem presented in the following chapters.

1.2. The European Common Agricultural Policy and its national implementation, voles and their predators...

1.2.1. *Establishment of practices and landscape*

The Jura Massif is known, among other things, for its cheese specialties with several emblematic protected designations of origin (PDO), such as Comté, Morbier, Mont-d'or and Bleu de Gex. From 67,154 tons in 2018, Comté production, for example, currently generates nearly 8,000 direct jobs. It accounts for 80% of the PDO cheeses in the Jura Massif and a quarter of the French PDO cheeses – but this was not always the case. In the 1950s, in this region of medium-altitude mountains (250–1,718 m), grassland covered between 20% and almost 75% of farmland, regardless of altitude (Figure 1.2). In the 1960s, common European and national agricultural policies encouraged the renovation of farms (consolidation and enlargement of plots, mechanization, etc.) and their specialization throughout France. Farmers above 500 m in the Jura have thus been encouraged to specialize in modernized dairy and cheese production. The shift from traditional mixed farming to exclusive livestock production led them to convert fields formerly ploughed for cereal or potato production into permanent grassland and, as a result, to increase the proportion of permanent grassland to 75%–100% farmland. This has also led to an increase in forage productivity from 2 to 3 tons of dry matter per hectare per year in the 1950s to 4 to 9 tons today. Milk production itself has increased from less than 2,000 l per cow per year in 1951 to an average of about 7,000 l today (the range is 5,400 –8,500 l).

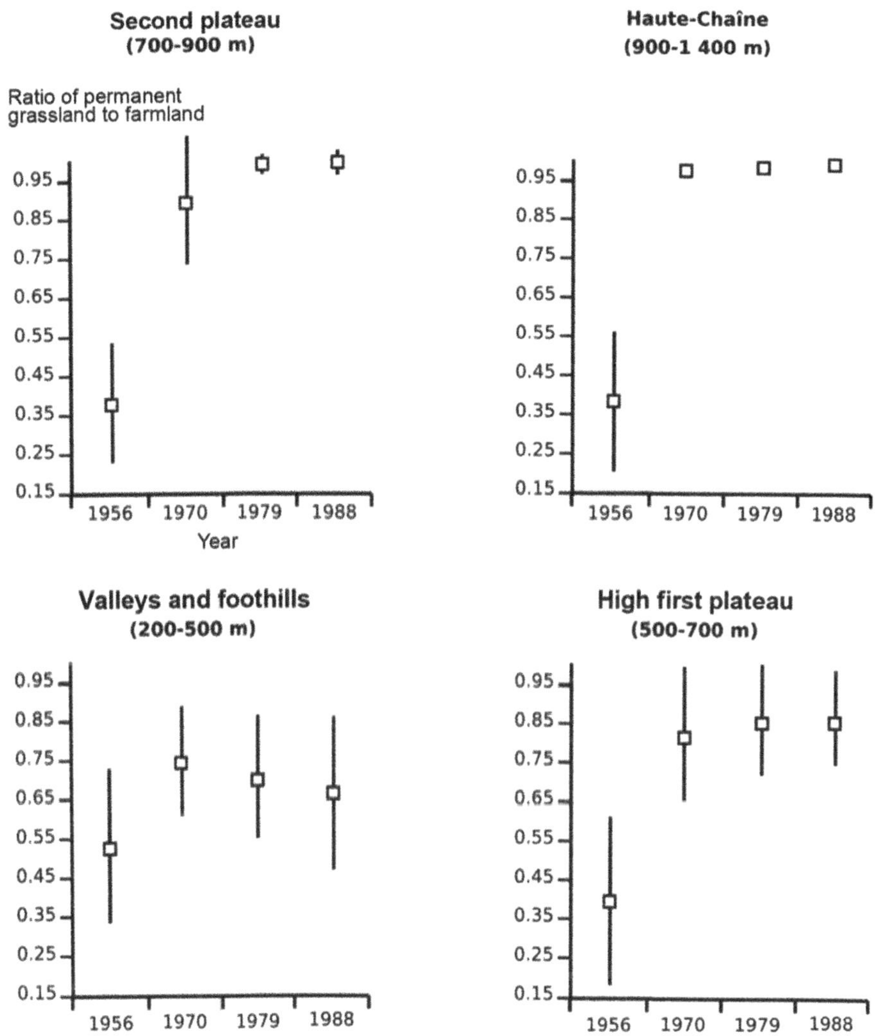

Figure 1.2. *Evolution of permanent grassland in the Doubs department, from* Giraudoux et al. *(1997). The squares represent the average of the communes and the vertical bars the standard deviation*

Massive outbreaks of a species of grassland vole, the water vole, *A. amphibius*, with cycles of 5–6 years spreading in the form of a traveling wave over all the mid-altitude grasslands of the Jura Massif, were reported by the Regional Plant Protection Service of the Ministry of Agriculture (*Service régional de protection des végétaux du ministère de l'Agriculture*, SRPV) as early as the 1970s. In the same area, outbreaks of another species of grassland vole, the common vole, *Microtus arvalis*, which was later found to be non-cyclic in the area (Giraudoux *et al.* 2019), were also observed, reaching, like the former species, several thousand individuals per hectare locally. For the past 50 years, the system seems to have stabilized around regular outbreaks of grassland voles, with peaks that can last locally for several years.

1.2.2. *Which measurement tools for which observations?*

Objective and standardized measurement of the phenomenon to be studied is a prerequisite for any observation, whatever the discipline. In the field of animal populations, there are various methods, but, in the absence of being able to count all individuals directly, those based on sampling, which are widely recognized, derive from two principles: either catch effort (this involves measuring the rate of decrease in capture success as a function of the effort needed for capture) or capture-mark-recapture (this involves measuring the dilution of individuals previously captured and marked or genetically identified in the population). Both methods generally aim to measure absolute abundances. For most species, these methods are logistically very cumbersome. In the case of voles, they require several days to implement for an estimate of only a few thousand square meters. From the late 1970s to the early 1990s, only these methods were calibrated and used, which reduced observation to local areas (a fraction of a farm plot) that were not related to the regional geographic extent of population variations. This put the researchers in the position of a meteorologist trying to understand how thunderstorms are triggered by staying in one place under an umbrella. However, empirical ratings based on the observation of surface activity indicators (earthen mounds, burrows, etc.), carried out by the SRPV in each municipality visited by car, already indicated that high densities of water voles affected very large areas over several tens of square kilometers, and covered all of the department's mid-altitude meadows in a few years (Pascal *et al.* 1985). We therefore decided, at the end of the 1980s, to calibrate these index methods on the standard methods of estimation by catch effort, first of the common vole (Delattre *et al.* 1990; Quéré *et al.* 2000), then of the water vole (Giraudoux *et al.* 1995). Since then, it has been possible to estimate population density variations by walking transects over several tens of kilometers, segmented into 10-step intervals, in which the presence or absence of signs of water vole, common vole or mole activity is noted. The SRPV communal scoring system was also standardized in the late 1980s (Habert 1988). For a long time, these methods met with reluctance from the editorial staff of international scientific journals on ecology, as they did not correspond to the commonly accepted standards of precision and demographic modeling. It was very difficult to convince the editors that the loss of local precision was acceptable in a system where the amplitude of variations in abundance is great (from zero to several thousand individuals per hectare), and that this loss was compensated for by a geographically broader vision than that offered by the other methods. Once this first obstacle had been removed, research could move from the plot scale to those that included the phenomenon observed at the regional scale.

1.2.3. *Landscapes and practices*

Examination of the SRPV communal scores confirmed the traveling wave phenomenon in the water vole population (Giraudoux *et al.* 1997; Berthier *et al.* 2014) (Figure 1.3) and the correlation between outbreak risk and landscape composition for the water vole and common vole (Figure 1.4). In addition, threshold effects were detected: for example, at the communal scale, for the common vole, a small increase in permanent grassland beyond the category of 50%–60% of farmland abruptly shifts from a low-risk system to a high-risk system (Figure 1.4(a)). The influence of semi-permanent legume crops (clover, alfalfa, etc.), which can considerably lower this threshold, has also been demonstrated (Delattre *et al.* 1992). In addition, index transects carried out over several hundred kilometers have demonstrated the coincidence between landscape type, as a spatial arrangement of habitats (grasslands, ploughs, hedgerows, forests, etc.), and the amplitude of variation of grassland vole populations, both for the water vole (Giraudoux *et al.* 1997; Duhamel *et al.* 2000; Fichet-Calvet *et al.* 2000; Morilhat *et al.* 2008) and the common vole (Delattre *et al.* 1992, 1996, 1999, 2009) (Figure 1.5). The proximity of wooded structures (hedges, forests) and dwellings, in particular, was generally related to lower average densities and lower density variations of both grassland species.

Finally, at the plot level, water vole population growth was found to be earlier with organic fertilization and in hay meadows. On the other hand, during the high-density phase, lower vole population abundance was observed in environments disturbed by grazing and ploughing (Morilhat *et al.* 2007). Finally, still on this plot scale, we have shown that the existence of networks of galleries dug by the mole (*Talpa europaea*) facilitated colonization by the water vole and accelerated the population growth phase (Delattre *et al.* 2006).

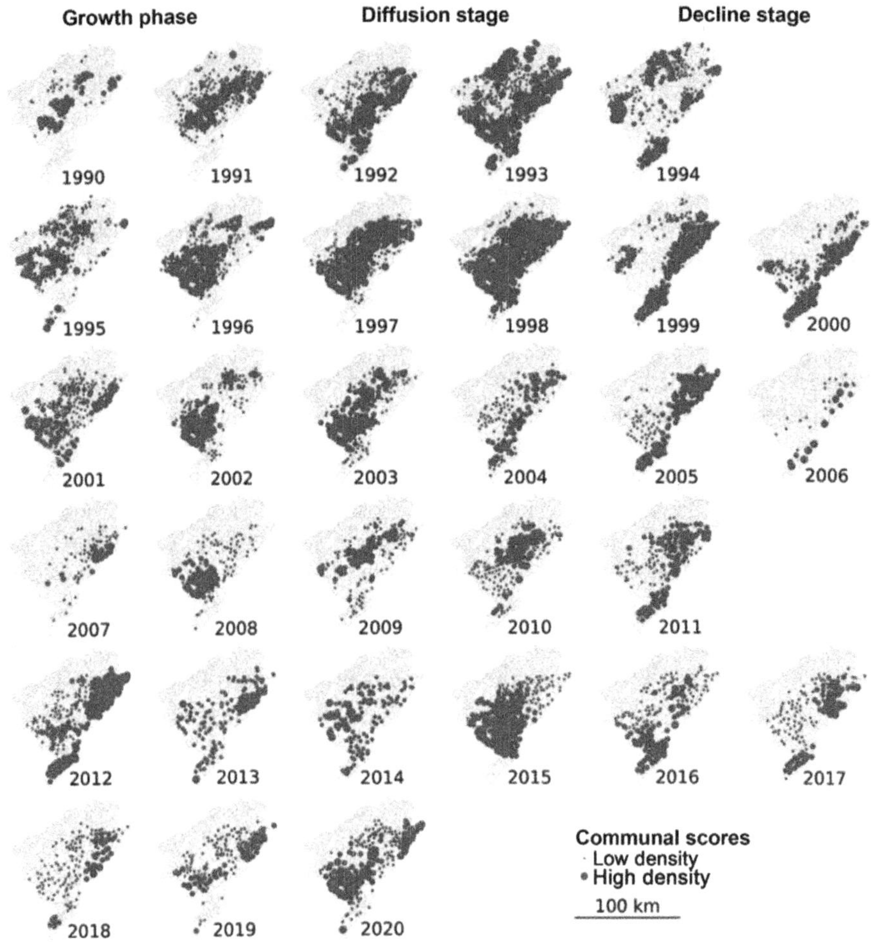

Figure 1.3. *Temporal evolution of water vole outbreaks in the Doubs department, during six demographic cycles (source: Fredon/DRAAF Bourgogne Franche-Comté). For a color version of this figure, see www.iste.co.uk/giraudoux/socioecosystems.zip*

All of these observations led to a hierarchy of factors correlated with the dynamics of vole populations at the scales where the correlations were observed (Table 1.1). However, the processes leading to these correlations remained hypothetical at this stage.

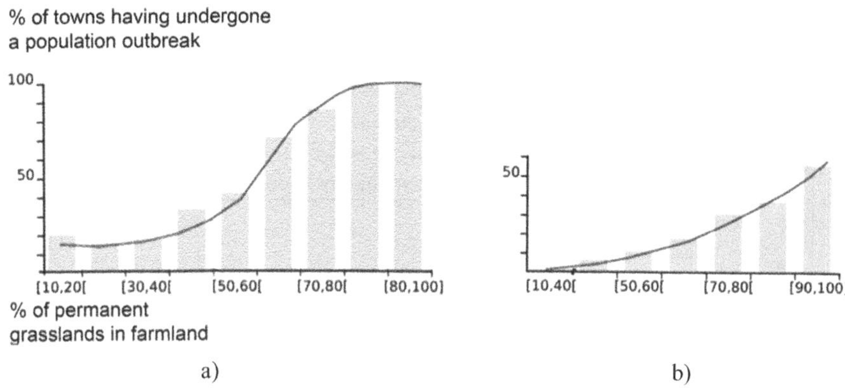

Figure 1.4. *Landscape composition and outbreak risk. a) common vole in Côte-d'Or (Delattre* et al. *1992). b) Water vole in Doubs (Giraudoux* et al. *1997)*

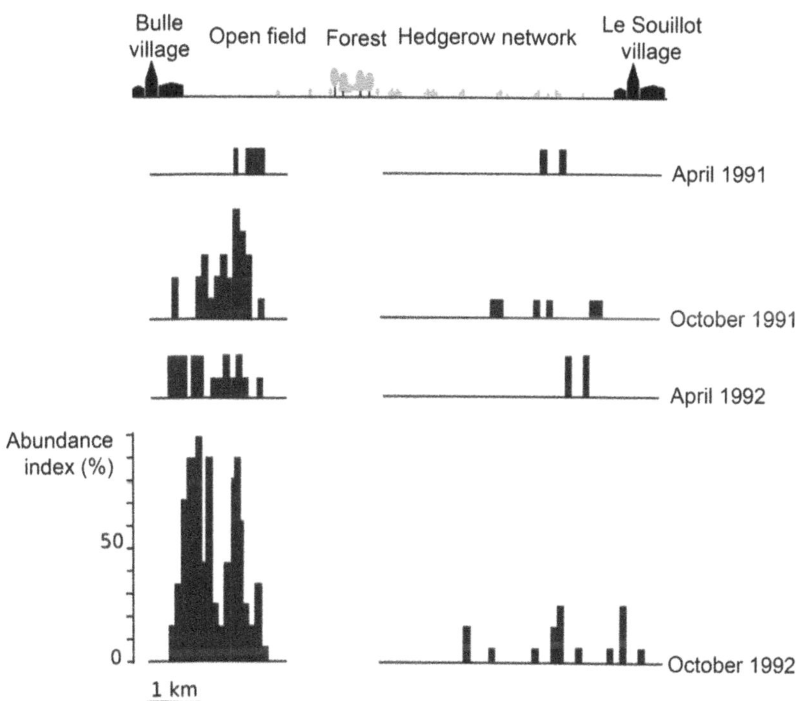

Figure 1.5. *Effect of landscape structure on common vole population dynamics. Open field areas occupied by permanent grassland show much greater variation in abundance than areas with hedgerow networks, according to Delattre* et al. *(1996)*

Spatial scale	Main results
Regional (2,500 km²)	Population dynamics and traveling wave correlated with landscape composition (proportion of permanent grassland).
Landscape (25 km²)	Magnitude of abundance fluctuations: – diminished by the proximity of hedges, forests and dwellings; – increased by semi-permanent legume crops.
Plot (0.01 km²)	Population growth: – increased by mole networks; – increased by grass height and fertilization; – decreased by ploughing, trampling by livestock and keeping in short grass (grazing).

Table 1.1. *Hierarchy of factors correlated with variations in montane water vole abundance, from Morilhat* et al. *(2008)*

1.2.4. *Synchronies*

From the early 1980s to the mid-1990s, population demographic monitoring of small mammals was carried out in the Septfontaines and Le Souillot area, covering nearly 20,000 hectares. This has shown, in particular, the concomitance of declines and phases of low population density of species as different as those living in wooded environments (hedges, forests) and grasslands (Figure 1.6) (Giraudoux *et al.* 1994). Furthermore, water vole declines were frequently observed synchronously from one plot to another, even in neighboring plots that were more sparsely colonized and where the vegetation was practically intact.

Synchronies were also sought between reproductive parameters, the age structure of common and water vole populations and their population dynamics. For the common vole, demographic parameters (sex, age, reproduction, etc.) were measured over 17 years and correlated with weather conditions (temperature and precipitation). Strong but complex relationships between female reproduction and meteorological variables were then revealed: female spring reproduction was decreased after cold winters. However, the rate of population growth was not correlated with weather conditions (at the time and up to 3 months earlier) or with female reproduction. The population could grow when reproduction was low or decrease when reproduction was at its peak. These results, combined with data on the age structure of the population, indicated that mortality (by predation, disease, stress or a combination of all three) was solely responsible for the observed multi-year population dynamics (Giraudoux *et al.* 2019).

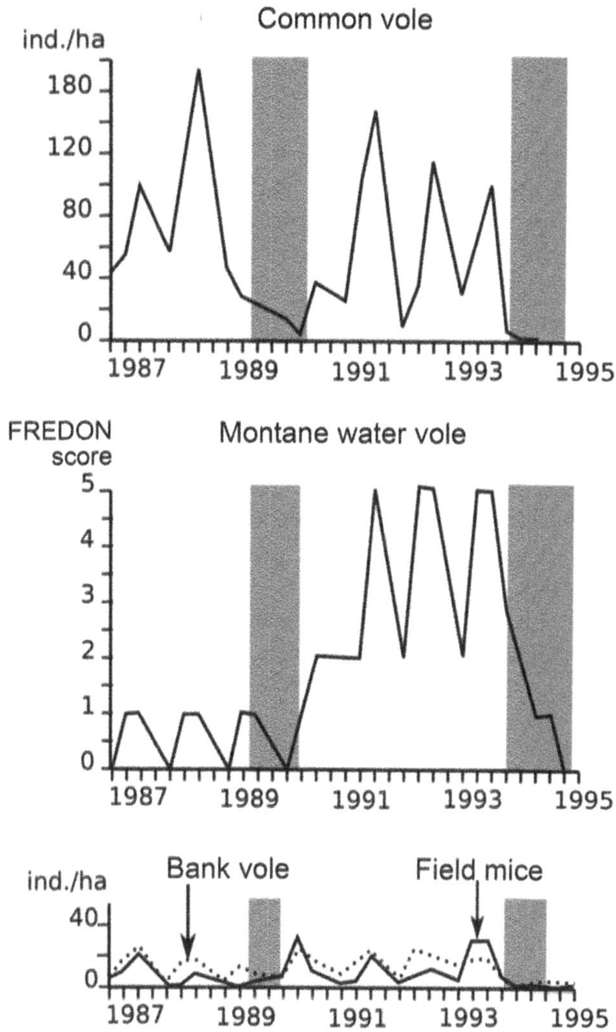

Figure 1.6. *Synchrony of decline and low-density phases (rectangles) of small grassland and woodland rodents, after Bernard* et al. *(2010). A Fredon score of 5 corresponds to densities of several hundred individuals per hectare (ind./ha)*

We have not been able to collect such a long time series of demographic parameters for the water vole. Studies conducted during two declines have shown contrasting results. In the decline observed in the study by Cerqueira *et al.* (2006), the population was reported to be aging (owing to a lack of recruitment of young) and reproduction of females slowed down. In the decline observed in the Villette *et al.* (2020) study, male body condition was poorer during the decline than during the peak of high density, without population aging, and female reproduction was reduced.

Although the approach adopted so far has been purely correlative, that is, it has focused on detecting correlations between variables in the system without being able to prove that they correspond to direct causalities, or to determine the underlying processes, these correlations or lack of correlations have already made it possible to eliminate a certain number of causalities. The existence of "traveling waves" observed among water voles, whose propagation is obviously totally independent of local weather conditions, ruled out the influence of meteorology on population dynamics on a regional scale. The local synchrony of the declines of this species, even in plots still rich in grass, showed that the hypothesis of regulation by the sole effect of the depletion of food resources was insufficient. The fact that the phases of low density were, moreover, synchronous whatever the species and their environment eliminated the action of genetic or other factors internal to each population. The factors responsible for the decline were therefore to be sought in additional non-meteorological causes likely to affect several species at the same time in different environments. The role of predation and diseases, whether combined or not, thus remained one of the most valid explanatory hypotheses.

1.2.5. *Predation*

It has been suggested that predation is a major driver of fluctuations in rodent populations. Theory predicts that specialist predators, such as weasels or stoats, that feed on a limited number of prey species may destabilize prey populations because they exert a delayed density-dependent mortality, whereas generalist predators, which feed on a wide variety of prey species, cause density-dependent mortality without delay, and thus stabilize prey populations (Andersson and Erlinge 1977). However, experimental tests of this prediction, for example, by predator removal, and comparative field studies provide evidence both supporting and rejecting this hypothesis (Korpimaki *et al.* 2002; Oli 2003; Gilg *et al.* 2006). Measuring the effect of predation on a prey population, when the system includes alternative prey and a large number of predator species capable of different diet or population variations (through breeding, movement or a combination of both), is virtually impossible with the current state of knowledge and techniques. However, it is possible in such systems to measure the functional (diet variation) and numerical (population variation by reproduction or movement) response of some predators to changes in prey abundance, and this has been attempted.

These studies showed that, except during their periods of low density, grassland voles constituted the majority and up to 90% of the prey remains found in the droppings or pellets of carnivores and raptors present in this system. The declines of the grassland voles, which occurred at the same time as the low-density phases of the other rodent species, also systematically showed a broadening of the food spectrum of the predators (Figure 1.7). The multiplicity of alternative prey in the field, however, led to complex responses, with the proportion of one species in the diet depending on both its abundance and the abundance and accessibility of other species (Bernard *et al.* 2010; Raoul *et al.* 2010; Baudrot *et al.* 2016b).

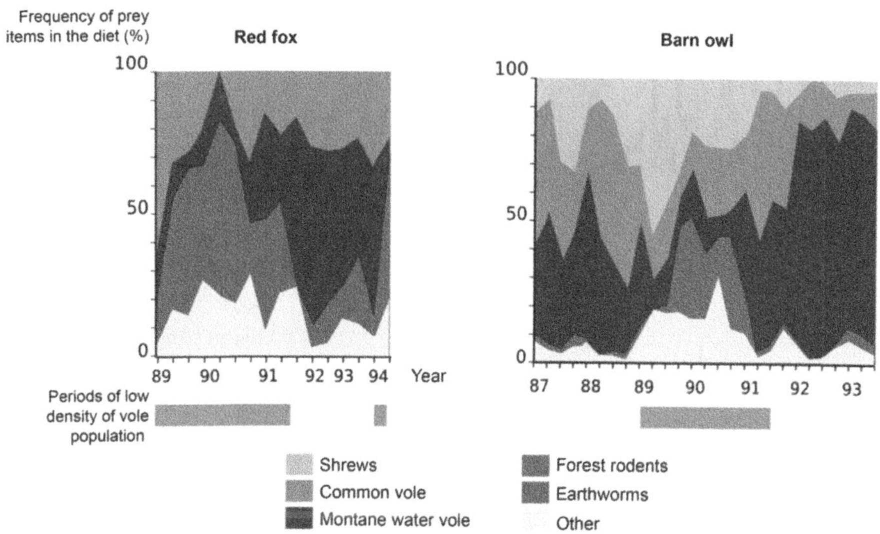

Figure 1.7. *Diet variations of foxes and barn owls. Note the diversification of diet during phases of low vole density, according to Quéré* et al. *(original data) for the red fox and Bernard* et al. *(2010) for the barn owl. For a color version of this figure, see www.iste.co.uk/giraudoux/socioecosystems.zip*

For 20 years, Fredon de Franche-Comté has been carrying out day and night roadside counts in an experimental vole control zone (*zone expérimentale de lutte anticampagnols*, ZELAC) in the Val des Usiers. However, this technique does not allow monitoring of small mustelid populations, which may play an important role in the regulation of vole populations. Nevertheless, they indicated that in such an ecosystem with large variations in grassland vole populations, although the proportion of each species of predator studied may change over time, the total number of predators present, as well as their total biomass and consumption, followed the outbreak cycles of grassland voles. For example, although the fox population increased tenfold during the first half of the study, the total number of predators present for a given vole density remained little changed throughout the study, indicating that populations of other predators declined in a compensatory manner. In addition, the higher predator densities observed during the grassland rodent peak were primarily owing to mobile birds of prey (Figure 1.8).

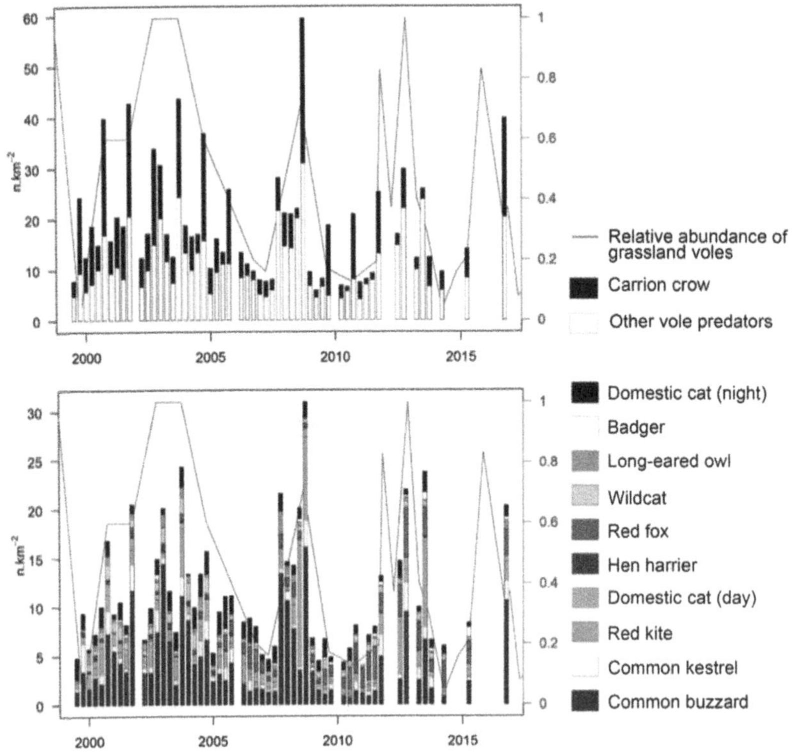

Figure 1.8. *Numerical response of vole predators observed by day and night roadside counts in the Val des Usiers, from Giraudoux* et al. *(2020). The broken line in the background and the y-axis on the right represent variations in the abundance of grassland rodents; the value 1 corresponds to the maximum value observed at the peak of the outbreak. For a color version of this figure, see www.iste.co.uk/giraudoux/socioecosystems.zip*

Calculations based on these estimates indicate that the predators observed during the low-density phase of the grassland rodent populations, alone, could slow the increase or even cause the extinction of rodent populations locally. However, the predator densities observed during outbreak peaks are insufficient to explain the decline in rodent populations observed after these peaks: this ecosystem periodically (permanently, on a large scale) offered an incredible biomass of several tens of kg/ha of easily accessible grassland voles to a large number of predator species, but consumption by the studied predators was multiplied by 7 at its peak, while the biomass of the small mammal population of grasslands was multiplied by thousands (Giraudoux *et al.* 2020). Furthermore, the concomitant increase in fox population with decreases in hare, wild cat and domestic cat suggested that foxes had a negative impact on the other species (Figure 1.9).

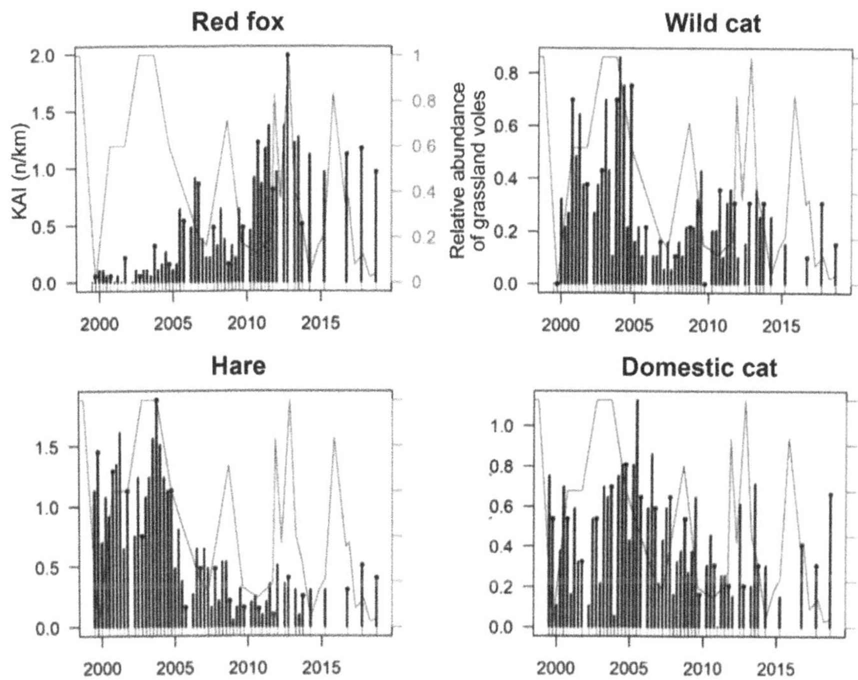

Figure 1.9. *Variations in the kilometric abundance index (KAI) of red fox, hare, wild cat and domestic cat, after Giraudoux* et al. *(2020). The broken line in the background and the y-axis on the right represent variations in abundance of grassland rodents and the scale is the same in all figures; the value 1 corresponds to the maximum value observed at peak outbreak*

Systems where a prey with high abundance dominates, permanently or periodically, make it possible to establish prey switches or even predation sinks, that is, situations where secondary prey undergoes strong predation that keeps them at low density, owing to the high density of one or more predators or the diversification of their diet during phases of lower density of the main prey. The concomitance between peaks in the abundance of water voles in the Risoux massif and the production of young capercaillie from 1976 to 1995 has been attributed to this type of process (Leclercq *et al.* 1997).

Finally, the ZELAC predator study indicates that domestic cat populations, concentrated around dwellings, could reach densities of 2.4–9.1 and, exceptionally, up to 18 individuals per km^2 within a 500 m radius, within which lower vole densities were generally observed (Figure 1.5) (Delattre *et al.* 1996, 1999).

1.2.6. *Bacteria and other vole parasites*

Predators alone do not appear to be able to trigger peak density declines in grassland voles, and these declines can occur even when food is abundant. Through successive eliminations, diseases caused by pathogens (viruses, bacteria, parasitic protozoa and metazoa), possibly combined with stress, were thus one of the remaining hypotheses to explain the declines. This hypothesis had not been fully tested in the past, as early studies had, for technical reasons, been limited to one or a few candidate infectious agents and proved inconclusive (Krebs 2013). More recent research emphasized the importance of considering pathogen communities as a whole, whose elements may act synergistically, neutrally or antagonistically (Telfer *et al.* 2010, 2011). Massive DNA sequencing techniques, popularized about a decade ago, with a reduction in their cost, make it possible to sequence all of the DNA contained in a sample and, by comparing it to sequences recorded in databases, to infer the taxa present. Nevertheless, the characterization of all the pathogenic organisms present still faces many difficulties: first, the existence of sufficiently informed and validated databases to make identifications possible and reliable; second, the proportion of genetic divergence to be retained in order to separate two species (one prefers to speak of an operational taxonomic unit, or OTU, rather than a species); third, the collection and conservation of samples, especially when it comes to RNA viruses, which are extremely fragile; and fourth, the differentiation between pathogens and commensals, as the functional barrier between the two is permeable. A pathogen that is harmless to a healthy organism can become a formidable pathogen in the case of immunosuppression caused by stress, for example.

Driven by an insistent request from Fredon de Franche-Comté, whose president at the time managed to find the necessary substantial funding over 5 years, outside the usual research funding circuit, we used high-throughput DNA sequencing to characterize bacterial communities during a peak and a decline in water vole populations, from 2014 to 2018. In addition, animal spleens collected by Nathalie Charbonnel's team (INRA) earlier, for other purposes, during another decline from 2003 to 2005, were preserved and frozen (Charbonnel *et al.* 2008), so we could compare bacterial communities from one decline to another (Villette 2018). The results can be summarized as follows:

– a surprisingly high number of 155 OTUs, representing at least 13 genera in 11 families, were detected, when only a few dozen bacterial species are known to infect the water vole so far;

– the number of OTUs per vole (individual richness) was higher during declines, with vole body condition lower and reproduction reduced;

– the residual OTUs in the low density phase were already present everywhere, from the beginning of the study and throughout the period, and the presence of certain OTUs was correlated with each other;

– the total number of OTUs in the local populations and the total population of voles did not show clear variations during the declines;

– a comparative study of samples that had been collected 10 years earlier, during a phase of decline, shows that the bacterial communities were not entirely the same at that time as those observed during the program.

Without being conclusive about the impact of bacterial communities on the dynamics of water vole populations, these results showed the previously underestimated complexity of bacterial communities and of the interactions between these communities and vole populations. They showed that such studies, in order to account for this complexity, should be conducted at several spatial scales: several times a year, over several years, and at local and long-distance vole dispersal scales (Villette *et al.* 2020).

Studies have also been conducted on other parasitic taxa, some of which are shared between the water vole and the common vole. The water vole hosts a variety of helminths in the region, including the cestodes *Echinococcus multilocularis*, *Taenia taeniaeformis*, *Taenia tenuicollis*, *Taenia crassiceps*, *Anoplocephala dentata*, *Paranoplocephaloides omphalodes*, *Aprostatandrya sp.* and the nematodes *Trichuris arvicuris*, *Capillaria murissilvatici* and *Heligmosomoides laevis*. Several viruses have also been detected, including *Puumala* virus, lymphocytic choriomeningitis virus and bovine pox virus (vaccinia). Research on these parasites focused on the correlation between their prevalence and the phases of the cycle, and its spatial distribution: *Trichuris arvicolae* was most prevalent in increasing and peaking populations and absent from declining populations, while the prevalence of *T. taeniaeformis* was higher in phases of low local small mammal host abundance (<0.1 km^2) than during peaks. Vaccinia virus prevalence was higher in late-peak than early-peak populations and seroprevalence was spatially aggregated, with some study sites showing high seroprevalences (>70%), while in others anti-vaccinia antibodies were completely absent. However, these relationships remain difficult to interpret, as their expression and hypothetical effects depend on the scale of the study and the variability of many other unknown ecological factors that could not be adequately monitored (health status, demographics and distribution of other hosts, other pathogen communities such as viruses and bacteria, etc.) (Charbonnel *et al.* 2009; Villette *et al.* 2020).

The ecology of *E. multilocularis* will be discussed in Chapter 4.

1.3. Controlling outbreaks and their consequences: from correlations to action

The research presented in section 1.2 of this chapter, thanks to data collection mechanisms on several scales of space and time, makes it possible to identify correlations between the variables of the system studied. It can therefore be posited that, being observational, they do not allow formal inference of causal links between these variables. These studies can be described as eco-epidemiological because, like epidemiological studies, they are based on a description of the components of a population system in space and time. Despite their intrinsic epistemological weakness in proving causality, compared to the hypothetico-deductive method chosen as the elective, if not exclusive, method of reductionist approaches (Popper 2006), observational studies are nevertheless powerful allies of the researcher (Ríos-Saldaña *et al.* 2018). They provide crucial data to better identify covariations in these systems and, through a comparative approach, to design tests and properly contextualized modeling at the appropriate locations, times and scales. They are also powerful allies for the manager as they inform decision-making and make adaptive management, necessarily linked to adaptive monitoring, possible (Lindenmayer *et al.* 2011). Adaptive management is a means of validating hypotheses, in line with a Popperian approach. We will demonstrate this here.

1.3.1. *Rodenticide treatments as a quasi-experiment*

Chemical control using bromadiolone, an anticoagulant, was advocated as early as the late 1970s to control the water vole. This application then became widespread during peak densities, at a time in the cycle when all categories of predators were concentrated in the areas concerned (see section 1.2.5 and Figure 1.8). From the outset, it resulted in massive mortality of vole predators through secondary poisoning, although this was only measured and recognized at a very late stage (Berny *et al.* 1997). The laboratory experiments conducted in the 1980s, which had concluded that the risk of secondary poisoning was low for buzzards and ermine, the only two wild species tested (Grolleau *et al.* 1989), were undermined by the reality of the field.

The half-life of the molecule in the soil seemed to be consistent with that observed in the laboratory, between 3 and 6 days (Giraudoux *et al.* 2006). However, Régis Défaut, a Fredon technician, observed in the field that water voles used to store poisoned bait (wheat grains) in underground caches of a few hundred cm^3.

Based on this observation, Sage *et al.* (2007), by creating artificial caches, showed that voles thus create conditions for the conservation of bromadiolone, multiplying the half-life of the molecule by a factor of 5–10, which can then reach 28–45 days in the spring and autumn, respectively. By reintroducing animals that had never been exposed to bromadiolone into tunnels abandoned after treatment, Sage (2008) also showed that voles recolonizing treated burrows could have access to bait stored by the population eliminated by the treatment and become intoxicated for at least 3 months following the treatment.

These findings, together with the observational studies described in the previous sections, led to a major revision of vole control strategies in the early 2000s, as discussed in section 1.3.2, including a significant reduction in bromadiolone use.

The massive mortality of predators observed in the 1990s made it possible to measure, almost experimentally, the effect of treatments on fox populations and the consequences of an increase in their density on other species living in the same ecosystem. The monitoring was carried out over the long term by the Fredon de Franche-Comté (ZELAC), the *Fédération départementale des chasseurs du Doubs* (472 segments, from 1 to 2 km, distributed throughout the Doubs department) and the researchers. Bromadiolone treatments during the water vole density peaks of the 1980s to 1990s had considerably reduced fox population densities by the early 2000s. The fox population then benefited from the change in control strategy implemented in the 2000s (see section 1.3.2), resulting in a steady increase in the fox population (Figure 1.10). The concomitance of these population changes with those of alternative prey populations, such as hare, wild cat and domestic cat, could be observed at that time, as long-term monitoring schemes were in place (Figure 1.9).

In addition, the negative effect of the considerable decrease in fox populations on the transmission intensity of a cestode parasite, *E. multilocularis* (see Chapter 4), could also be measured during the 1990s (Raoul *et al.* 2003).

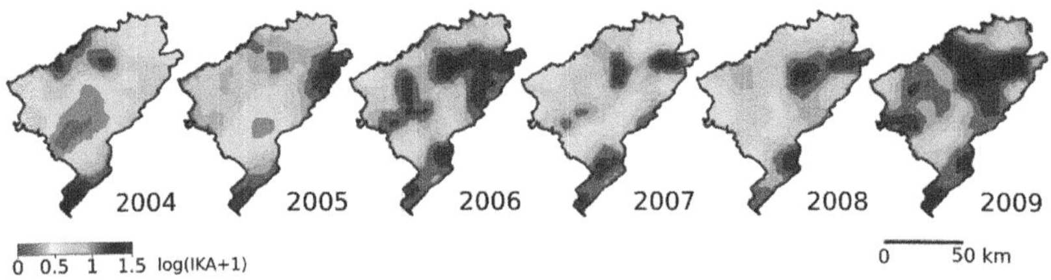

Figure 1.10. *Changes in the relative abundance of foxes in Doubs department following the decrease in the quantities of bromadiolone used in vole control operations from Jacquot* et al. *(2013). For a color version of this figure, see www.iste.co.uk/giraudoux/socioecosystems.zip*

1.3.2. *The toolkit*

It took about 20 years to move from a curative, entirely chemical control, often carried out too late during the peak or even shortly before the spontaneous decline of a water vole outbreak, to a preventive control aimed at preventing the densities from exceeding levels unbearable for the farmer. This has led to a change from the administration of a maximum of 40 kg bromadiolone bait per ha in the form of carrots in the 1980s to 20 kg/ha in the late 1990s, and then, by implementing research-based concepts, to an average of no more than 1–2 kg/ha in 2019–2020, compared with the maximum permitted dose of 7.5 kg/ha at that time (Figure 1.11), and to increasingly smaller areas. In practice, this approach has resulted in a local reduction of bromadiolone use by more than 44 times (Giraudoux *et al.* 2020).

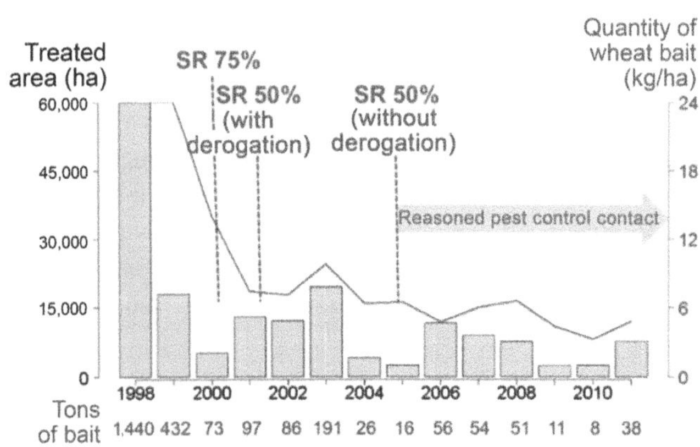

Figure 1.11. *Evolution of bromadiolone use in water vole control. RT, regulatory threshold (% of positive intervals in a transect, see section 2.1) from Cœurdassier* et al. *(2014a). Above this threshold, treatment is prohibited. The wheat bait contains 50 mg bromadiolone/kg*

It has also significantly reduced the unintended effects of treatments on non-target wildlife (Figure 1.12).

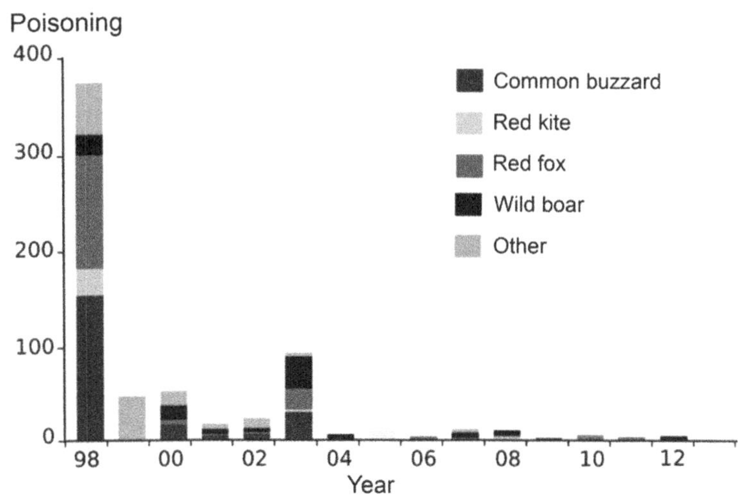

Figure 1.12. *Evolution of wildlife mortality with confirmed or suspected death by anticoagulant intoxication in the Doubs department from 1998 to 2013. SAGIR data – ONCFSFNC-FDC network, per Cœurdassier* et al. *(2014b). For a color version of this figure, see www.iste.co.uk/giraudoux/socioecosystems.zip*

The principle of control was based on selecting the correlations identified in the observational studies as possible causes of the observed effects. Since none of the identified variables could, according to the studies, statistically explain the variations in abundance, the multifactoriality of these causes was accepted as a likely hypothesis. Furthermore, the demography of grassland voles is, in the absence of any limitation, exponential (a pair of voles by interlocking generations can produce more than a hundred in a single breeding season). As a result of this reproductive dynamic, late control attempts were failures and it became clear that, to be effective, control had to be early, preventive and carried out on a population that was still numerically weak, in refuge areas considered "sanctuaries" for the population, even though they were still small in size. We will come back to this notion in Chapter 7, in connection with cholera. A toolkit was therefore progressively set up in experimental zones bringing together a few dozen volunteer farmers, first in the Val des Usiers (ZELAC, experimental vole control zone) in 1999, then in the Charquemont area (CLAC, *Comité de lutte anticampagnol de Charquemont*) in 2005, with the technical support of the Fredon de Franche-Comté and researchers. We will come back later on to the human and political conditions that made this association possible over a sufficiently long period of time to produce results, which is necessarily longer than several vole cycles.

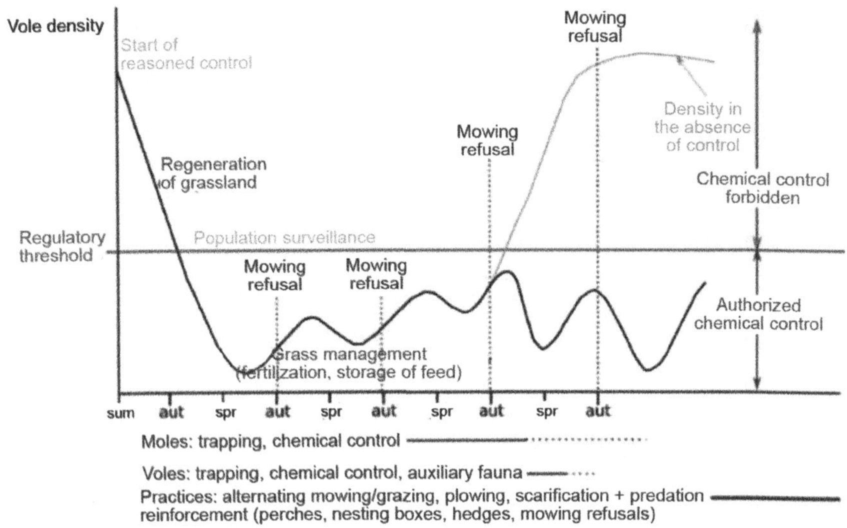

Figure 1.13. *The different proposals in the toolkit over the course of a 5-year control contract, from Couval and Truchetet (2014)*

Based on the research literature, the actions implemented, according to a reasoned chronology (Figure 1.13), were grouped under the triptych "observation, collective engagement, combination of methods" (Couval and Truchetet 2014):

– observation refers to the need for permanent but reinforced surveillance in the low-density phase, in order to identify the first colonies and implement preventive actions before the outbreak. The observation is organized in a regional surveillance network, composed of volunteer farmers and technicians, which feeds a specific plant health bulletin. Distributed to as many people as possible, it allows each farmer to contextualize his or her own surveillance on a more local scale;

– collective commitment is made necessary by the spatial dimension of the outbreak, which can cover areas of several thousand hectares. The speed of colonization of isolated plots of land can therefore defeat attempts at control that are too local, for example, on the scale of a single agricultural plot of a few hectares;

– the combination of methods drawn from a "toolkit" (Table 1.2) includes direct control (trapping and chemical control as soon as the first burrows appear) and indirect control (which acts on the rodents' habitat and thus on the supposed causes of outbreaks). They can be implemented at different spatial scales (plots, groups of plots, municipalities) depending on the constraints linked to the farm (size of the plots, fragmentation, working time, size of the herd, agricultural speculation, etc.).

Method of control	Practical applications
Vole trapping	Effective if used at very low density but requires significant human resources.
Chemical control of vole	In the burrow, at low density, using small quantities of bromadiolone* bait, zinc phosphide and, according to a regulatory framework defined by interministerial order of May 14, 2014, hydrogen phosphide generators by certified applicators.
Mole control	Hydrogen phosphide traps or generators by licensed applicators.
Crop and grazing management techniques	Objective: to hinder the installation or re-installation of the water vole, to decrease the proportion of favorable habitats and to increase the visibility of the first colonies: – tillage with shallow or deep tools (ploughing). Two scales are possible – on a parcel basis to remove old galleries/burrows or on a rotational basis on a farm scale to reduce the ratio of grassland to farmland; – alternating mowing and grazing to cause the collapse of the underground burrows by the trampling of livestock.
Protection and enhancement of predator action	– Establishment of hedge and grove networks; – installation of perches, nesting boxes, re-opening of bell towers and maintenance of shelters favorable to predators (stone walls between plots, called *murgers* in the Jura Massif, etc.); – regulatory measures for the specific protection of predators.

Table 1.2. *Toolkit used in integrated pest management contracts between Fredon and volunteer farmers. *Bromadiolone has been forbidden since May 2021*

This set of measures allowed, from then on, the control of vole outbreaks to the farmers who applied them, with lesser unintentional environmental impact (Figure 1.14).

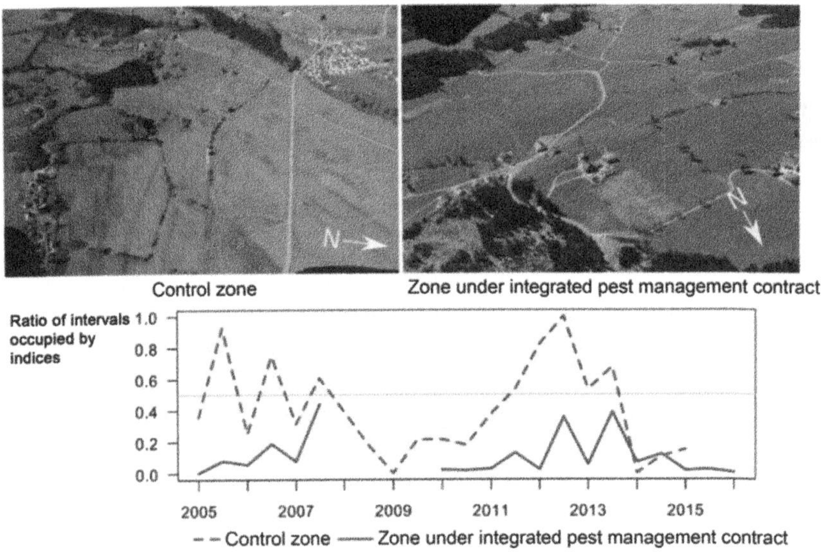

Figure 1.14. *Comparison of an area under an integrated pest management contract with a neighboring area in the Charquemont region, Doubs, from Giraudoux* et al. *(2017). The photos show the difference in vegetation condition in April 2013, at the peak of the outbreak (photos Dominique Mougin), and the graph shows the population dynamics of the water vole. For a color version of this figure, see www.iste.co.uk/giraudoux/socioecosystems.zip*

This result proves that, even if not everything is yet understood in the decline processes, the systemic conditions promoting outbreaks are sufficiently described and known to be actionable, in contrast to reductionist studies and single-factor methods. To date, and despite nearly a hundred years of research, none, to our knowledge, has led to the prevention and sustainable control of a pest outbreak cycle and its effects, let alone "with less environmental damage", but together they have led to a long list of conclusions, all correct, about what does not explain them (Lidicker 1988; Oli 2019) without leading to coherent management recommendations. The effectiveness of the herders' action constitutes, downstream of the eco-epidemiological approach, a quasi-experiment validating the overall systemic scheme and the spatial and temporal hierarchies that generate it (see Figure 1.16), without validating the details of the processes underlying each of the identified correlations, which is out of technical reach, alas, with the current state of methods. Furthermore, this long-term experimentation and multisectoral collaboration served as the basis for national regulations on the use of rodenticides (Legifrance 2014) and for the development of decision-making support tools.

1.3.3. *Anthropology to the rescue of the application*

Since the end of the 1970s, farmers' demands have always been expressed through professional organizations and the press, presenting the vole outbreaks in the areas of the Jura Massif producing PDO cheeses as a scourge that threatens the economy of many farms. In fact, without adaptation, it appears that on an intensive farm, 1 year of outbreaks can lead to a loss of gross surplus of up to 10,000 euros per labor unit (Schouwey *et al.* 2014). It is therefore surprising to note that, despite this vivid expression and the objective measurement of the losses due to outbreaks and the gains made by controlling them, among 2,600 family farms in the PDO Comté area (Doubs and Jura), only 126, or less than 5%, subscribed to a control contract (as of June 1, 2020). There is therefore still an unbridgeable gap between, on the one hand, the results of research and their application in experimental areas run by a minority of a few dozen particularly mobilized farmers who are anxious to tackle the problem through a research-action approach (which is ultimately successful) and, on the other hand, a large number of farmers who revolted until the early 2000s ("Get rid of it right away! You have to choose between buzzards and farmers! What is research doing?"), and were then resigned to the losses incurred since.

An anthropological approach, based on long interviews with farmers in Franche-Comté, has allowed us to better understand how the water vole was perceived by the farmers who suffered from its damage (Michelin *et al.* 2014). It combines an assessment of their technical leeway in the management of their livestock and an anthropological analysis of their perception of the vole and control methods. For some farmers, the vole is considered an intruder that must be eradicated. For others, it is a disruption of the ecosystem that must be dealt with. These two opposing conceptions lead them to react differently, preferring certain solutions and rejecting others, even though they are technically effective within the range of possibilities offered by the toolkit (Morlans and Michelin 2014). Michelin *et al.* (2014) infer four classes of farms associated with different profiles of response to outbreaks. Other studies conducted in Auvergne, in comparable ecosystems, reveal three types of perception, according to Yves Michelin and Shantala Morlans, in Couval *et al.* (2013). We are quoting the terms and conclusions of this analysis in full as they seem informative and difficult to reformulate in any other way without misrepresenting them:

> On the surface: the vole is perceived through the clods of earth it produces, which do not "look clean" and harm the quality of fodder and milk. The important thing is to make the soil disappear. The fact that a few individuals remain is not a problem if they do not cause any

damage. As a result, during low density periods, these farmers do not consider taking action, although this is when it is most effective.

In depth, it is seen as an animal, with its specific behaviors but also often with a strong symbolism: “There are rats under my meadow that even came and ate my boots when I was in my field the other time.” Some farmers will fight to the last animal, even if it is not necessary.

In dynamics, it is seen as a population, likely to pullulate and invade a production area. The actions carried out then integrate the whole cycle with low density interventions to prevent or at least slow down the start of outbreaks. To do this, the proponents of this approach have developed empirical knowledge of population level diagnosis that helps them to anticipate events.

As a result, whether a farmer fights or not depends of course on the technical or economic impact of the outbreak on the production system, but also on the idea she or he has of it and her or his values: some overestimate the damage while others, preoccupied by other problems, pay little attention to it. The decision to act is based first and foremost on what the farmer observes on her or his plots and on the consequences she or he draws from them regarding the impact on her or his future production. However, not everyone has the same acuity of vision or the same knowledge of possible means of control. Moreover, and above all, they do not attribute the same values (technical, moral, ethical) to the actions to be carried out, in particular to chemical control. Finally, other factors play an important role, such as the image that each farmer intends to give of her or his work and of herself or himself in the eyes of others. This leads some to fight actively to “keep things clean”, because “we are not pigs”, and others to refuse all use of chemical products because “we do not want to be seen as those who sow death in nature”. The structure of the collective is also essential. Depending on whether farmers are used to discussing and working together or whether they are very isolated, it will be more or less easy to take a collective approach.

Finally, other stakeholders such as hunters and naturalists can encourage certain methods (e.g. systematizing trapping in Natura 2000 areas) or move towards confrontation. This indirectly pushes some farmers to stop chemical control, even at low density, following birds of prey or game being poisoned. The farmer therefore always has reasons to justify her or his decision or refusal to use certain techniques, even if these decisions are not always understood by others. In order to promote the dissemination of integrated pest management methods, it is therefore essential to understand how a vole disrupts the functioning of the production system and to understand how the farmer reacts and on what basis. This is necessary in order to work with the farmer to find the modalities that are best adapted to the constraints of her or his system and way of dealing with the problem (Table 1.3).

<table>
<tr><th>Each group represents itself the problem differently</th><th>A target = a speech = one solution: it does not work</th><th>There is an interconnection of several systems of thought</th><th>The solutions are to be co-constructed</th></tr>
<tr><td>Cattle breeders:
the water vole is a pest of grasslands</td><td rowspan="3">There is no such thing as a universal good practice because it has to be adapted to the local context:

– production systems;
– of the network of local actors;
– ecological processes related to environmental characteristics.</td><td rowspan="3">The different ways of thinking of the different actors must be taken into account
The management of the problem must be collective and not only agricultural

↓

Change of posture</td><td rowspan="2">– Listen to all parts;

– analyze the constraints of each;

– understand the motivations;

– facilitate dialog between stakeholders.</td></tr>
<tr><td>Wrestling technicians:
the water vole is invasive and must be regulated</td></tr>
<tr><td>Researchers:
the water vole is an agricultural problem and prey for predators within an ecosystem that must be managed in its complexity</td><td>– Build on knowledge gained through research;
– develop an expertise of the phenomenon through observation;
– acquire mediation skills.</td></tr>
</table>

Table 1.3. *Rethinking advice to better meet needs, from Couval* et al. *(2013)*

The vole outbreak is primarily a problem of competition for grass resources between farmers and rodents. However, as it is a discreet burrowing species, it is only seen when the populations have already reached such a level that little can be done. When they are not yet visible, it is difficult to motivate farmers to undertake time-consuming and sometimes costly practices. The "vole" phenomenon is therefore a source of anxiety. Often, it is added to other factors that the farmer does not control: increase in the price of food supplements, drop in the price of milk, drought...

1.3.4. *The status and contribution of models*

"All models are wrong, some are useful." This popular aphorism invites us to situate the place of modeling in the approach presented in this chapter. Statistical modeling and inference, whether spatial or not, have underpinned the validation of the basic observations of the eco-epidemiological approach from the outset and on a permanent basis. This type of modeling is therefore intrinsically linked to the observational approach in ecology.

Other types of modeling were called upon as needs and collaborations arose, and as convenient tools emerged to implement them. The R[1] development platform, which became popular in the early 2000s, has certainly boosted not only the statistical approaches mentioned above, but also the possibilities of interfacing biologist-statisticians with mathematicians and computer scientists. The first non-statistical models were essentially conceptual, and mathematical-computational tools were only put to good use for process-oriented modeling after eco-epidemiological approaches had sufficiently contextualized the subject in terms of the variables and assumptions about the processes involved, and the hierarchy of spatiotemporal scales of application.

1.3.5. *Conceptual models*

Apart from the dozens that were scribbled on a corner of a table and conference presentations of preliminary versions (Delattre *et al.* 1986, 1988), the first conceptual model that guided the research was proposed internationally in Delattre *et al.* (1992). The aim was to provide a typology of systems linking landscape composition, predation and common vole population dynamics (Figure 1.15). The weasel component of this model could never be validly tested afterward, but the link established between landscape and population dynamics, with all the variants described in the chapter, has still not been invalidated and is still used as a basis for current studies.

1 https://www.r-project.org.

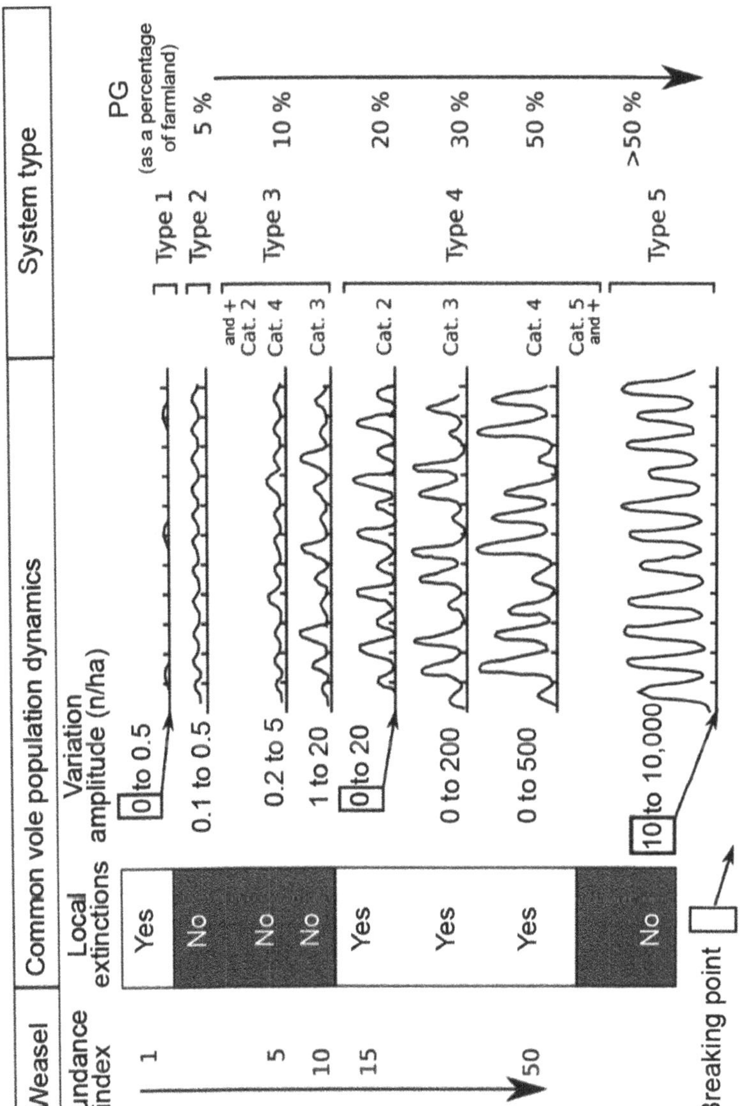

Figure 1.15. *Typology of common vole population dynamics according to environmental variables, from* Delattre et al. *(1992). PG, Permanent grassland*

In addition to the hierarchical classification of the system's control variables (Table 1.1), it was not until 2014 that a global synthetic conceptual model of the socioecosystem studied was finalized, focusing on the factors inducing fluctuations in the abundance of water vole populations (Figure 1.16). This allows the conflicts of interest and synergies between the processes and the corresponding agronomic, health and conservation biology domains to be visualized. It also shows the possible linkages with other elements of the system not studied here, but which will be the subject of Chapters 2, 3, 4 and 5.

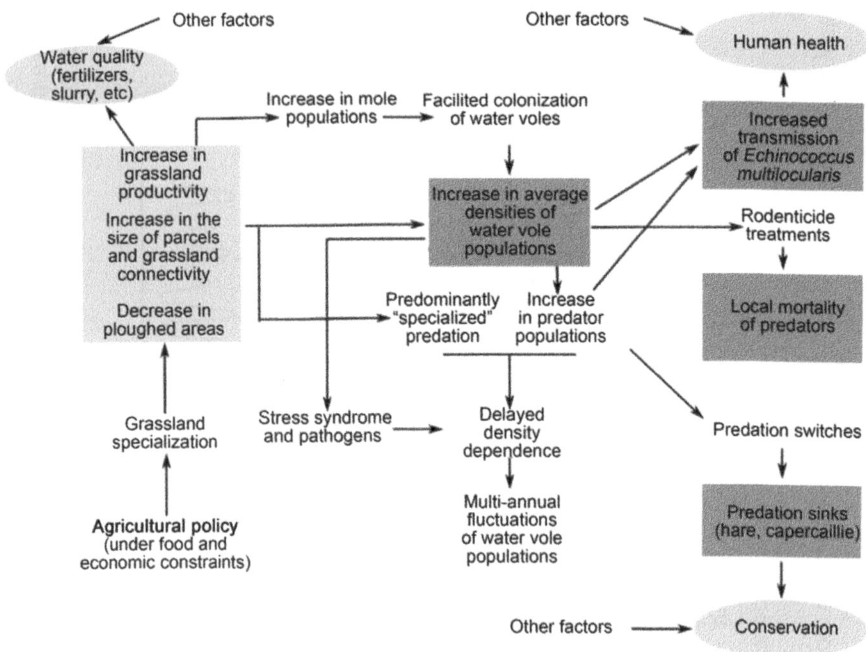

Figure 1.16. *Cascade of consequences of grassland specialization on water vole populations, their predators, alternative prey and transmission of Echinococcus multilocularis, from Giraudoux (2014). The green box groups the proximal outcomes of grassland agricultural specialization and the red boxes the undesired effects. The blue ovals indicate the distal non-agricultural domains affected in such a system. For a color version of this figure, see www.iste.co.uk/giraudoux/socioecosystems.zip*

1.3.6. *Quantitative models*

Incorporating the spatial heterogeneity of real landscapes into population dynamics remains extremely difficult and is a critical factor at a fundamental level (what is the outcome of interactions of spatial mosaics of interacting communities?), once the challenge is to advocate management actions. Based on Graphab software (Foltête *et al.* 2012), the first explicitly spatial modeling of water vole outbreaks confirmed that landscape modeling using graph theory can produce parameters consistent with field observations and thus pave the way for simulating the effect of landscape modification on population dynamics (Foltête and Giraudoux 2012). This approach made it possible to compare several strategies for disconnecting grassland networks at the level of the Doubs department and to detect the main theoretical points of weakness in the network (Foltête *et al.* 2016). The external validation of such regional models is still currently hampered by the logistical impossibility of conducting adequate experiments at the required spatial scales and over the required duration.

An explicitly spatial, process-oriented modeling based on the combination of mathematical equations and computer agents was designed to simulate a wave of water vole colonization that had been monitored for 13 years in the Haute-Romanche, an Alpine valley (Marilleau *et al.* 2018), in parallel with and inspired by studies conducted in the Jura Massif. It was made necessary by the fact that equation-based models could not sufficiently consider spatial heterogeneity, and that the number of voles involved, if they had all been modeled as individuals, exceeded the capacity of the clusters of the computing centers. This approach shows that among the 6,059 random simulations of six demographic variables, only 27 combinations of values, which are very different from one another, account for the observed speed of colonization. Compensation effects are therefore possible between the variables, and the same effect (colonization speed) is then observed for different causes. Moreover, the observed colonization speed cannot be reached without a dispersion of at least 1,100 m, concerning between 0.2% and 20% of individuals in the population, thus underlining the theoretical importance of this category of the population in the speed of propagation of outbreaks.

Several non-spatial equation models were explored to account for specific processes. For example, data collected on the diet of barn owls and foxes were used to test a new general multispecies functional response model, disentangling predator preference for prey categories as a simultaneous function of variation in prey frequency and total density, and proposing a new concept of "rank switching" (Baudrot *et al.* 2016b). This approach led to theoretical thinking informed by observational data collected on the parasite *E. multilocularis*, addressing the dynamics of the parasite in a context of multiple intermediate hosts differentiated by their competence, and complex predator foraging behavior (Baudrot *et al.* 2016a).

Finally, based on observational data, a non-spatial equation model was used to explore the theoretical impact of alternating regulation of voles by predation or chemical treatment in a four-component system: a population of voles, foxes, small mustelids (weasels or stoats) and a chemical treatment likely to poison voles and also, indirectly, predators consuming poisoned voles (Baudrot *et al.* 2020). This model reveals the need to maintain refuges with enough unpoisoned voles to support the mustelids, in order to maintain the predator community, unless we switch to a system where chemical treatments would become permanent owing to the lack of predators.

"[...] some are useful." A few comments on the usefulness of the non-statistical models presented here are in order. The first is that they can only be designed after observational data have been collected at scales appropriate to justify them – the minimum time periods exceeding those of a pest outbreak cycle, some extending over decades. Early attempts, based on insufficient observation time, sometimes at inadequate spatial scales and with naïve assumptions, which did not take space into account, have all failed to produce useful knowledge in the meantime. Non-statistical modeling therefore accompanies downstream thinking, rather than preceding it, and its main utility is to be able to explore theoretically and elegantly (mathematics has an intrinsic beauty) certain sub-systems in a context framed by long-term observations. One then tests whether or not the small number of assumptions and variables (relative to their number in the real world) in each model is sufficient to explain (in theory) the observations. Although much more favorably received in the major ecological journals than the observational studies that motivate them – we can testify to this – their contribution to knowledge of the system is still marginal at this stage, because of their poor capacity to predict the functioning of the theorized sub-system in the more complex real system with which it interacts. Added to this is the impossibility of testing the predictions when, for example, only logistically and ethically impossible landscape or population manipulations on the appropriate space and time scales allow it. Some of the conclusions may also be disappointingly trivial: for example, concluding that predators in the system must be able to survive in order not to be forced into permanent use of rodenticides does not seem to be a scoop to a practitioner, even if the elegance of the mathematical demonstration provides access to one of the most reputable journals of applied ecology. It should be noted, however, that the inverse resolutions allowed by certain simulations or analytical resolutions, when the population parameters are unknown, make it possible to point out the potential importance of certain variables. This is the case, for example, with regard to the speed of propagation of an outbreak, of the theoretically important role of long-distance dispersers, which only constitute a small fraction of the population. However, this simulation finding refers to the field and to appropriate methods, perhaps genetic, to study rare events in a population. In conclusion, yes, models are locally useful, and we do not hesitate to invest time in them, but much less than their overvaluation in the ecological literature in recent years has suggested (Ríos-Saldaña *et al.* 2018), to the detriment of descriptive, eco-epidemiological studies, which are the indispensable, but more costly to acquire, basis of knowledge of socioecosystems, which are, alas, generally idiosyncratic in their details.

1.3.7. *Toward new questions...*

1.3.7.1. *Consequences of global warming and the ecological transition*

Global changes are of course having an impact on the Jura Massif, starting with the reduction in the duration of snow cover, which determines the duration of tourist operations in ski resorts and is thus an important part of the economy of higher altitude areas. For its part, the grassland production system is increasingly subject to spring and summer heat waves, which considerably modify the phenology of grass production (Figure 1.17).

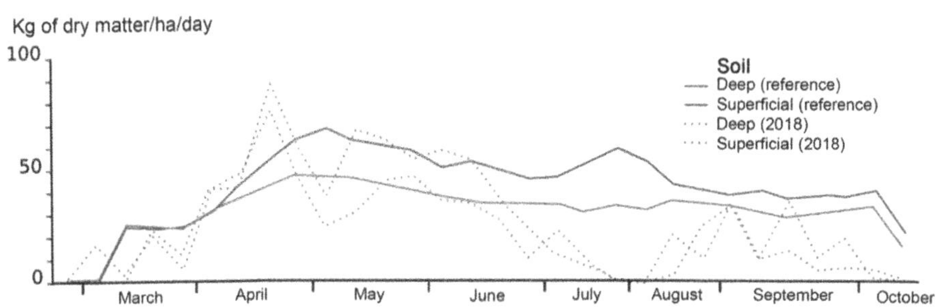

Figure 1.17. *Grass growth on the Jura plateaus comparing the standard reference to that observed during the 2018 heat wave, from Jean-Marie Curtil, Interdepartmental Chamber of Agriculture of Doubs, Belfort region. For a color version of this figure, see www.iste.co.uk/giraudoux/socioecosystems.zip*

Farmers are currently considering the technical and economic adaptations that will be necessary. At present, earlier grass production at a time when rainfall makes hay difficult to harvest, followed by prolonged summer droughts that block it during the summer months, suggests an overall decrease in forage production, which is all the more noticeable when the production system is intensive. The response of vole populations to such changes in phenology remains an enigma. It may directly affect the length of the breeding season (possibly earlier and later owing to the milder climate, but possibly also slowed or interrupted for further periods owing to reduced summer vegetation growth). The evolution of the economy of the farms can itself integrate more or less ploughing with the evolution of the price of cereals, within the limits of the specifications of the PDO, which would then result in a variation in the proportion of favorable habitats, which we have seen to have direct consequences on the amplitude and frequency of the peaks of vole abundance. It is therefore the current stability of the socioecosystem that may be affected, to an as yet unappreciated degree, potentially with the same force as it was at the time of the 1960s, which saw massive regional outbreaks of grassland voles at the base of today's complex food webs involving their predators and parasites. This illustrates that regional socioecosystems are not stable in the long term and follow constrained trajectories. Their adaptability to change requires constant monitoring and innovation.

Another category of foreseeable change concerns the ecological transition from productivist agriculture that pays little attention to its environmental externalities (soil preservation, water management, etc.) to conservation agriculture, also known as agroecology. In "large-scale farming" (cereals, oleaginous plants, protein crops and legumes), which takes us away from the Jura Massif and its farmers, simplified cultivation techniques (SCTs), for example, are cultivation methods that avoid ploughing. The advantage is that they maintain vegetation cover throughout the year, and thus limit erosion. However, this is at the cost of increased herbicide use at planting time (which is problematic); they preserve the structure and vertical organization of the soil, its biodiversity, and thus improve carbon storage while reducing input requirements (fertilizer, fuel, etc.). Permanent and very productive vegetation cover is, however, favorable, as we have seen in this chapter, to the development of the common vole (Delattre *et al.* 1992; Delattre and Giraudoux 2005). In "ploughed" systems, which leave an inhospitable, bare and overturned soil throughout the winter, common vole populations are limited to herbaceous margins and semi-permanent legume crops. They only spill over into annual crops at times and seasonally during peak population periods. A new frontier is therefore opening up in action research, namely, that of knowing enough about the ecology of this species, the consequences of future outbreaks in this new socioecosystem, and the margins for modifying production itineraries and their environment, in order to avoid them without deleterious ecological consequences (reduction of biodiversity, use of pesticides, etc.).

1.3.7.2. *Consequences of the environmental drawbacks of a commercial success*

"Why is the Comté being accused of polluting Franche-Comté's meadows and rivers?" This was the question posed by the media, following a damning scientific report produced in February 2020, "Study of the health status of karst rivers in relation to anthropogenic pressures on their basins" (Badot and Degiorgi 2020). The studies on which this report is based sought to explain the algal blooms in the rivers of the Jura plateaus, which have been observed over the last 20 years. They are accompanied by a rarefaction and, in some cases, an extinction of aquatic species considered sensitive, such as large plecopterans, the white-clawed crayfish (*Austropotamobius pallipes*), large mayflies and trichopterans, and, as a result, a fivefold decrease in salmonid biomass in less than 20 years. It has become apparent that the main cause of these losses is the increase in nitrate and phosphate concentration, as well as multiple contaminations by phytosanitary products, biocides and active substances from veterinary medicines. In fact, the intensification of agricultural practices over the last few decades is blamed, even if it is inaccurate to attribute it to the Comté alone. Examination of the figures in detail shows that it is in fact an overall increase in cheese production, in which, since the early 1990s, a shift away from Emmental production has been more than compensated for by an increase in the production of Comté, Morbier and Mont-d'Or PDOs (Brischoux *et al.* 2015). This is also the consequence of the switch from a "manure" to a "slurry" effluent treatment system, the latter, owing to its high ammonia fraction, behaving as a mineral fertilizer. We have seen in previous sections that the increase in the dairy performance of the herd (genetics, number of cows, etc.), from the 1960s to the 1970s, led to an increase in grass production and allowed the development of large populations of water voles and common voles. It is therefore very likely that the measures that will be taken to reduce the ecological externalities linked to current cheese production will have consequences on, on the one hand, the technical itineraries of grass production (reduction of manure spreading, attention to biodiversity, limitation of livestock density, etc.) and, on the other hand, the means of controlling grassland vole populations (supervision of soil turning, which is part of the "toolkit" but leads to the rapid mineralization of organic matter and therefore, if no precautions are taken, can contribute to the removal of carbon from the soil and to the eutrophication of rivers, etc.). It is not yet known how vole populations will respond to these changes, the extent of which is currently difficult to predict. However, the necessary ecological transition, which will lead to greater attention being paid to the protection of water, which is itself becoming scarcer as a result of climate change, may once again lead to profound changes in the characteristics of the socioecosystems of the Jura within a few years.

1.4. What methodological lessons can be drawn from this experience?

It has taken about 30 years to describe and understand the factors that determine vole outbreaks in the Jura Massif in PDO grassland areas and to establish the basis for controlling them. Most of this work (research and experimental validation) was completed in 2006, at the end of a State-Region planning contract to which we will return. One may ask whether this period could have been shortened and, if so, what were the obstacles that prevented it. To answer this question, it is necessary to return to the organization of the research and the applications of its results, as described above.

The first incompressible point is the duration of an outbreak cycle. From 5 to 7 years for the water vole, non-cyclic fluctuations for the common vole, but with declines synchronized with that of the water vole and whose outbreaks, even if they can last several years, appear chaotic (Giraudoux *et al.* 2019). Thirty years therefore only offer the opportunity, locally, in longitudinal monitoring, of about five observation replicates. This can be compensated for by a synchronic approach: comparing at a given time locations that are at known stages of a cycle. This approach, which has in fact been implemented, assumes that adequate information is available to make the comparisons. It is therefore necessary that a regional observation network exists that has gathered this information during at least one outbreak cycle, and that the next one (the one during which comparisons are planned) is perfectly identical to the first. This is almost never the case. It is therefore not possible to ensure full comparability of observations (the landscapes and the history of the populations compared are never exactly the same). Moreover, given the logistical burden (trapping, intervention in the environment, etc.) and the limited human and financial resources allocated to this type of study, this reduced comparability cannot be compensated for by the acquisition of large statistical series for each category. Ecology does not benefit from the same volume of funding as that allocated to nuclear physics, astronomy or the conquest of space.

The second point, which is difficult to compress, concerns the sociological evolution of the research network, the formation of which we will return to later. Three periods can be distinguished:

– 1979–1996 (18 years!), a period during which the agricultural demand was, it has to be said, to wait for some kind of magic bullet that would have relieved the farmer of the problem, through a minimum of operations that did not involve major changes in practices. The leitmotif, not always kind to the researchers, was: “What is research doing?” The 1980s was also the period of collection of the first observational data on protocols, which were essential to objectively answer the questions that arose. From 1992 to 1996, a regional contract (Franche-Comté regional council), entitled “Rodents and the management of rural areas”, finally allowed the implementation of action-research protocols bringing together the various parties and the publication of the first results.

– 1997–2000 was a period when the leitmotifs were, with regard to the deleterious effects of chemical treatments on non-target fauna: “You have to choose between the buzzards and the farmers”, “You can’t make an omelette without breaking eggs” and, with regard to the results of research: “The results don’t suit us; one has to find something else”. Consequently, these results were little or not taken into account, and the chemical treatments, too late, in the peak of the outbreak, were extended to even larger areas. The foreseeable crisis generated by these treatments, with a massacre of hundreds of rodent predators (buzzards, foxes, etc.) and animals consuming poisoned bait (wild boar, etc.) in 1998, made headlines in the national media, mobilized nature protection organizations and, locally, departmental hunters' federations. This situation led to a vigorous response from the Prefect and the regional director of agriculture and forestry at the time, followed by a change in the leadership of the technical agricultural bodies in charge of the issue, resulting in the establishment of a State-Region Planning Contract (CPER) from 2001 to 2006 (i.e. one outbreak cycle) and a steering committee coordinating the funding of research.

– From 2001 onward, therefore, the CPER brought the conflicting parties together on the basis of jointly constructed research projects and the creation of experimental zones (Delattre and Giraudoux 2009). This period of joint work, with the results obtained and their validation in the experimental zones being taken into account, quickly led to a new positioning of the representatives of the agricultural world. Charles Schelle, president of Fredon Franche-Comté, summed it up in May 2015: "We must keep in mind our threefold requirement for our agriculture to be successful: the social aspect, the economic aspect and the environmental issue must be part of our requirements. These requirements are all the more important as the fight against the vole is never-ending. This is why the work done with research is fundamental, because researchers and farmers have a shared destiny."

It is therefore certain that at least 10 years could have been gained over the 30 years if financial support for research had been less strictly linked to crises, less discontinuous and even less often interrupted at each interval between crises. This perennial precariousness systematically leads to opportunistic dispersion and thematic acrobatics, and thus to an absurd waste of energy in order to adapt conjuncturally to the objectives of short-term calls for tenders available on other issues (according to the thematic trends of the moment), almost irrespective of the content, while trying to keep the guiding line as best as possible over the long term (Figure 1.18). It is impossible to find funding for this alone: "To answer the questions asked, data have been collected for years and it is necessary to continue to do so for years to come."

It is possible, however, that long periods of time and crises are essential for ideas to percolate, for actors to question their practices with stasis and acceleration, and for consensus to finally be established and lead to action. "Social" time, the time necessary for the evolution of each party's positions, is perhaps not one of the compressible variables. One of the major difficulties in this type of action research is to understand and accept that the time of the vole (and its predators), the researcher and the cattle breeder are not the same (Figure 1.19).

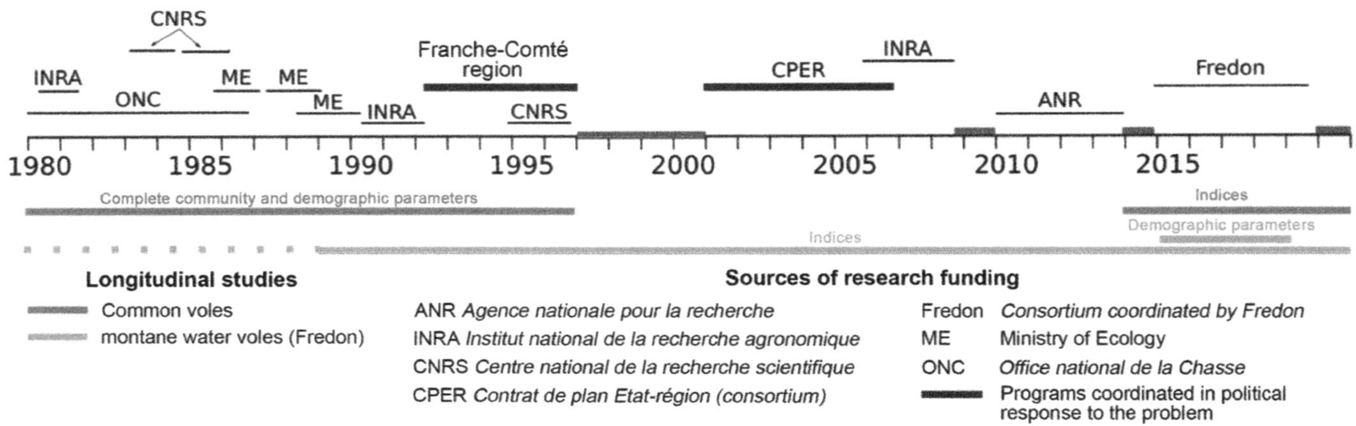

Figure 1.18. *Discontinuity, dispersion and short-term sources of research funding, despite the fact that research was conducted over the long term. The red segments mark the periods without any funding. For a color version of this figure, see www.iste.co.uk/giraudoux/socioecosystems.zip*

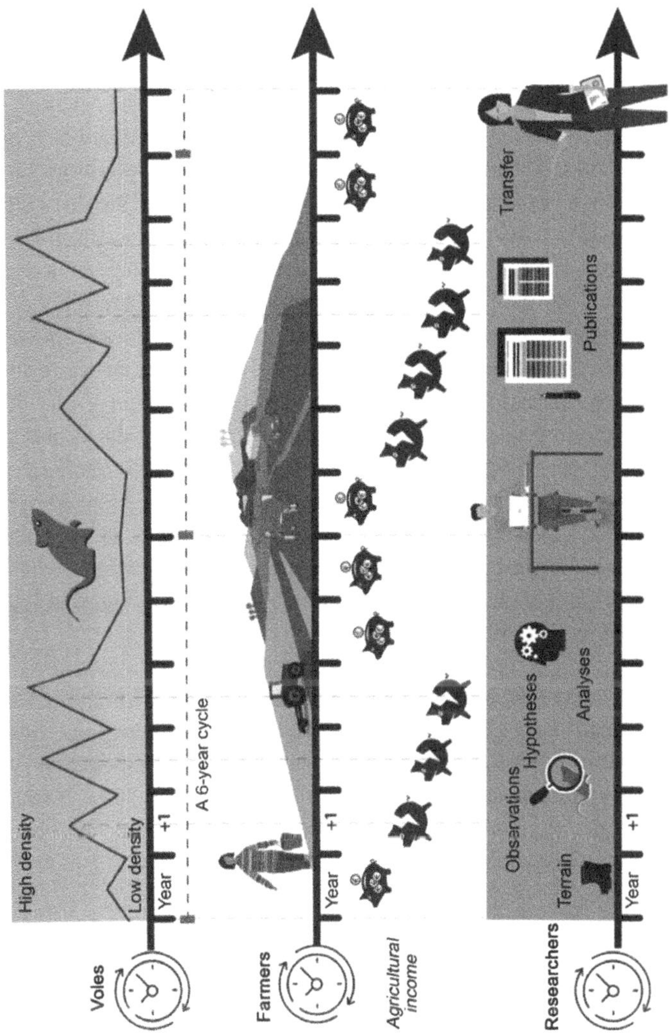

Figure 1.19. *The time of the voles, the farmer and the researcher (infographics by Estelle Franc). For a color version of this figure, see www.iste.co.uk/giraudoux/socioecosystems.zip*

Was this research multidisciplinary? There is no doubt that it was, but to varying degrees. It is customary to distinguish between "major" and "minor" multidisciplinarity. The "major" multidisciplinarity refers to that which brings together disciplines that are far remote from one another, such as the humanities and the so-called "hard" sciences. The "minor" multidisciplinarity was almost permanent, bringing together ecologists, health scientists, geographers, modelers (mathematicians, computer scientists) and tools ranging, to use a cliché, from satellites (space imagery) to microsatellites (molecular genetics). The "major" was also present with the socio-anthropological studies presented in section 1.3.3, but it must be admitted that, like the others, this was an opportunistic addition "in the course of research" to answer the emerging question of the origin of the difficulty of transferring results to cattle breeders. The incorporation of socio-anthropological and economic questions, from the outset, would probably have helped to better anticipate this difficulty, as well as to better understand the functioning of the socioecosystem, as we shall see in Chapter 2. Honesty compels us to add, however, that despite the institutional exhortations to multidisciplinarity, no structure in the field in question has the means to mobilize a sufficient budget to carry out all the studies that would be required in the long term. Figure 1.18 clearly shows that over the nearly 40 years of the timeline, there have been only two institutionally coordinated multidisciplinary and multisectoral actions with a specific budget, both in response to crises, and covering a total of only 10 years. Neither incorporated a socio-anthropological or economic component as such. However, beyond what has been already presented in this chapter, the value of integral socio-anthropological studies is also discussed in Chapter 2 and Chapter 4.

The research was also, and equally critically, multisectoral, involving professions interested in the results, but not necessarily tasked with conducting research (Figure 1.20). It was during the periods when research and implementation were coordinated by a specific multisectoral consortium (1992–1996, then 2001–2006) that the most important results were achieved, results that cannot, however, be obtained without the long-term continuity of pre- and post-program observations.

These *ad hoc* umbrella consortia are indeed essential to coordinating the mobilization of the necessary funds, both for research and for application, and thus reflect the overall integration of all elements of the system (Figure 1.16). Without this, the researcher is faced with an exhausting and sterile game of ping-pong between potential funders, with the environment administration passing on to agriculture the burden of dealing with what it sees as an agricultural problem, and agriculture passing on to the environment what it sees as a species protection problem, or to health what it sees as a public health problem, and the public health sector offloading what it considers to be primarily an agricultural problem, all of them considering that they do not have to finance research because there is a Ministry of Research responsible for it – a Ministry of Research that is of the view that it is not its vocation to finance the operation of environmental observatories, because it does not have the financial means to do so anyway. However, a consortium must not become a bureaucracy and must itself be supportive and adaptive. Multisectoriality is also imposed by the fact that research labs almost never have the human and material resources to maintain long-term observation systems in ecology, especially in the field of animal populations. On the other hand, long-term agronomic monitoring (agricultural pests), sanitary monitoring (diseases) and monitoring related to the knowledge and protection of biodiversity (fauna and flora conservatories and associations) are a mission for many regional organizations, with which intersectoral collaborative research is possible.

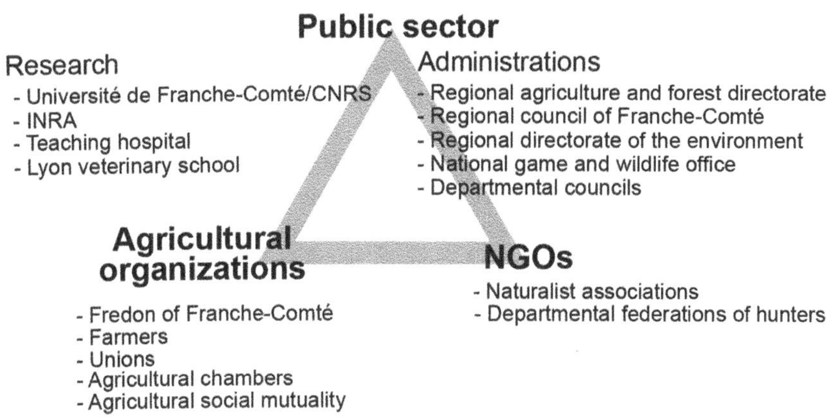

Figure 1.20. *Professional sectors involved in the research*

Since the will to last is not guaranteed by institutions, the critical question of persistence over time arises. It is, therefore, based on the question of the humans that pursue it. This structural problem was already raised in the last century: "Most long-term studies are still from Europe and the USSR[2] while North American series cover mainly the few years needed for a doctoral thesis" (Hansson and Henttonen 1988). We cannot provide any examples of long-term continuity provided by a succession of short contracts. The accelerated turnover of positions in administrations, short fixed-term contracts and the annualization of budgets, for example, make program continuity beyond a few years problematic. Duration here implies the personal commitment of several generations of researchers and their partners over several decades, and their tiling is essential. We cannot avoid

2 Author's note: where permanent positions were the most common situation.

providing here a few decisive *ad hominem* examples of statutory positions whose persistence and synergy over the long term have allowed the continuity of the work. Pierre Delattre was the initiator of the research in 1979, especially the studies on the common vole and the community of small mammals of similar size. As a researcher at INRAE, he had a permanent position that enabled him to collect basic data for more than 10 years, without publishing on these data (the first international article was published in 1992). He was thus able to lead part of the research, ensure the continuity of the fieldwork and seek funding until his retirement in 2006 (28 years). Patrick Giraudoux, then a high school biology-geology teacher, joined the team in 1987 and was recruited as an associate professor at the University of Burgundy in 1992, thus with permanent status. He retired in 2019. Trained in Pierre Delattre's team as a PhD student, he took charge, from the beginning, of the scientific animation of a complementary part of the long-term research and continues to do so under the status of emeritus professor (35 years as of 2021). Michel Habert was director of the Regional Plant Protection Service of Franche-Comté (Ministry of Agriculture) until 2002 (over 30 years) and his successor, Philippe Guillemard, until 2017 (15 years). Denis Truchetet, agricultural engineer, was trained in 1989 by Pierre Delattre and Patrick Giraudoux on the subject, and joined the Plant Protection Service from 1995 to 2015 (20 years). During this period, he was the vertebrate pest expert for the Ministry of Agriculture. Régis Defaut, technician, joined Fredon in 1989 and was the main point of contact between this organization and the researchers and the leading figure behind experiments conducted with the cattle breeders, until his death in 2008 (20 years). Geoffroy Couval joined Fredon in 2006 as an environmental engineer (16 years to date) and has been, since 2014, in a liaison position (time shared between Fredon and university) and technical reference point for voles for the Fredon France network. All of these people, who regularly met in the field for joint work and to compare their experiences, formed the core group that made it possible to continue collecting data and conducting research over the long term, even during periods of funding shortfall.

KEY POINTS. –

The composition and structure of the landscape, determined by the socio-economic constraints on the socioecosystem, determine the dynamics of vole populations.

The details of the causes of fluctuations in abundance are still poorly understood, but they appear to be multifactorial, embedded in a systemic conceptual model, and hierarchical across space and time scales. They include the proportion of optimal habitats in the landscape, predation and disease, possibly combined with stress.

Changes in practices, based on systemic hypotheses derived from long-term observational (eco-epidemiological) studies, constitute quasi-experiments, that is tests of the validity of these hypotheses. This eco-epidemiological approach is at the heart of the adaptive management of vole populations and their effects on other components of the system.

Non-conceptual modeling does not allow the system to be modeled in its entirety. It does, however, make it possible to explore certain local sub-systems when they are framed (in space and time) by observational studies, to verify the theoretical sufficiency of certain hypotheses, to make certain hidden assumptions explicit and to estimate the possible values of certain non-measurable variables by inverse resolution.

The relevant space and time scales (several tens of thousands of square kilometers and several decades) for understanding the combined functioning of prey and predator populations and human societies are beyond the means of observation of researchers alone. Multisectoral partnerships make it possible to study such systems. They are facilitated by the establishment of consortia that coordinate sectors (including government agencies) around management issues. The selection and long-term professional stability of people motivated by such studies and multidisciplinarity are indispensable.

Multifactoriality requires multidisciplinarity to varying degrees. It can only be coherent if a shared question is answered first. Here: what are the mechanisms and consequences (proximal and ultimate) of regional outbreaks of grassland rodents?

1.5. Acknowledgments

Thanks to Geoffroy Couval, Estelle Franc, Marc Montadert and Francis Raoul for their review of this chapter.

1.6. References

Andersson, M. and Erlinge, S. (1977). Influence of predation on rodent populations. *Oikos*, 29, 591–597.

Badot, P.-M. and Degiorgi, F. (2020). 2012–2020 : Bilan d'étude sur l'état de santé des rivières karstiques. Report, UMR 6249 CNRS-UFC usc INRA Chrono-environ-nement/université de Franche-Comté, Besançon [Online]. Available at: https://chrono-environnement.univ-fcomte.fr/recherche/themes-actions-et-projets/article/2012-2020-bilan-d-etude-sur-l-etat-de-sante-des-rivieres-karstiques [Accessed August 30, 2020].

Baudrot, V., Perasso, A., Fritsch, C., Raoul, F. (2016a). Competence of hosts and complex foraging behavior are two cornerstones in the dynamics of trophically transmitted parasites. *Journal of Theoretical Biology*, 397, 158–168.

Baudrot, V., Perasso, A., Fritsch, C., Giraudoux, P., Raoul, F. (2016b). The adaptation of generalist predators' diet in a multi-prey context: Insights from new functional responses. *Ecology*, 97(7), 1832–1841.

Baudrot, V., Fernandez-de-Simon, J., Cœurdassier, M., Couval, G., Giraudoux, P., Lambin, X. (2020). Trophic transfer of pesticides: The fine line between predator-prey regulation and pesticide-pest regulation. *Journal of Applied Ecology*, 57(4), 806–818.

Bernard, N., Michelat, D., Raoul, F., Quéré, J.-P., Delattre, P., Giraudoux, P. (2010). Dietary response of Barn Owls (*Tyto alba*) to large variations in populations of common voles (*Microtus arvalis*) and European water voles (*Arvicola terrestris*). *Canadian Journal of Zoology*, 88(4), 416–426.

Berny, P.J., Buronfosse, T., Buronfosse, F., Lamarque, F., Lorgue, G. (1997). Field evidence of secondary poisoning of foxes (*Vulpes vulpes*) and buzzards (*Buteo buteo*) by bromadiolone, a 4-year survey. *Chemosphere*, 35(8), 1817–1829.

Berthier, K., Piry, S., Cosson, J.-F., Giraudoux, P., Foltête, J.-C., Defaut, R., Truchetet, D., Lambin, X. (2014). Dispersal, landscape and travelling waves in cyclic vole populations. *Ecology Letters*, 17(1), 53–64.

Brischoux, A., Gurtner, J.-P., Gurtner, M., Matton, M. (2015). Il est temps de "comté" la fruitière. Ville de Pontarlier, Ornans [Online]. Available at: https://catalogue.bnf.fr/ark:/12148/cb44446257h.

Buckle, A.P. and Smith, R.H. (eds) (2015). *Rodent Pests and their Control*, 2nd edition. CAB International, Wallingford/Boston.

Cerqueira, D., De Sousa, B., Gabrion, C., Giraudoux, P., Quéré, J.-P., Delattre, P. (2006). Cyclic changes in the population structure and reproductive pattern of the water vole, Arvicola terrestris Linnaeus, 1758. *Mammalian Biology*, 71(4), 193–202.

Charbonnel, N., Chaval, Y., Berthier, K., Deter, J., Morand, S., Palme, R., Cosson, J.-F. (2008). Stress and demographic decline: A potential effect mediated by impairment of reproduction and immune function in cyclic vole populations. *Physiological and Biochemical Zoology*, 81(1), 63–73.

Charbonnel, N., Deter, J., Gabrion, C. (2009). Parasitisme. In *Le Campagnol terrestre : prévention et contrôle des pullulations*. QAE, Versailles.

Chevret, P., Renaud, S., Helvaci, Z., Ulrich, R.G., Quéré, J.-P., Michaux, J.-R. (2020). Genetic structure, ecological versatility, and skull shape differentiation in Arvicola water voles (Rodentia, Cricetidae). *Journal of Zoological Systematics and Evolutionary Research* [Online]. Available at: http://onlinelibrary.wiley.com/doi/abs/10.1111/jzs.12384 [Accessed May 15, 2020].

Cœurdassier, M., Berny, P., Couval, G., Decors, A., Jacquot, M., Queffelec, S., Quintaine, T., Giraudoux, P. (2014a). Évolutions des effets non intentionnels de la lutte chimique contre le campagnol terrestre sur la faune sauvage et domestique. *Fourrages*, December, 327–335.

Cœurdassier, M., Riols, R., Decors, A., Mionnet, A., David, F., Quintaine, T., Truchetet, D., Scheifler, R., Giraudoux, P. (2014b). Unintentional wildlife poisoning and proposals for sustainable management of rodents. *Conserv. Biol.*, 28(2), 315–321.

Couval, G. and Truchetet, D. (2014). Le concept de lutte raisonnée : combiner des méthodes collectives contre le campagnol terrestre afin de conserver une autonomie fourragère. *Fourrages*, December, 343–347.

Couval, G., Truchetet, D., Cœurdassier, M., Michelin, Y., Jacquot, M., Giraudoux, P., Berny, P., Decors, A., Morlans, S., Quintaine, T., Renaude, R. (2013). La lutte raisonnée contre le campagnol terrestre. *Phytoma*, 33–36.

Delattre, P. and Giraudoux, P. (2005). Enjeux phytosanitaires pour l'agriculture et l'environnement : pesticides et biopesticides, agriculture durable. In *Le contrôle des rongeurs non commensaux : impasse du tout chimique et perspectives de lutte intégrée*, Regnault-Roger, C., Fabre, G., Philogène, B. (eds). Tec & Doc, Cachan.

Delattre, P. and Giraudoux, P. (eds) (2009). *Le campagnol terrestre. Prévention et contrôle des pullulations*. QUAE, Versailles.

Delattre, P., Damange, J.-P., Roger, M. (1986). Analyse comparative des fluctuations de populations de rongeurs. Rôle de la prédation et de la structure des paysages. In *Colloque CNRS de biologie des populations*. CNRS.

Delattre, P., Damange, J.-P., Pascal, M., Habert, M. (1988). Rôle de la prédation et influence de la structure du paysage agraire sur le développement des cycles d'abondance d'Arvicola terrestris scherman. *EPPO Bulletin*, 18, 415–422.

Delattre, P., Giraudoux, P., Damange, J.-P., Quéré, J.-P. (1990). Recherche d'un indicateur de la cinétique démographique des populations du campagnol des champs (*Microtus arvalis*). *Revue d'écologie – Terre et vie*, 45, 375–384.

Delattre, P., Giraudoux, P., Baudry, J., Musard, P., Toussaint, M., Truchetet, D., Stahl, P., Poule, M.-L., Artois, M., Damange, J.-P., Quéré, J.-P. (1992). Land use patterns and types of common vole (*Microtus arvalis*) population kinetics. *Agriculture, Ecosystems & Environment*, 39(3), 153–168.

Delattre, P., Giraudoux, P., Baudry, J., Quéré, J.-P., Fichet, E. (1996). Effect of landscape structure on Common Vole (*Microtus arvalis*) distribution and abundance at several space scales. *Landscape Ecology*, 11(5), 279–288.

Delattre, P., De Sousa, B., Fichet, E., Quéré, J.-P., Giraudoux, P. (1999). Vole outbreaks in a landscape context: Evidence from a six-year study of *Microtus arvalis*. *Landscape Ecology*, 14, 401–412.

Delattre, P., Clarac, R., Melis, J.P., Pleydell, D.R.J., Giraudoux, P. (2006). How moles contribute to colonization success of water voles in grassland: Implications for control. *Journal of Applied Ecology*, 43(2), 353–359.

Delattre, P., Morellet, N., Codreanu, P., Miot, S., Quéré, J.-P., Sennedot, F., Baudry, J. (2009). Influence of edge effects on common vole population abundance in an agricultural landscape of eastern France. *Acta Theriologica*, 54(1), 51–60.

Duhamel, R., Quéré, J.-P., Delattre, P., Giraudoux, P. (2000). Landscape effects on the population dynamics of the fossorial form of the water vole (Arvicola terrestris sherman). *Landscape Ecology*, 15, 89–98.

Elton, C. (1942). *Voles, Mice and Lemmings: Problems in Population Dynamics*. Clarendon Press, Oxford [Online]. Available at: https://www.cambridge.org/core/journals/polar-record/article/voles-mice-and-lemmings-problems-in-population-dynamics-by-charles-elton-clarendon-press-oxford-1942-496-pp-64-tables-22-figures-frontispiece-30s/B069B5114C46BEFF295609F9E3974672 [Accessed June 4, 2020].

Fichet-Calvet, E., Pradier, B., Quéré, J.-P., Giraudoux, P., Delattre, P. (2000). Landscape composition and vole outbreaks: Evidence from an eight-year study of Arvicola terrestris scherman. *Ecography*, 23, 659–668.

Foltête, J.-C. and Giraudoux, P. (2012). A graph-based approach to investigating the influence of the landscape on population spread processes. *Ecological Indicators*, 18, 684–692.

Foltête, J.-C., Clauzel, C., Vuidel, G. (2012). A software tool dedicated to the modelling of landscape networks. *Environmental Modeling & Software*, 38, 316–327.

Foltête, J.-C., Couval, G., Fontanier, M., Vuidel, G., Giraudoux, P. (2016). A graph-based approach to defend agro-ecological systems against water vole outbreaks. *Ecological Indicators*, 71, 87–98.

Gilg, O., Sittler, B., Sabard, B., Hurstel, A., Sane, R., Delattre, P., Hanski, L. (2006). Functional and numerical responses of four lemming predators in high arctic Greenland. *Oikos*, 113(2), 193–216.

Giraudoux, P. (2014). Équilibre écologique et santé des écosystèmes : entre mythe biologique et consensus social. In *Nature ou culture*, *Les colloques de l'Institut Universitaire de France*. Université de Saint-Étienne, Saint-Étienne.

Giraudoux, P., Delattre, P., Quéré, J.-P., Damange, J.-P. (1994). Structure and kinetics of rodent populations in a region under agricultural land abandonment. *Acta Oecoligica*, 15(4), 385–400.

Giraudoux, P., Pradier, B., Delattre, P., Deblay, S., Salvi, D., Defaut, R. (1995). Estimation of water vole abundance by using surface indices. *Acta Theriologica*, 40(1), 77–96.

Giraudoux, P., Delattre, P., Habert, M., Quéré, J.-P., Deblay, S., Defaut, R., Duhamel, R., Moissenet, M.-F., Salvi, D., Truchetet, D. (1997). Population dynamics of fossorial water vole (Arvicola terrestris scherman): A land usage and landscape perspective. *Agriculture Ecosystems and Environment*, 66, 47–60.

Giraudoux, P., Tremollières, C., Barbier, B., Defaut, R., Rieffel, D., Bernard, N., Lucota, E., Berny, P. (2006). Persistence of bromadiolone anticoagulant rodenticide in Arvicola terrestris populations after field control. *Environmental Research*, 102(3), 291–298.

Giraudoux, P., Couval, G., Levret, A., Mougin, D., Delavelle, A. (2017). Suivi à long terme d'une zone de pullulation cyclique de campagnols terrestres : le contrôle raisonné des populations est possible ! *Fourrages*, 169–176.

Giraudoux, P., Villette, P., Quéré, J.-P., Damange, J.-P., Delattre, P. (2019). Weather influences *M. arvalis* reproduction but not population dynamics in a 17-year time series. *Scientific Reports*, 9(1), 13942.

Giraudoux, P., Levret, A., Afonso, E., Coeurdassier, M., Couval, G. (2020). Numerical response of predators to large variations of grassland vole abundance and long-term community changes. *Ecology and Evolution*, 10(24), 14221–14246.

Grolleau, G., Lorgue, G., Nahas, K. (1989). Toxicité secondaire, en laboratoire, d'un rodenticide anticoagulant (bromadiolone) pour prédateurs de rongeurs champêtres : buses variable (Buteo buteo) et hermine (Mustela erminea). *Bulletin OEPP*, 19, 633–648.

Habert, M. (1988). Fondement des réseaux d'alerte et évolution des densités du campagnol terrestre en France. *EPPO Bulletin*, 18, 423–427.

Hansson, L. and Henttonen, H. (1988). Rodent dynamics as community processes. *Trends in Ecology and Evolution*, 3(8), 195–200.

Jacquot, M., Cœurdassier, M., Couval, G., Renaude, R., Pleydell, D., Truchetet, D., Raoul, F., Giraudoux, P. (2013). Using long-term monitoring of red fox populations to assess changes in rodent control practices. *Journal of Applied Ecology*, 50(6), 1406–1414.

Korpimaki, E., Norrdahl, K., Klemola, T., Pettersen, T., Stenseth, N.C. (2002). Dynamic effects of predators on cyclic voles: Field experimentation and model extrapolation. *Proceedings of the Royal Society B:Biological Sciences*, 269(1495), 991–997.

Krebs, C.J. (2013). *Population Fluctuations in Rodents*. The University of Chicago Press, Chicago/London.

Leclercq, B., Desbrosses, R., Giraudoux, P. (1997). Cycles démographiques du campagnol terrestre (*Arvicola terrestris*), et reproduction et cinétique des Tetraonidae du massif du Risoux, Jura, France. *Gibier et faune sauvage*, 14, 31–47.

Legifrance (2014). Arrêté du 14 mai 2014 relatif au contrôle des populations de campagnols nuisibles aux cultures ainsi qu'aux conditions d'emploi des produits phytopharmaceutiques contenant de la bromadiolone [Online]. Available at: https://www.legifrance.gouv.fr/affichTexte.do?cidTexte=JORFTEXT000029039908 [Accessed January 26, 2019].

Lidicker, W.Z.J. (1988). Solving the enigma of microtine "cycles". *Journal of Mammology*, 69, 225–235.

Lindenmayer, D., Likens, J., Haywood, A., Miczis, L. (2011). Adaptive monitoring in the real world: Proof of concept. *Trends in Ecology and Evolution*, 26(12), 641–646.

Marilleau, N., Lang, C., Giraudoux, P. (2018). Coupling agent-based with equation-based models to study spatially explicit megapopulation dynamics. *Ecological Modeling*, 384, 34–42.

Michelin, Y., Coulaud, F., Morlans, S., Ingrand, S. (2014). Pullulations de campagnols terrestres : perception du phénomène, impact sur les systèmes bovins laitiers de Franche-Comté et perspectives pour l'action. *Fourrages*, 220, 285–290.

Morilhat, C., Bernard, N., Bournais, C., Meyer, C., Lamboley, C., Giraudoux, P. (2007). Responses of Arvicola terrestris scherman populations to agricultural practices, and to Talpa europaea abundance in eastern France. *Agriculture Ecosystems and Environment*, 122(3), 392–398.

Morilhat, C., Bernard, N., Foltête, J.-C., Giraudoux, P. (2008). Neighbourhood landscape effect on population kinetics of the fossorial water vole (Arvicola terrestris scherman). *Landscape Ecology*, 23(5), 569–579.

Morlans, S. and Michelin, Y. (2014). Comment limiter la lutte chimique contre les campagnols terrestres ? Enseignements d'une analyse agro-anthropologique dans 4 territoires montagneux français. *Fourrages*, 357–363.

Oli, M.K. (2003). Population cycles of small rodents are caused by specialist predators: Or are they? *Trends in Ecology & Evolution*, 18(3), 105–107.

Oli, M.K. (2019). Population cycles in voles and lemmings: State of the science and future directions. *Mammal Review*, 49(3), 226–239.

Pascal, M., Pradier, B., Habert, M., Vallat, H. (1985). Le Campagnol terrestre : premiers résultats des réseaux d'observation des régions Franche-Comté et Auvergne-Limousin (1983–1984). *La Défense des végétaux*, 231, 1–8.

Popper, K.R. (2006). *Conjectures et réfutations : la croissance du savoir scientifique*. Payot, Paris.

Quéré, J.-P., Raoul, F., Giraudoux, P., Delattre, P. (2000). An index method applicable at landscape scale to estimate relative population densities of the common vole (*Microtus arvalis*). *Revue d'écologie – Terre et vie*, 55, 25–32.

Raoul, F., Michelat, D., Ordinaire, M., Decote, Y., Aubert, M., Delattre, P., Deplazes, P., Giraudoux, P. (2003). *Echinococcus multilocularis*: Secondary poisoning of fox population during a vole outbreak reduces environmental contamination in a high endemicity area. *International Journal for Parasitology*, 33(9), 945–954.

Raoul, F., Deplazes, P., Rieffel, D., Lambert, J.-C., Giraudoux, P. (2010). Predator dietary response to prey density variation and consequences for cestode transmission. *Oecologia*, 164(1), 129–139.

Ríos-Saldaña, C.A., Delibes-Mateos, M., Ferreira, C.C. (2018). Are fieldwork studies being relegated to second place in conservation science? *Global Ecology and Conservation*, 14, e00389.

Sage, M. (2008). Bromadiolone transfer (baits/soils – grassland rodents – foxes): Environmental study of the persistence and non-invasive method for exposure monitoring. PhD Thesis, Université de Franche-Comté, Besançon [Online]. Available at: https://tel.archives-ouvertes.fr/tel-00404261 [Accessed January 4, 2022].

Sage, M., Cœurdassier, M., Defaut, R., Lucot, E., Barbier, B., Rieffel, D., Berny, P., Giraudoux, P. (2007). How environment and vole behaviour may impact rodenticide bromadiolone persistence in wheat baits after field controls of Arvicola terrestris? *Environmental Pollution*, 148(1), 372–379.

Schouwey, B., Cassez, M., Couval, G., Fontanier, M., Michelin, Y. (2014). Campagnol terrestre et lutte raisonnée : quels impacts économiques sur les exploitations en AOP Comté ? *Fourrages*, December, 297–302.

Telfer, S., Lambin, X., Birtles, R., Beldomenico, P., Burthe, S., Paterson, S., Begon, M. (2010). Species interactions in a parasite community drive infection risk in a wildlife population. *Science*, 330(6001), 243–246.

Telfer, S., Lambin, X., Birtles, R., Beldomenico, P., Burthe, S., Paterson, S., Begon, M. (2011). Microbe interactions undermine predictions – Response. *Science*, 331(6014) [Online]. Available at: https://science.sciencemag.org/content/331/6014/145.1 [Accessed January 4, 2022].

Villette, P. (2018). Spatial and temporal characteristics of bacterial parasite communities in outbreaking fossorial water vole (*Arvicola terrestris*) populations: Static uniformity or dynamic heterogeneity? PhD Thesis, Université de Franche-Comté, Besançon.

Villette, P., Afonso, E., Couval, G., Levret, A., Galan, M., Goydadin, A.-C., Cosson, J.-F., Giraudoux, P. (2020). Spatio-temporal trends in richness and persistence of bacterial communities in decline-phase water vole populations. *Scientific Reports*, 10(1), 9506.

2

The Pollution of a River: A Sociological Investigation of Knowledge and Expertise

Simon CALLA

Agroecology, Université Bourgogne Franche-Comté, Besançon/Dijon, France

GUIDING QUESTIONS. –

What are the contributions of socio-anthropology to the study of so-called "pollution" problems?

What are the different types of actors and knowledge involved in the discovery of cases of fish mortality?

How do these different actors and knowledge participate in the "settlement" and redefinition of the problematic situation?

How are the relations between the different groups of investigators working on the pollution of the Loue organized?

2.1. Introduction

Between 2009 and 2015, several major rivers in the Doubs department of France – the Loue, the Doubs and the Dessoubre (Figure 2.1) – regularly carried fish corpses. This was particularly true of trout and grayling. The situation soon gave rise to many questions. Individuals taking on the role of river defenders wondered who was responsible for this "river pollution", and scientists sought to understand the causes of this fish mortality. In addition, there were always differences in the

Socioecosystems,
coordinated by Patrick GIRAUDOUX.

location and the nature of the problem, as well as in the way the problem was dealt with. Moreover, always situated and differentiated according to the individuals' belonging to a "social world"[1] (Becker 2010; Latour 2010a), the different attempts to define the problem and its solutions led to the emergence of a dispute, or even a conflictual situation, between river defenders and farmers or local authority representatives. Faced with the exacerbation of tensions and seeking to control the social effects of these fish mortality events, representatives of the General Council and the Prefecture of the Doubs department organized public meetings entitled *Conférence départementale de la Loue et des rivières comtoises* (The Departmental Conference of the Loue and the Franche-Comté Rivers) (Figure 2.2). These meetings, which were intended to provide a peace offering (Monnerie 2018) between the various proponents, also involved representatives of a scientific group commissioned by government representatives to investigate the situation.

Figure 2.1. *The Loue River at Mouthier-Haute-Pierre (photo J.-P. Grandmont)*

In November 2013, while I was working on a doctoral research project, Dominique Jacques-Jouvenot, who was then director of the Laboratory of Sociology and Anthropology (LaSA-UFC/EA3189), asked me to accompany her to a meeting at the General Council where the issue of river pollution in the department would be

1 Think, for example, of the worlds of "fishing", "science", "administration", etc. We might add that Bruno Latour defines a "world" as a set of beings "that a particular culture binds together in practical forms of life" (Latour 2010a). We can add to Becker that these beings together form networks of chains of cooperation (Becker 2010).

discussed. Working for many years on the rural world and the agricultural profession (Jacques-Jouvenot 1989, 2001), she had been contacted by a hydrogeologist of the *Laboratoire Chrono-environnement* (CE-UFC-CNRS/UMR6249) to take part in the work carried out by the scientific group associated with the *Conférence départementale de la Loue et des rivières comtoises.* Particularly interested in what is called "environmental issues" and still hoping to obtain funding for the completion of my doctorate, it was without really knowing what to expect, but with pleasure, that I accepted her invitation.

Figure 2.2. *Second Conférence départementale de la Loue et des rivières comtoises, Fort de Beauregard, April 11, 2014, Besançon (photo S. Calla)*

Whether on foot, by bicycle or by car, I regularly passed the headquarters of the Doubs General Council, but this was the first time I was invited to enter. When we arrived, about 15 people had taken their seats in a meeting room on the first floor. They were all gathered to discuss the phenomenon of large-scale fish deaths:

– *the spokesman for a river defense collective* spoke of "nutrients", nitrates and phosphates that he accused of contributing to the eutrophication of rivers. He also pointed to farmers and public authorities as being responsible for this "deplorable situation";

– *the representative of the Chamber of Agriculture* recalled the steps taken to encourage farmers to move toward a "more environmental" activity: the reduction in the quantities of fertilizers imported into the territory and investments to modernize livestock buildings, etc.;

– *the head of the "Environment and rural space" department of the Doubs General Council* mentioned a project for the creation of new spreading plans in order to better manage agricultural practices and the proper functioning of most of the wastewater treatment plants in the watershed;

– *the head of the "Water, risks, nature, forest" service of the Departmental Directorate of the Territories* reminded us of the standards defining the quality of the watercourses and reminded us that, even if fish died there, the Loue was considered to be in "good ecological condition" by the European Water Framework Directive (WFD);

– *the university ecologist* spoke of cyanobacteria, beings which had just been cleared but which had yet to be monitored, for although their danger to animals had been demonstrated, their action on humans was still unknown. It was necessary to be able to prevent their return;

– *the university hydrobiologist* confirmed what many fishermen had said: invertebrate and fish communities had deteriorated. He also mentioned the fact that the absence or non-standardization of data had not enabled the evolution of the river to be traced over time;

– *the university toxicologist* was talking about micropollutants from domestic or agricultural activities. If the quantities found in the water flow were low, it could be because of the transformation of the molecules used. According to him, it was necessary to change the methods and look in the river sediments;

– *the university hydrogeologist* spoke of the "karstic" character of the Jura Massif. Here, the subsoil was a "gruyère" made up of numerous underground circulations that needed to be better understood in order to know what was reaching the watercourses.

While observing this scene, I could not but notice the fact that many actors with different characteristics were concerned by the situation of the Franche-Comté rivers. Moreover, I noticed the diversity of points of view and the multiplicity of resources on which the participants in the meeting relied to qualify it. On this basis, the research project that I proposed to the scientific group for transmission to the scientific council of the Water Agency aimed at questioning the existence of pollution among different groups of stakeholders: fishermen, ecologists, farmers, local elected officials, industrialists, residents, experts, etc.

Unfortunately, no further action was taken on this research project. During a brief exchange with a member of the scientific council of the Water Agency, I learned that the absence of an economic dimension in the project proposed was one reason why it had not been retained. On the other hand, this person explained to me that he had previous experiences of working with anthropologists which, according to him, "had not led to anything interesting". Was I to understand that the analysis of environmental problems by the social sciences was only of interest insofar as it proposed an accounting and anthropocentric reduction of what Western common sense calls "nature"? Was I also to understand that the personal opinions of the members of scientific councils were admissible arguments to justify the refusal to fund a scientific project?

I was not, however, discouraged and undertook my work without funding. Although precarious, this situation allowed me a certain freedom in the conduct of my research. Moreover, the refusal to fund my project and, above all, the arguments that were returned to me strengthened my conviction that I needed to carry out a research project showing that sociology and anthropology had something to say about pollution and that this subject of research was not the property of so-called "nature scientists". If this was self-evident to sociologists and anthropologists (Douglas 2005; Van Staëvel 2006; Charvolin *et al.* 2015), it did not seem to be so to those in other disciplines, and I imagined – rightly, as evidenced by the present book and the research prospects following my defense – that the less narrow-minded might find it of interest.

Focusing on the emergence of a situation described as "a pollution problem", my thesis led me to the following conclusion: pollution is not an entity already present, waiting to be discovered and denounced for its action, but a political solution to a disorder faced in scattered order and in sometimes conflicting relationships by actors who are forced to engage in order to preserve their interests (Calla 2019). The fieldwork and the reflections constituting this work are also at the root of the discussion developed in this chapter. By paying attention to the way in which certain actors were concerned by the episodes of fish mortality in the Loue and their ways of acting, I will show that the diversity of knowledge and relations questions their capacity to engage in a collective activity of defining the problem and its solutions.

2.2. Different types of knowledge to qualify the situation

2.2.1. *At the water's edge, indexical knowledge*

Patrice, a fly fisherman, had decided to go to the banks of the Loue on this morning in May 2009. He wanted to take advantage of the calm of the dawn to "get a head start". What a surprise when he arrived at the water's edge. One, two, three, four, five... In total, he counted about 15 dead trout on the bank. By preventing Patrice from practicing his sport, the irruption of the corpses in his field of vision constituted a "trouble" that broke the course of action (Dewey 1967). "What has happened? Something is wrong," thought the fisherman. He took out a phone from his pocket and called the agent of the French National Office for Water and Aquatic Environments (*Office national de l'eau et des milieux aquatiques*, Onema[2]) in charge of the sector he was in. About an hour later, he arrived on the scene to observe the facts.

2.2.1.1. *Mycosis on the body of fish*

As their ancestors would have done thousands of years ago, the two men examined the scene for signs of what might have happened and what the situation was (Ginzburg 1980). Looking closely at the stream, they found that the fish were not only dead but also had a kind of white foam on their bodies. Those that were still alive looked the same. They stood out more from the bottom of the river because this "thing" broke the camouflage of their scales. "It's scary walking along the river. Every time you look, you see a white spot that moves," said the president of a fishing association later.

Figure 2.3. *Trout affected by Saprolegnia. The white filaments and spots correspond to the development of a mold of the genus Saprolegnia (photo T. Poulleau, Fédération de pêche du Doubs). For a color version of this figure, see www.iste.co.uk/giraudoux/ socioecosystems.zip*

2 Also known as Onema, the *Office national de l'eau et des milieux aquatiques* was, at the time of this work, the public body responsible for implementing and verifying the application of public water policy in France. Its informal name was also the "water police". It was merged in 2016 into the *Agence française pour la biodiversité*, and then in 2020 into the *Office français de la biodiversité*.

Apart from the presence of dead fish, the appearance of the corpses and those still alive became a source of concern for those who frequented the banks of the Loue. “[Some] are completely eaten away on all sides. It’s awful to look at,” said one. However, despite its spectacular nature, this situation was not “abnormal”, according to the Onema agent. The phenomenon and the name of the parasite that caused it were well known. It was saprolegniasis, or a disease caused by *Saprolegnia*.

“They are everywhere in the environment,” he explained to a journalist who came to interview him. For him, there was no reason to worry: “These are mycoses present in the water as microbes are present in the air; and as they are weakened by spawning, the fish contract them and develop them more easily.”[3]

The agent therefore considered the presence of this white foam on the bodies of the fish in the Loue to be “normal”. Nevertheless, if we think about the fact that “health is the luxury of being able to fall ill and to recover from it [and that] any illness is, on the contrary, the reduction of the power to overcome others” (Canguilhem 2009), it remained that this white foam could suggest the existence of a potential health problem in the fish. This was also the opinion of a fly fisherman for whom the presence of *Saprolegnia* was to be taken as a sign of river dysfunction.

> In a river where there is no water quality problem, there is never *Saprolegnia*. After spawning, there are some fish mortality, but they don’t have the fungus. If you see *Saprolegnia* growing like that, there’s a problem in the river.[4]

The presence of *Saprolegnia* in the Loue could then be interpreted as a sign of “pathological living standards” (Canguilhem 2009) corresponding to a lower quality of living environment.

2.2.1.2. *The number of dead fish*

While the presence of *Saprolegnia* drew attention to the fish, overall it did not seem to cause any particular concern. Some fishermen had observed the phenomenon on several occasions and placed it in the particular context of the end of the spawning season. Nevertheless, the number of dead fish was another sign that the “normality” of the situation was being questioned. “I had never seen sick fish on

3 “Troubles sur la Loue”, *L’Est républicain*, May 6, 2009.

4 Interview conducted with an academic geochemist and member of a river defense collective, October 30, 2015.

such a scale,"[5] lamented the president of a fishing association. The characterization of fish mortality episodes therefore also required a quantitative dimension.

When he went to the site to observe the reported facts, the Onema agent did not consider the situation "abnormal". Contrary to what Patrice thought, he thought that the presence of dead fish was not caused by the discharge of a product into the river. There was a significant number of dead fish in the water, but it was nothing like the hundreds of dead fish he had seen in the past following such events. Moreover, none of his colleagues had heard of an "acute pollution" event occurring. According to him, it was an episode of post-spawn mortality and the action of *Saprolegnia* on weakened individuals. "That said, it is true that this year the number of dead fish is higher than usual,"[6] he told a journalist before explaining the reasons that led him to this conclusion.

> The waters are low, quite cold and perhaps more polluted than in other years, since we had a long and harsh winter, the salting of the roads was sustained. So many salts and chlorides that drained residues on the roadways and constituted a pollutant flow in the water table. So it's a whole, but there is no particular pollution episode.[7]

According to the Onema agent, the situation could therefore be explained by poorer water quality, itself linked to the salting of roads during the past winter. The fish were, in a way, the collateral victims of the human reaction to a climatic hazard. However, the situation remained "normal" and did not correspond to "a particular pollution episode".

Nevertheless, more than just the presence of dead bodies in the water, it was the sheer number of dead fish that raised questions. "It's also a question of quantity," explained a hydrobiologist from the Doubs Fishing Federation. His colleague, a development officer, stated that although fishermen had long been expressing their concerns about the state of the Loue, it was "[when] the fish started to die rapidly [that] it awakened people's conscience a little, saying to themselves: 'Shit, I see dead fish everywhere'".[8] So the high number of dead fish in a small space and time was what raised questions about what was going on in the river.

---

5 Interview conducted with the president of a fishing association, October 21, 2015.

6 "Troubles sur la Loue", *L'Est républicain*, May 6, 2009.

7 Ibid.

8 Interview conducted with a development officer from the Doubs Fishing Federation, November 20, 2015.

However, the number of dead fish observed was not enough to consider the phenomenon "abnormal". According to an engineer from Onema, "saying 'I saw nine dead fish at such and such a place, on such and such a day' is meaningless, it is uninterpretable". If it was only done by counting the number of corpses from the bank, as Patrice had done, the determination of the situation remained confused.

> You can count them [dead fish]. But you can't see all the fish that are doing well. You can see them, but they don't behave the same way. A sick fish is easy to see because it's wandering around: it's weak, it's not going to hide, it's not lively. In addition, it has fungus that makes white spots, so you can see it.[9]

While it is therefore possible, even easy, to observe a phenomenon of fish mortality from a simple observation made at a precise location from the bank, it is, on the other hand, impossible to say how many have died at this point in the river and thus to qualify the phenomenon. Characterizing the phenomenon was difficult, if not impossible, "because I don't know where these dead fish come from," explained the Onema engineer.

> Over a month, for example: every day I count ten dead fish. They come in and I take them out. Basically, for one month, I had the same level of mortality. But if during that same month, I have a flood: I have more water coming in and fish that will possibly come from further away. What am I going to say, "I have a spike in mortality at that time"? No. There was no spike in mortality, but there was a flood that brought more fish.[10]

The only way to characterize the situation from the number of cases observed was to calculate the prevalence of the phenomenon, "that is, the ratio of the number of sick fish to the total number of fish," he says. This study would have given us the percentage of diseased fish in the river. But conducting it – and therefore characterizing the situation – was difficult because it required a significant protocol to determine the total number of fish in the river. If, in addition to the visibility of the dead fish, their number increased the suspicion of an "abnormal" situation, uncertainty persisted.

9 Interview conducted with an engineer from the regional delegation of Onema, October 9, 2015.

10 Ibid.

2.2.1.3. *The identity of the victims*

At the end of April 2010, the owner of a pub on the banks of the Loue called the agents of the Doubs Fishing Federation. "I went down to the banks of the Loue this morning and I saw more dead fish. Fifteen today, twenty three days ago… It's weird," he explained. An agent went to the site and accompanied the owner of the lodge to the water's edge. "It's been several days since we've had a lot of calls to the federation. I'm going to take a closer look," said the agent. He put on his waders and headed for the floating corpses. At first glance, they were mostly trout and grayling. "Another post-spawn mortality episode," he thought. However, as he observed the scene more closely, he noticed the presence of small individuals. Back on the shore, he called his hydrobiologist colleague.

> Shit, this sucks. I see dead loaches and sculpins too!
>
> – Indeed, it sucks! We hadn't found any dead bodies among these species yet.
>
> – This is definitely not a typical case of post-spawning mortality. The sculpin finished spawning a month ago and the loach has not started yet.[11]

The observation of the sculpin and loach corpses was interpreted as a sign of an "abnormal" situation for two reasons: these fish were not coming out of a reproduction period and were therefore "not supposed to die at that moment"; and, because the loach is "a fish that is relatively resistant to organic aggression," says the agent of the Fishing Federation. It can indeed live in rivers where the water quality is uncertain and does not have the same "ecological requirements" as trout, which need living, fresh and oxygenated water. Because their "gait" is likely to provide – to those who know how to look at it – information on the "norms of an environment" (Canguilhem 2009), fish are "indicator species" of water quality (Beeby 2001). Better still, they can play the role of witnesses in the attempt to characterize the phenomenon of fish mortality.

2.2.2. *In the laboratories, scientific knowledge*

Although he recognized that the experience and observation skills of some fishermen can enable them to interpret the functioning of rivers accurately, a university hydrobiologist felt that "their deductions are sometimes terribly wrong".

---

11 Interview conducted with two agents of the *Fédération de pêche du Doubs* on November 20, 2015.

Indeed, according to him, they could be biased by the representations or particular attachments to a certain species, a certain river, and so on. This was also the position of the representatives of the State, in particular the Prefecture and the Doubs Departmental Directorate of Territories, who called upon analysis laboratories to qualify the situation of the Loue. They needed certainty in order to respond to the questions raised by local residents, fishermen, elected officials and environmental protection associations.

2.2.2.1. *60 km from the Loue, in a veterinary analysis laboratory*

In May 2010, at the request of the Departmental Directorate of Territories, several Onema agents returned to the banks of the Loue. They had to take moribund fish and send them to a veterinary analysis laboratory located about 60 km away, in the Jura department. Why this one and not another? The choice of the service provider was not contingent, it was linked to the fact that this laboratory not only had an "animal health sector", offering services for "aquaculture pathologies", but also accreditations allowing it to carry out "official" analyses on behalf of the State services. Moreover, this transport was not just a geographical move. By moving from the Loue to the laboratory bench, the fish changed their status. They were no longer just near-dead animals, they had become "samples", representatives, even "spokespersons" (Latour 2010b) of their fellow creatures that the veterinarians would interview in order to better understand the causes of the mortality episodes. Through the establishment of a "conversation" (Mead 2006), the bodies of these non-humans could indeed provide answers to various questions posed by the veterinarian: was it a "natural" death? Had the individuals been killed? If so, how and by whom?

The veterinarian carried out a sanitary assessment of the fish that had been sent to her by the Onema agents. This operation included an autopsy and parasitic, virological and bacteriological examinations. It consisted of "searching for all pathogens – bacteria, viruses and fungi – likely to cause an alteration in the state of health of the fish," explained the veterinarian. If the operation made it possible to look at the fish with a close microscopic focus (which observations from the banks did not allow), the veterinarian nevertheless invited others to be wary of the quality of the representation of these spokesmen. Indeed, several biases could alter the quality of the answers obtained and therefore the diagnosis. These included the attitude of the samplers, who could focus on "extraordinary" specimens and "force things towards sampling sick fish, so it's not representative of the overall population at the time," said the veterinarian. Second, "the freshness of the fish", because "once they are dead, the problem is that other things come along and we could say anything," explained an agent from the Fishing Federation. Finally, the absence of contextual data (oxygen, temperature) on the state of the river at the time of sampling, which could make it difficult to determine the causes of fish mortality.

Nevertheless, this precaution as to the representativeness of the fish that passed through the laboratory benches, and therefore the possibility of generalizing these observations to all the fish of the Loue, did not prevent the veterinarian from transmitting a veterinary report, an extract of which can be seen below:

> The examinations revealed the presence of various parasites in massive quantities, indicating a probable alteration in the immune capacities of the fish, in particular:
>
> – the presence of worms of the genus *Gyrodactylus sp.* in massive quantities;
>
> – Acanthocephalans, trematodes, nematodes, either as larvae encysted in organs (liver, digestive wall) or adults and free in the digestive tract;
>
> – cestodes related to the digestive tract;
>
> – and also protozoa of the genus *Trichodina sp.*, *Apiosoma sp.*, *Chilodonella sp.* on the skin of fario trout.
>
> Virology did not reveal any virus.
>
> Bacteriology allowed the isolation of saprophytic bacteria from the aquatic environment (*Pseudomonas fluorescens*, *Aeromonas hydrophila*, *Shewanella putrefaciens*, *Aeromonas sobria*), without an invasive or septicemic character.
>
> Skin lesions with a white down or velvet of mycelial tangles led to the isolation of *Saprolegnia parasitica* […] but other *Saprolegniaceae* have also been isolated from these lesions.[12]

This report, written after the examination of the fish, was the result of practical activities that produced new conditions of visibility of the situation and at the same time "revealed new objects" (Boltanski and Thévenot 1991). From then on, it was no longer just a question of the presence of trout and grayling corpses in the Loue, but of a dozen other beings settled on/in the bodies of the fish – including a pathogen, *Saprolegnia*, which would continue to attract attention.

12 Expert report on fish kills and cyanobacteria blooms in the Loue River, Onema and Bioemco, March 2012, p. 6.

The latter was known to fishermen and had already been recognized from the banks of the Loue. However, the veterinarian could not confirm that it was the cause of the fish mortality. It had already been spotted in several rivers in the region, but its presence had not given rise to such mortality episodes. "Why does *Saprolegnia* manage to kill so many fish this year when in other years it kills two or three?" wondered a university ecologist who is a member of a river defense collective. However, experiments conducted by the veterinarian showed that the strain of *Saprolegnia* present in the fish was not "highly pathogenic". The pathogen could therefore not have killed the fish. However, the veterinarian said that since "the presence of a pathogen is often associated with a triggering factor", a change in the river was now to be investigated. The question became, "What in the river has weakened the fish to the point where *Saprolegnia* can take hold?"

By ruling out the possibility that it was directly involved in the death of the fish, the veterinarian was inviting a reorganization of the way the situation was viewed. In order to find a satisfactory explanation, it was no longer necessary to look only at the action of the pathogen, but to shift the focus to the environment in which the fish lived. More importantly, she felt it was necessary to look at something that was, in her words, "a completely obvious fact" at the time the fish mortality occurred: the presence of cyanobacterial blooms.

2.2.2.2. *400 km from the Loue, in a plant biology analysis laboratory*

"With the bibliography that I have made on this topic and the various people that I have asked about the potential role of cyanobacteria in fish, I have maintained that there was a capacity for toxinogenesis with a possible sensitivity of fish," says the veterinarian. This statement was the result of mobilizing resources and expertise that she did not have. Since even under the microscope the fish were not enough to make sense of the situation, she looked for scientific publications produced far away from other cases and contacted various specialists in fish pathology. Once again, the progress of the investigation into the death of the fish in the Loue river required distancing oneself from the precise locations where it had been observed. It was necessary to get out of the river to understand what was happening there. Moreover, it was by linking a plurality of visions on different elements of the situation and thus by building a "library of cases" (Damien 1995; Tripier 2009) that the veterinarian was able to make sense of the cases of fish mortality.

However, despite the contribution of this led to the progress of the investigation, the veterinarian was not able to test her hypothesis. "The 'cyanobacteria' criterion is not part of our services," she said, recalling that testing laboratories sell a limited range of services and are businesses placed in specific markets. "We [therefore] offered and strongly encouraged people who are interested in environmental quality to do that research or to have it done at a competent lab." The reorganization of the view of the situation and its settlement by new actors appeared to be dependent on the skills of other mobilized actors and modified the form of the network built through the survey. In order to continue the research on fish mortality, it was necessary to find competent actors to follow the cyanobacteria trail and question the new definition of the situation. The services of the Departmental Directorate of Territories then turned to a research office located in the Haute-Vienne (about 400 km from the Loue) and specialized in the determination of cyanobacteria.

Having received the necessary recommendations for carrying out this work, the Onema agents returned to the banks of the Loue. Rather than looking at the river again, they had to "look at it differently". Since the new explanation for fish mortality was the presence of cyanobacteria in the aquatic environment, they no longer had to take "fish samples", but "river samples". In other words, water, sand or plants where cyanobacteria might be found. To do this, the officers had to get the right equipment.

> We had to find a circle that was 15 centimeters in diameter so that we'd have about the equivalent of ten by ten, something like that. To have a hundredth of a square meter. And then we needed a net behind it, which we were never able to get because the French Regional Health Agency (*Agence régionale de santé*, ARS) had it and they never wanted to give us 600 euros to pay for it, so I had to manage to recover and rub the stones.[13]

As with the pedocomparator used by the scientists Bruno Latour followed at the edge of the Amazon, "it is through this invention that the world of things [was] to become a sign" (Latour 2007). It was thanks to this artifact "cobbled together" by the Onema agent that the cyanobacteria could be detached from the many other entities inhabiting the river. The operation had to follow a defined protocol: "The samples are taken by scraping the substrate on a known surface. We put it in a bottle, we rinse it with water, and then we find ourselves with something suspended in the water," explained a hydrobiologist from the fishing federation. Each time, the agents recovered "raw water" from the water column; "coarse sand substrate" and "filamentous algae substrate" from the river bottom. The procedure was carried out on the "right bank" and "left bank", at locations chosen according to the presence of corpses or their frequentation for water sports. "Then you send it to the lab and they say, 'There are so many cyanobacteria cells per milliliter of water'," said an agent

13 Interview conducted with an agent from the departmental service of Onema, November 6, 2015.

from Onema. Again, there was a removal and containment of the situation for investigation.

In the same way that the list of pathogens observed in fish sent to the veterinarian was translated into "entries" in the laboratory instruments, so were the water and substrate samples. They took the form of a double entry table from which it became possible to read "cyano-bacteria concentrations". From then on, it was no longer a question of observations from the shore or even under the microscope, but of measuring practices aimed at achieving accuracy. The laboratory step was thus a passage "from the world of approximation to the universe of precision" (Koyré 1986). Moreover, these inscriptions were the trace – if not the proof – of the presence of cyanobacteria and invited the production of discourses on the situation (Callon *et al.* 2001).

> The cyanobacteria analyzed at three points in the Loue: Lods, Ornans and Rurey, were mainly benthic cyanobacteria, fixed or attached to substrates (pebbles), hence the large differences between the concentrations found in the water layer: respectively 48,000, 20,000 and 2,500 cyanobacteria cells cc/ml of water and those found from the "substrate" samples: 205,000, 506,000 and 1,800,000 cc/ml. A neurotoxin (toxoid A) was also found in the "substrate" sample taken at Ornans.[14]

The above extract is taken from the document published by the Doubs Prefecture at the end of June 2010. At that time, it was no longer just a question of a river in which individuals had observed dead fish, but of a watercourse in which the presence of cyanobacteria was now proven. Moreover, they were now quantifiable, locatable and characterizable: mainly located in the substrate and in the bottom of the watercourse, they were qualified as "benthic"; their development along the Loue was not homogeneous; and their presence was associated with the presence of toxins in the substrate. The analysis report then became the object of interpretation, so that "we never see anything but a continuous series of interlocking elements, each of which acts as a sign for the previous one and as a factor for the next" (Latour 2007).

2.2.3. *In the administrations, legal knowledge*

"Set out to conquer the oceans across the board around 3,400–3,200 Ma, then without any atmospheric oxygen" (Boeuf 2014), cyanobacteria were present at the

14 Meeting of MISE (*mission interservices de l'eau*, French Interdepartmental Water Mission) "Loue special", Doubs Prefecture, June 25, 2010.

beginning of life on the Earth. Moreover, according to an academic hydrobiologist, some futuristic biologists would estimate that "they will surely be the last to disappear because they are able to find favorable conditions for their development in extremely different environments". Cyanobacteria were also described as "opportunistic, adaptable and able to compete with other species in the midst of the environment".[15] Moreover, like those found in the water of the Loue, some "are capable of emitting toxins to poison their friends and prevent them from grazing, thus fighting against their own predation," added the hydrobiologist. This characteristic, which was also raised by the veterinarian, could turn cyanobacteria into ideal suspects, but for the moment there was nothing to indicate that they were necessarily involved in the death of the Loue fish. It was still necessary to show their sensitivity to these toxins.

Moreover, this new lead was reinforced by the discovery of a dead roe deer on the banks of the Loue in early June 2010, whose death was attributed to "intoxication due to cyanobacteria of the *Oscillatoria* type contained in the water of the Loue".[16] The death of the animal added a sense of concern, as a new risk had to be integrated. While the episodes of fish mortality only led to questions about the toxicity of cyanobacteria for fish, the death of the deer led to the consideration of a potential risk for humans. The situation then shifted into a new dimension because "it is not the same biopolitics that is applied to animals and humans" (Keck 2013).

2.2.3.1. *Fear of a health risk*

"What would happen if a child was playing in the river and swallowed water?" If their mobilization was born from the occurrence of fish mortality episodes, this new dimension of the situation, characterized by the fear of a health risk linked to the use of the Loue, increased their determination and their capacity to mobilize. For their part, the agents from the Prefecture did not take the news lightly. They could not afford to do so, as health security was one of their obligations. If nothing was done and if people developed pathologies – or even died – after having been in contact with the water of the Loue, their responsibility and that of the State would be called into question. The members of the Prefect's office therefore had to respond as quickly as possible to these events. Thus, they consulted the services of the ARS at the end of May 2010:

> What is the risk of cyanobacteria in water?

15 Meeting of MISE (*mission interservices de l'eau*, French Interdepartmental Water Mission) "Loue special", Doubs Prefecture, June 25, 2010.

16 "Un animal tué par l'eau de la Loue," *L'Est républicain*, June 4, 2010.

– No human deaths have been observed to date. The only deaths mentioned in the international scientific literature are those of animals, dogs, sheep or cows that drank from certain ponds, rivers or lakes, replied a representative of the Regional Health Agency.

– What about the effects on human populations? asked a cabinet member.

– Several surveys have shown that there is a possibility of itching, acute gastroenteritis, neurological damage, liver damage and reproductive disorders in the event of chronic exposure. This is recorded in the regional public health plan for the period 2006–2008. I am sending it to you.[17]

Although the sensitivity of fish in the Loue to the toxins produced by cyanobacteria had not yet been demonstrated, one thing was certain for scientists and representatives of the health authorities: there was indeed a sensitivity in mammals.

Several cases of animal poisoning observed in the past had in common their presence in the water ingested by the victims. There were 15 dogs and 20 sheep near a lake in Great Britain in 1985; 100 cows in Switzerland between 1970 and 1995; and three dogs on the banks of the Loue River in 2003. Precedents were added to the veterinarian's library of cases and the Loue situation was now populated by cases of animals that had died at other times and in other countries. Its assessment by public officials involved interrogating the memory of previous crises (Keck 2013).

The information available to the representatives of the ARS and now the Prefecture on the activity of cyanobacteria did not, however, indicate a lethal risk to human populations. Nevertheless, as the Loue River is not only a location for numerous leisure activities such as fishing, swimming or canoeing, but also and above all a source of water for some communes in the valley – including Besançon, the regional capital, it was used daily by several thousand people. This is why certain precautionary measures appeared necessary in the eyes of the agents from the Prefecture.

But what could be done to limit the risks related to the use of the Loue water? Bathing sites and water distributed in the water supply networks were the subject of additional analyses; letters were sent to doctors, pharmacists, elected officials and

17 Dialog reconstructed from field observations.

tourist offices to inform them of the presence of cyanobacteria and their potential effects; recommendations were made via a press release relayed by the various titles of the local press[18]; and prefectoral decrees were published to regulate the use of the Loue. If, until June 2010, the agents from the Prefecture seemed to be in the background regarding the situation of the Loue, the spectacular aspect of the events, the work of mobilization carried out by the defenders of the environment and, above all, the uncertainty concerning the existence of a health risk linked to the use of the water of the Loue led them to take a stand, to focus their attention more on the river.

2.2.3.2. *Prefectoral decrees in response to the risk*

After having consulted the agents of other State services and more particularly the Departmental Directorate of Territories, the Water Agency, the Regional Health Agency and the French Food Safety Agency (AFSSA), the members of the Prefect's office proposed a response to the situation. It took the form of prefectoral decrees published on June 4, 2010:

– the first one was a "ban on using the water of the Loue for the watering of the cattle from its source to the dam of Quingey"[19];

– the second was a "ban on consuming all species of fish caught and an obligation to release fish caught in the Loue from its source to the Quingey dam".[20]

More than administrative documents, they were legal instruments whose analysis allowed for a better understanding of how the authorities responded to a situation characterized by a certain degree of uncertainty.

This answer was elaborated on a particular situation whose problematic aspect was presented in two ways: "The significant fish mortality observed since the second half of April on the Loue river from its source to the Quingey dam" and "the results of cyanobacteria and cyanotoxin analyses of the samples taken on May 4, 2010 in the Loue river". It was, moreover, built from the reference to legal and regulatory texts framing the action of the public authorities. Regulation EC

18 In particular "to avoid ingesting water, to take a careful shower after swimming or water sports activities and to clean water sports equipment", to ensure that young children "do not ingest substrate or water, nor play in areas of stagnant water", and to "not allow domestic animals to approach the river bank". See "Interdictions et précautions, suite aux conclusions des analyses pratiquées sur le chevreuil mort...", *L'Est républicain*, June 4, 2010.

19 Prefectoral decree No. 2010-0406-01991 prohibiting the use of water from the Loue for livestock watering from its source to the Quingey dam, June 4, 2010.

20 Prefectoral decree No. 2010-0406-01993 prohibiting the consumption of all species of fish caught and requiring the release of fish caught in the Loue river from its source to the Quingey dam, June 4, 2010.

No. 183/2005 of the European Parliament laying down requirements for food hygiene, the Charter of the Environment, the General Code of Local Authorities, the Public Health Code and the Administrative Justice Code stipulated the terms. These references, which in legal language are called "Visas" and which are preceded by the locution "VU", served to justify the regularity of an action which was based on the formulation of "bans".

The orders were therefore binding and made manifest "the State's power of command" (Delpeuch *et al.* 2014). However, the latter had to be justified and the Prefect had to explain the reasons, in fact or in law, that supported his decision by setting out elements likely to prove its "validity" – what legal language calls the "Recitals". These included, for example, the fact that "significant concentrations of toxin-producing cyanobacteria [had] been detected in the so-called 'substrate' water samples in the Loue"; and that a "probable health risk linked to the contamination of foodstuffs by cyanobacterial toxins" had been mentioned by Afssa and Afsset[21] in a 2006 report. The proven presence of toxin-producing organisms and the potential health risk were therefore the main reasons for the restrictions on the use of the Loue.

The reaction of state representatives and the legal decisions that resulted from it thus appeared to be the result of an activity to qualify the situation through the selection of elements deemed relevant and based on a collective production involving heterogeneous actors and instruments (Weller 2007).

Examining the singular case of the Loue in the light of a collection of disparate elements describing similar cases and the legal and regulatory principles to which it belonged, the agents from the Prefecture thus implemented an argumentation based on casuistry. The objective of this approach was to resolve cases of conscience by applying theoretical principles to real-life situations, and it allowed them to justify restricting the use of the river and imposing measures deemed necessary. Because they are made up of a set of heterogeneous elements intended to allow for the government of the entities involved in a situation, prefectoral decrees appeared to be "devices" (Foucault 1994).

21 The *Agence française de sécurité sanitaire des aliments* (AFSSA) and the *Agence française de sécurité sanitaire de l'environnement et du travail* (AFSSET) merged into the *Agence nationale chargée de la sécurité sanitaire de l'alimentation, de l'environnement et du travail* (ANSES) in July 2010.

2.3. Several groups of scientists investigating the same situation

2.3.1. *The construction of the content of expert reports*

On April 20, 2011, a meeting of the Interdepartmental Water Mission[22] was organized at the Doubs Prefecture. Its objective was to review the situation of the Loue and to present the work carried out to better understand the phenomenon of fish mortality. About 15 people attended, including several researchers from different disciplines (ecology, hydrobiology, toxicology, etc.) and institutions (university, CNRS, Onema, Jura department analysis laboratory, etc.). In addition, these researchers belonged to two groups: the "national experts" and the "local experts".

Spokespersons for each of them were asked to present the work being undertaken and the objectives pursued. The "national experts" thus presented their study on *fish mortality in the Haute-Loue and the possible links with cyanobacteria proliferation*; and the "local experts" presented their research program on *the state of health of the karstic rivers of the Franche-Comté region in relation to anthropic pressures on their catchment areas*. At that moment, the content of the expert reports could be considered to have stabilized. Being an outside observer, I only saw constructed research objects and proposals for investigations already underway, or even first results. However, analysis of the situation showed that the involvement of experts was linked to the existence of "windows of opportunity" (Kingdon 1984) and that their content was the result of negotiation between sponsors and experts.

2.3.1.1. *"Local" expertise and interest in toxics*

Before becoming a hydrobiologist at the University of Franche-Comté and for a research consultancy, the spokesman for the local experts was an engineer for the *Conseil supérieur de la pêche.*[23] As part of his activities, he had the opportunity to carry out several studies that led him to alert the representatives of the public authorities to the state of the Loue as early as 2002. "There is a big problem with the

22 Created in 1993, the Interdepartmental Water Mission (called MISE, *mission interservices de l'eau*) is a grouping of different State services and public establishments carrying out missions in the field of water. It "is likely to ensure a better coherence of the interventions of the State and to allow a global approach of the questions relating to water" (circular of January 22, 1993). It was replaced in 2010 by the Interdepartmental Mission for Water and Nature (*mission interservices de l'eau et de la nature*, MISEN).

23 Created in 1948, the *Conseil supérieur de la pêche* (or CSP) was a body co-managed by the State and the Fishing Federation to carry out water policing and river quality control missions (Bouleau 2016). It was replaced in 2006 by Onema (see note 2).

Loue. It is sinking more and more. It's not going well, there are tufa problems, algae problems and probably problems with contaminants, micropollutants,"[24] he said.

Concerned about the state of the river, the representatives of a mixed syndicate of the Loue contacted him in 2008 in order to make this river a "pole of environmental excellence". Together with his toxicologist colleague, the hydrobiologist proposed to conduct research on the toxic elements and micropollutants present in the river. At the time, this was a subject that was still not widely discussed, but which, according to the researchers at the University of Franche-Comté, could explain the problems that were then being observed in the Loue. However, this proposal to work on toxic substances did not attract the attention of environmental protection associations and fishermen, who remained focused on agricultural discharges. "For them, it was only because of agricultural discharges... it was only 'nitrogen, phosphorus, carbon', so the classic pollution that can cause problems. They were afraid that the fact that they were going to focus on toxic substances would prevent them from dealing with this pollution,"[25] says the hydrobiologist. The constitution of a pole of environmental excellence on the Loue and the realization of a study on toxics and micropollutants were suspended because of this "refusal of the transaction".

When episodes of fish mortality were observed in the Loue in 2010, the agents of the Franche-Comté regional council wanted to investigate. They contacted a university ecologist with whom the then director of the agriculture and environment department had collaborated in the context of work on vole outbreaks (see Chapter 1). He directed them to the hydrobiologist and his ecotoxicologist colleague.

> The region remembered that we had said that there were problems and so it told us: "Make us a program, co-construct it with the local actors, the consular chambers, the farmers, the residents, the mayors so that the solutions don't come from above and from academic knowledge, only to fall flat in reality."[26]

This request appeared to be a godsend for the two researchers. As requested by the representatives of the environment, water and energy department of the regional council, they set up a research program described as "integrative" or "holistic" (without, however, involving the social sciences) and were able to include their hypotheses, which had been left in abeyance, on the impact of toxic substances and micropollutants on the quality of the water of the Loue.

---

24 Interview with university hydrobiologist, November 18, 2015.

25 Ibid.

26 Ibid.

2.3.1.2. *"National" expertise and interest in cyanobacteria*

At about the same time, the general management of Onema proposed to the Prefect of Doubs to conduct an expert assessment of the situation of the Loue. This was following the completion of inventory fisheries in July 2010. The Prefect replied to the director general of Onema in a letter dated September 21, 2010 and expressed his expectations regarding this proposal:

> Mr. Executive Director,
>
> I would like to thank you for the responsiveness of your public establishment in carrying out the fish inventory of the Upper Loue Valley […].
>
> Moreover, you have shown, from the very first exchanges with my services, your interest in conducting, from the case of the Loue, national expertise on the phenomenon of fish mortality, which would tend to develop, in particular in aquatic environments known for their quality. The phenomenon observed on the Loue – degradation of the health status of fish, cyanobacteria efflorescence, and piscicultural mortalities – was thus noted this season on other rivers of the region (Doubs Franco-Swiss; Ain), which it is thus advisable to include in the perimeter of reflection.
>
> I can confirm the importance I attach to this expertise, the challenge of which will be to define the relevant, targeted and concrete action levers to prevent this type of episode in the coming years. In terms of a timetable, it would be necessary for us to be able to define a – preventive – monitoring strategy as early as February, and for us to have a first level of recommendations for targeted actions in March […].[27]

The Prefect of Doubs therefore wanted the assessment to study several aspects of the situation, to cover other rivers in the department in addition to the Loue and to provide levers to prevent similar episodes from occurring. However, the director general of Onema responded to the Prefect's "generalist" request with a more "specific" response. Arguing that the multiplicity of probable causes and the lack of data would make it impossible to determine precisely the origin of the cases of fish

27 Letter sent by the Prefect of the Doubs to the Director General of Onema, September 21, 2010. Expert report on fish mortality and cyanobacteria blooms in the Loue. Onema and Bioemco, March 2012, p. 37.

mortality, he stated that the expertise coordinated by his services would focus solely on cyanobacteria blooms.[28] The stakes were high for Onema, as little data were available on this phenomenon, even though it was observed in several places in the territory. The situation of the Loue therefore appeared to be an opportunity to study it.

2.3.2. *The cohabitation of two groups of experts*

From April 2011 onward, two expert reports were conducted on the Loue case. Starting from a similar troubled situation, the "local" and "national" expert reports proposed two investigations with different focuses. The river thus became a research field for these two groups of researchers (we will see later that the river defenders constitute a third). While the meetings I attended initially suggested that there was scientific collaboration, on closer examination it became clear that there were tensions between the two groups of experts.

2.3.2.1. *Recognition of expert status*

These tensions were partly related to the establishment of a hierarchy between the group of "local experts" and the group of "national experts", who were named and identified as such in official and working meetings. Regretting that the latter group "wanted to be in charge", a member of the former group refuted such a distinction, explaining that it did not reflect their place in the scientific community. "We feel that there is a hierarchy. We want to be 'local', but... we consider ourselves 'international'. In any case, we write in English". Admittedly, these scientists from the University of Franche-Comté were "local" because of their proximity to the region's rivers, but they belonged to a research unit that conducted work in several countries around the world and published in English-language journals. For the hydrobiologist, the group of "local experts" was therefore no less national or international than the one convened by the general management of Onema and mandated by the Prefect of Doubs.

Moreover, the fact that the general coordination of the surveys was entrusted to the president of the group of "national experts" was looked on badly by these researchers from the University of Franche-Comté, who had been working on the Franche-Comté rivers for more than 30 years and who had been students and/or colleagues of Jean Verneaux, the pioneer of hydrobiology in France (Verneaux

28 Letter sent by the Director General of Onema to the Prefect of the Doubs, October 11, 2010. Expert report on fish mortality and cyanobacteria blooms in the Loue. Onema and Bioemco, March 2012, p. 39.

1973). "I think they were very offended that the president of the scientific group was not a local and that we went for a Parisian," explained a geochemist belonging to the same laboratory. This choice would have been interpreted as a disavowal of their skills. "We are a team. We have our feet in it, we have measurements. Even when it was no longer funded, we continued to make some measurements. We have methods, etc.," says the hydrobiologist coordinating the local expertise. For him, if experts had to be appointed to supervise the surveys on the water quality of the Loue, it had to be them because they knew the field perfectly.

Finally, tensions also arose over the assignment of roles to each group of experts, which was included in a slideshow presented by representatives of Onema's general management at the April 2011 meeting of MISE.

> The president is assisted by a scientific and technical facilitator... who is supported by:
>
> – the head of expertise at Onema to provide the secretariat;
>
> – a project team made up of Onema project managers for scientific and technical aspects (data analysis, bibliographic notes, etc.);
>
> – the group of local experts, led by the Rhone-Mediterranean and Corsica Water Agency (Besançon delegation), whose objectives are:
>
> > - to organize and make available the knowledge elements (technical reports, in particular) and the data available on the studied catchment areas,
> >
> > - to provide the national group of experts with detailed knowledge of the terrain, the history of the functioning of the Loue and human activities, in the form of notes or audits,
> >
> > - to develop proposals for action, particularly for pre-emptive monitoring.[29]

The group of "local experts" was presented as "led by" the Water Agency. It was therefore not seen as being in sole command of its investigation, but guided by the staff of another State service. In addition, it was responsible for organizing and making available to the "national" elements of field knowledge. In doing so, the group formed by the researchers from the University of Franche-Comté was deprived of what characterizes expertise: analysis of the material collected and the production of a judgment on a situation (Trépos 1996). In the end, Onema managers

29 Expertise nationale relative aux mortalités de poissons dans la Haute-Loue et les liens éventuels avec les proliférations de cyanobactérie. Loue Special Issue, April 20, 2011.

asked them to do work that a master's student could do, and the locals became like assistants to the nationals. This role assignment was perceived as an affront by the locals, as it did not correspond to their position as producers of scientific knowledge and called into question their status as experts.

2.3.2.2. *The division of expertise work*

Although a form of division of labor was established between the national and local experts, each of these groups also benefited from the work carried out by others. Indeed, the expert reports were provided with research contracts and fed by the acquisition work carried out by certain agents of the regional delegation and the departmental service of Onema. As a result, when the national expertise was set up, an agent from the departmental service thought that he would have a role to play. After all, the Loue had been part of his sector for more than 20 years and the body for which he worked was at the initiative of this survey. Invited to attend a meeting at the Doubs Prefecture, he went with a colleague, but things did not turn out as they had imagined.

> There was the university, there was the Onema "regional delegation", there was the DDT [Departmental Directorate of Territories], there were the associations, we felt a bit isolated. We saw the national experts, they came. I saw someone who was appointed. It cost about 100,000 euros. We were struggling to make ends meet and we never saw this person again.[30]

This was not his world. The officer was not a scientist or a government administrator. He felt that he was in the middle and could not participate in the meeting – except in bits and pieces, when someone asked his opinion because of his presence, almost as a diplomatic gesture. The officer hoped to be able to help by bringing in his experience, but the "national experts did their own thing," he says. Feeling sidelined by his own institution, he took a critical stance toward this expertise and felt that it would be out of touch with reality.

> There is only one who called me. He said to me: "Sir, you are on the Loue, what do you think? How do you see it?" "So I gave him a little bit of the stuff, but he's the only one. And that's serious for me. They're going to come up with cobbling together some stuff that won't fit anything. They're going to do some reading, they're going to

30 Interview conducted with an agent from the departmental service of Onema, November 6, 2015.

> take some elements, but they're not going to get the feeling of the base".[31]

Although the Loue was also part of his sector and he was ready to provide technical assistance to the national experts, the engineer of the regional delegation also felt that he could not participate in the survey.

> I was waiting for the national experts to say, "OK, we don't know what's going on. However, in order to find answers to the questions, we need to work on these compartments, to recover this type of data"; or even something a little vague, to say: "It would be good to dig up these leads, so we'll recover these data, and then we scientists will work on them, we'll look for them, and either that's not what we're looking for and we'll look somewhere else, or it's a lead and we'll dig".[32]

The agents of Onema's departmental and regional services therefore expected the national experts to rely on their skills, their knowledge of the field and the data they produced throughout the year. However, this was not the case. The "nationals" called on them very little, if at all, and the agents had the impression that they were being sidelined and, gradually, they dropped out.

2.3.3. *River advocates and experts*

On June 19, 2010, local elected representatives of *Europe Écologie-Les Verts* (the national main green party) organized a demonstration in Ornans. This event was also the starting point for the constitution of a collective for the defense of the Franche-Comté rivers. Initially led by a few fly fishermen and environmental activists, it was a response to the feeling of inaction of the representatives of the public authorities and the powerlessness of the actors in charge of the management of the rivers. From the beginning, the newly created collective was characterized by a constant search for efficiency in action and by a commitment in the scientific sphere.[33]

31 Interview conducted with an engineer from the regional delegation of Onema, October 9, 2015.

32 Ibid.

33 They also engaged in the legal and public spheres but this will not be discussed here. For more information on this subject, see Calla (2019).

2.3.3.1. *Activist expertise*

From his experience as an activist, the founder of the river defense collective had learned that the balance of power with the stakeholders in a dispute had to be established through "knowledge of the case". In order to be listened to by the representatives of the public authorities, he felt it was necessary to bring elements that could "carry weight". The scientific arguments, unlike those evoking for example an attachment to the river or a heritage to be preserved, were part of these. Also, the fact of counting "trained" or "professional" scientists among the members of the collective had the advantage of being able to produce "serious discussions" (Dreyfus and Rabinow 1992).

A young ecologist member of this group – now a lecturer – showed in particular that "the fine monitoring carried out upstream and downstream of Ornans made it possible to highlight a pronounced enrichment of the interstitial water for nitrogen and especially phosphorus" (Frossard 2006). His work then led him to assert that, in order to avoid algal growths prompting the appearance of cases of fish mortality, it was necessary to limit the contribution of these nutritive elements to the watercourse, but it was still essential to know where these elements came from.

The members of the collective had different opinions on this subject. Some said that "river pollution is 70% sewage problems" while others – including the founder, who had specialized in agricultural issues during his activist career – argued that "it's at least the other way around, at least 70% agriculture". They therefore decided to end the debate by conducting research based on a scientific approach and using only articles published in the international literature as a source of information.

A university geochemist and fly fisherman conducted this study in 2011. He may not have had the necessary knowledge of sanitation or agriculture, but he knew and used the scientific method in his professional practice. After analyzing the data produced by a sewage treatment plant located on the banks of the Loue, he told the founder of the collective: "In the results I arrive at, it is absolutely not the sewage treatment plants that are the major problem. I assure you that the figures I find are from agriculture".[34] Satisfied to note that his assertions were confirmed by a scientist, the founder of the collective did not fail to mobilize these results resulting from a study "rigorously carried out" at the time of the various demonstrations in which he took part.

34 Eutrophisation de la Loue : sources principales des nutriments, azote (N) et phosphore (P). Société d'histoire naturelle du Doubs, May and June 2013.

He also wanted to enter the scientific arena in a more sustainable way by working to secure funding for a research project led by the university ecologist and member of the river advocacy collective.

> David has been sampling invertebrates, algae and fish for four years. He also did a core sample of a ditch in the Loue. He looked at the cadastral map and saw that it had existed for at least two centuries. He only got negative answers from his bosses [the people in charge of the research laboratory at the University of Franche-Comté, ed.] So at one point I said, "I'll get you some money."[35]

This study, which consists of carrying out isotopic signatures of nitrogen in order to determine its trace in animals, plants or in the soil, would never have been carried out by the researchers of the University of Franche-Comté. If, according to the founder of the collective, it was because "the University of Besançon works a lot for the comté and there are conflicts of interest", it is rather because his local expert colleagues "consider [toxic substances] more important," said the ecologist-activist.

2.3.3.2. *A political commitment considered problematic*

However, the involvement of "professional" scientists in the river advocacy group was seen as problematic by their colleagues who were at the same time in the position of "local" experts.

> We think they're messing around because when they give lectures, they advertise themselves as academics from the lab. Basically, we're already in an intellectual-scientific dispute with the other expert groups and, within the university, there are people who are agitating. It's very difficult to manage. They're friends, but I've already told them several times that I want them to keep things in perspective. For example, on agriculture, if I'm asked my opinion as a citizen, I say: "Farmers, they have to make huge efforts and that's it." But at the scientific level, I say: "That's the impact they have. It's not for me to decide as a scientist, it's for the manager to decide." So there you have it, I think there's a confusion on their part and that bothers me.[36]

"It is the eternal question of the place of the scientist in the city," says the university ecologist and member of the river defense collective – of the relationship between *the scientist and the politician* (Weber 1963). A dispute emerged between

35 Interview conducted with the founder of a collective for the defense of rivers, June 19, 2015.

36 Interview with university hydrobiologist, November 18, 2015.

the researchers of the University of Franche-Comté and thus revealed the existence of different ethical and political positions. Indeed, while the university hydrobiologist and expert argued for a separation between the scientist's and the citizen's position (which, according to him, should not give rise to the production of the same statements), the university ecologist and activist explained that he had a different conception of the role of the scientist.

> Paul [the hydrobiologist] I like him, he does a good job, he has a very good knowledge and I think he is going in the right direction. Afterwards, what diverges a bit is that they say – well, it's mostly Guy [the toxicologist] who says that – that the scientist is not there to make judgments. He's there in the formidable, shining neutrality of the scientist's ivory tower. I don't have that approach. I am paid by all the people who work, I am paid by their taxes, so once I know something, I say it in the public arena. That's why I'm happy to have public debates. Sometimes, there are farmers who insult us, but I have the impression that I am more in touch with reality; and that what I know, paid by the city, well… I try to give it back to the city, but with arguments.[37]

He felt that he could neither stand back from public life nor place himself behind the separation between science and society because of the existence of a debt owed to "citizens". This idea of debt was also perceptible in the discourse of the academic and expert hydrobiologist, but in a different way:

> We are sometimes reproached, saying: "You scientists are in your ivory tower"; but we are obliged, at the deontological level, as well as at the concrete level, as well as at the financial level, to have distance. I get on well with the members of the collective, we have good discussions. I don't withhold information, but sometimes I tell them: "I can't give you information about a research program that is funded before I give it to the funders. They are public funders, so they have to give you the information. It's not up to me to give it, to short-circuit." Moreover, when we short-circuit, in general, it annihilates the effect because it becomes controversial, the funders get angry with us and it is not very honest either.[38]

---

37 Interview with a university ecologist and member of a collective for the defense of rivers, June 24, 2015.

38 Interview with a university hydrobiologist, November 18, 2015.

Unlike the ecologist-activist, the expert hydrobiologist did not feel that he owed a debt to French citizens, but to the funders of his research program. It was, according to him, to the latter that he should be accountable and leave the results to the political leaders to act. Nevertheless, this position was considered by the activist ecologist to be problematic, because it was a position of dependence on, and even cronyism with, the political powers that be. He was also critical of another colleague, a university toxicologist and local expert.

> "The scientist is not there to militate, the scientist is there to take a step back." All those who talk to you like that are in bed with the government and are state scientists. They are there to say that "if we have to say that it is warmer in Besançon, we will find data that will say so". That's not my thing, maybe because I'm younger and don't have much to lose. Guy, on the other hand, has so many networks that I think he's pretty much got it covered everywhere. But I don't want to worry about that because I wanted to do my duty as a person who knows three things.[39]

However, the ecologist-activist was no less sympathetic toward this local expert and explained their different stances by the fact that they did not occupy the same position within the "scientific field" (Bourdieu 1976). On the one hand, there was a university professor, whose accumulation of that "particular kind of social capital" that is scientific authority led him to have a dominant position and to interact with the dominant figures in the "political field", and, on the other hand, an ecologist who had recently completed his doctorate, a "newcomer" to the field, who claimed to have nothing to lose and to want to accomplish what he considered his "duty".

Moreover, this question of the place of the scientist in the city – of the relationship between science and politics – was at odds with a discourse on the objectivity of facts and discourse. "When you are involved, you are told that you are subjective and that you are in a militant approach that is outside science," deplored the academic and ecologist activist, while the professor from the same laboratory kept reminding us of the principle of validation by peers. "We are scientists and we confront our data to the validation by peers before publishing them," he said during an interview given to France 3 Franche-Comté (a local TV channel).[40] In other words, even if the representatives of the river defense collective published the results of their investigations, these could not be considered scientific and objective facts because they had not been validated by other scientists who are experts in the

39 Interview with a university ecologist and member of a collective for the defense of rivers, June 24, 2015.

40 Filmed interview, France 3 Franche-Comté, April 22, 2015.

fields in question. However, this was not a problem for the academic ecologist and activist, who claimed to apply scientific rigor to his commitment: "I only have scientific data. There would be a problem if I claimed to have scientific knowledge and said anything and people would say: 'Holy crap, he's a joke, he says it's science, but he's cracking up…'".[41]

2.4. Conclusion

This contribution is based on the work done for my doctoral thesis in sociology on the problem of river pollution in Franche-Comté (Calla 2019). Unlike other texts that make up this volume, it does not look back on an experience of multidisciplinary research (notably because of the youth of my career and, as explained in the introduction, a lack of interest on the part of potential funders), but aims to give an account of a situation whose attempt to resolve it involved a diversity and plurality of actors and knowledge.

In the first part, we were able to note that the turmoil generated by the discovery of large-scale episodes of fish mortality in the Loue from 2009 onward had led to the implementation of three types of knowledge (Calla 2017): "ordinary" knowledge, based on the observation of environmental details by local residents in order to understand a situation and decide on the actions to be taken; "scientific" knowledge, seeking to objectify the initial interpretations of the situation based on a detour through laboratories and their measuring instruments; and, finally, "legal" knowledge, whose casuistic reasoning was to make it possible to find the right response to be proposed with regard to the situation.

Furthermore, it has been shown that this knowledge – and the surveys with which it was updated – was achieved through practical operations (observation, measurement, comparison, etc.) consisting of implementing a real "work of discernment" (Guinchard and Ogorzelec 2018), which should make it possible to qualify the situation and adjust the actions to be taken accordingly. Indeed, whether it is to characterize episodes of fish mortality, to objectify observations made from the banks or to legislate because of a potential health risk, this knowledge was based on a logic that consisted of examining the situation in terms of its components and comparing it with other existing cases.

Finally, we were able to notice that this knowledge was not all there – that is, integrated into the situation – at the time of the discovery of the dead fish in the

41 Interview with a university ecologist and member of a collective for the defense of rivers, June 24, 2015.

Loue. On the contrary, they were successively summoned (before coexisting) because of a process of ever denser settlement of the situation (leading to increasing complexity) and the need to respond to new constraints. Indeed, while at the beginning it was only about 15 dead fish, the situation was then characterized by the presence of bacteria producing toxins, then by the potential health risk for human populations.

In the second part of this contribution, it first appeared that this situation, characterized by the discovery of dead fish in the Loue, had led to the constitution of three different groups of investigators: "national" and "local" experts, respectively mandated by the services of the Doubs Prefecture and the Franche-Comté regional council, as well as members of a river defense collective coordinated by environmental activists. The analysis of the situation could have led to distinguishing other groups or sub-groups, but these appeared to be the main ones to invest the scientific arena that occupied us in the framework of this contribution.

In looking at the engagement of these three groups, we first noticed that while they were concerned with a similar problematic situation, their investigations were focused on different objects. While the "national" experts investigated cyanobacteria blooms, the "local" ones questioned the presence of toxic molecules, while the members of the river defense collective focused their investigations on the origin of the nutrients leading to the eutrophication of the river. How, then, can we understand these different focuses applied to the same situation? From this point of view, the analysis showed that the research objects are not already there in the situation, but are the result not only of a construction involving the particular interests of the researchers or the sponsors, but also of the existence of a "window of opportunity" to engage in work that has already been envisaged in the past and/or that can be a source of funding for young researchers.

Moreover, if sociology is a science that is interested in "social groups", it is also a science of relations and interactions between the different groups studied. From this point of view, we were forced to note that tensions had emerged between the national experts, the local experts and the members of the river defense collective. More precisely, we noticed that these tensions were caused by: the recognition of the status of expert and the legitimacy of the words spoken; the establishment of a division of labor and therefore of a hierarchy; and the conception of the role and place of the scientist in the city. Also, in general, we observed the existence of difficulties for these groups composed of professional scientists to get along and to deploy a real collective work in order to solve the problematic situation of the Loue.

In conclusion, the confrontation of local residents with massive episodes of fish mortality in the Loue and the constitution of this phenomenon as a problem to be solved mobilized a plurality of actors with different interests and knowledge. Its complexity and the diversity of its stakes exceeded the capacity and the intention of each actor to solve it individually, which is why the situation required a collective investigation to define the problem and its solutions. The investigation showed that this was not the case and it seems that, in order to do so, it is necessary to recognize the distributed nature of knowledge (Cicourel 1994), as much as the need to set up cooperative research practices.

KEY POINTS.–

Fish mortality events have effects that extend beyond the location of their occurrence and the riverbed, and they have effects that deserve to be studied by socio-anthropology.

Residents, scientists and government representatives are different groups of stakeholders concerned by fish mortality events. They do not belong to the same "world", do not have the same interests and do not have the same knowledge about how to solve this problem.

Indexical, scientific and legal knowledge do not deal with the same elements of the situation and apply different focuses. What they have in common, however, is that they all carry out the work of discernment to qualify episodes of fish mortality.

The relationships between the various groups of investigators working on cases of fish mortality are influenced by dimensions related to the recognition of expert status, the establishment of a division of labor and the existence of different ethical and political positions.

The complexity of the phenomenon of fish mortality exceeds the capacity of each actor to solve it individually. It requires a collective activity to define the problem and its solutions.

2.5. References

Becker, H. (2010). *Les Mondes de l'art.* Flammarion, Paris.

Beeby, A. (2001). What do sentinels stand for? *Environmental Pollution*, 112(2), 285–298.

Boeuf, G. (2014). *La Biodiversité, de l'océan à la cité.* Collège de France, Paris.

Boltanski, L. and Thévenot, L. (1991). *De la justification. Les économies de la grandeur.* Gallimard, Paris.

Bouleau, G. (2016). Pourquoi chercher la petite bête ? Les enjeux politiques de l'indice biotique en France (1964–1969). *VertigO*, 16(2).

Bourdieu, P. (1976). Le champ scientifique. *Actes de la recherche en sciences sociales*, 2(2), 88–104.

Calla, S. (2017). Des poissons morts, des enquêtes et des rivières différentes. *Développement durable et territoires*, 8(1) [Online]. Available at: http://journals.openedition.org/developpementdurable/11549.

Calla, S. (2019). Des poissons, des hommes et des rivières. Sociologie d'un problème de pollution en Franche-Comté. PhD Thesis, Université de Bourgogne Franche-Comté, Besançon.

Callon, M., Lascoumes, P., Barthe, Y. (2001). *Agir dans un monde incertain. Essai sur la démocratie technique.* Le Seuil, Paris.

Canguilhem, G. (2009). *La Connaissance de la vie.* Vrin, Paris.

Charvolin, F., Frioux, S., Kamoun, L., Roussel, I., Mélard, F. (2015). *Un air familier ? Sociohistoire des pollutions atmosphériques.* Presse des Mines, Paris.

Cicourel, A.V. (1994). La connaissance distribuée dans le diagnostic médical. Joseph, P. and Conein, B. (transl.). *Sociologie du travail*, 36(4), 427–449.

Damien, R. (1995). *Bibliothèque et État. Naissance d'une raison politique dans la France du XVIIe siècle.* Presses universitaires de France, Paris.

Delpeuch, T., Dumoulin, L., de Galembert, C. (2014). *Sociologie du droit et de la justice.* Armand Colin, Paris.

Dewey, J. (1967). *Logique. La Théorie de l'enquête.* Presses universitaires de France, Paris.

Douglas, M. (2005). *De la souillure. Essai sur les notions de pollution et de tabou.* La Découverte, Paris.

Dreyfus, H.L. and Rabinow, P. (eds) (1992). *Michel Foucault. Un parcours de l'objectivité et de la subjectivité.* Gallimard, Paris.

Foucault, M. (1994). *Dits et écrits : 1954–1988. Tome 3 : 1976–1979.* Gallimard, Paris.

Frossard, V. (2006). Étude des proliférations algales sur la Loue. Master's Thesis, Université de Franche-Comté, Besançon.

Ginzburg, C. (1980). Signes, traces, pistes. *Le Débat*, 6(6), 3–44.

Guinchard, C. and Ogorzelec, L. (2018). Discerner des singularités. De l'embellissement des façades et des jardins à la construction des vertus dans une ancienne cité minière d'Alsace. *Tracés*, 65–81.

Jacques-Jouvenot, D. (1989). *Voyage au pays des montbéliardes.* Textel, Lyon.

Jacques-Jouvenot, D. (2001). Expertise, technique et tradition culturelle. La culture paysanne. In *L'expertise*, Damien, R. (ed.). Presses universitaires de Franche-Comté, Besançon.

Keck, F. (2013). Surveiller les animaux, préparer les humains. In *Le Gouvernement des catastrophes*. Karthala, Paris.

Kingdon, J.W. (1984). *Agendas, Alternatives, and Public Policies.* Little Brown, Boston.

Koyré, A. (1986). *Études d'histoire de la pensée philosophique.* Gallimard, Paris.

Latour, B. (2007). *Petites leçons de sociologie des sciences.* La Découverte, Paris.

Latour, B. (2010a). *Cogitamus. Six lettres sur les humanités scientifiques.* La Découverte, Paris.

Latour, B. (2010b). *La science en action.* La Découverte/Poche, Paris.

Mead, G.H. (2006). *L'Esprit, le soi et la société.* Presses universitaires de France, Paris.

Monnerie, D. (2018). Échanges ou élaboration des relations ? Les cérémonies kanak (Kanaky Nouvelle-Calédonie. Océanie). *Revue du MAUSS*, 277.

Trépos, J.-Y. (1996). *La Sociologie de l'expertise.* Presses universitaires de France, Paris.

Tripier, P. (2009). Sociologie et management éclairés par les arts de la guerre ? Les grammaires d'action. In *Je connais la situation mais comment agir ?*, De Coninck, F. (ed.). L'Harmattan, Paris.

Van Staëvel, E. (2006). *La Pollution sauvage.* Presses universitaires de France, Paris.

Verneaux, J. (1973). Cours d'eau de Franche-Comté (massif du Jura). Recherches écologiques sur le réseau hydrographique du Doubs : essai de biotypologie. Université de Besançon, Besançon.

Weber, M. (1963). *Le Savant et le politique.* Union Générale d'Éditions, Paris.

Weller, J.-M. (2007). La disparition des bœufs du père Verdon. *Droit et société*, 713–755.

3

Farm Environment, Raw Milk and Immunity: A "Field" Study of Tolerance Learning

Dominique Angèle VUITTON[1], Jean-Jacques LAPLANTE[2] and Amandine DIVARET-CHAUVEAU[3]

[1] *Université Bourgogne Franche-Comté, Besançon, France*

[2] *ASEPT Franche-Comté/Bourgogne, Besançon, France*

[3] *Chrono-environnement UFC-CNRS, Université Bourgogne Franche-Comté, Besançon, France*

GUIDING QUESTIONS. –

Is the farm environment really involved in protecting children from atopic allergic diseases?

If so, what are the elements determining protection and the crucial periods of their intervention?

What are the mechanisms of the environmental and dietary protective effect?

How does the microbiological biodiversity of the environment and food intervene?

Are non-microbial elements also at work in the modulation of immunity by the environment?

Socioecosystems,
coordinated by Patrick GIRAUDOUX. © ISTE Ltd 2022.

3.1. Introduction: from farm disease to farm protection, a rural environmental story

It is difficult to approach this chapter without going back to the birth of the "Health and Rural Environnement-Franche-Comté University" (*Santé-Environnement Rural-Université de Franche-Comté*, SERF, EA 2276) research team and the intrinsic multidisciplinarity that presided over the creation of this atypical team dedicated, in a School of Medicine, to a topic defined by "the environment". At the end of the 1980s, there was simply no unit, team or institute in any of the major research organizations – Inserm, CNRS, Inra and even in French universities – specifically dedicated to this type of research. The proof: in 1991, these major organizations came to find us[1] to fill in a large blank page in a report on the "environment and health" theme that had been commissioned by the directors of Inserm and CNRS.[2] It happened to be precisely the one that concerned "health and the rural environment". This request came at a good time: 1990 had been the year of the first contracting of university research teams to the French Ministry of Health; in the "health" field of our university, few "disciplinary" proposals had received the ministerial seal of approval, despite a level of scientific publication of which we had nothing to be ashamed?[3] A certain frustration on the part of the local players quickly turned into determination and, on the occasion of the thesis defense of a pneumologist colleague, we had a kind of breakthrough: we had to build on what we had each built on our own on rural pathologies, cemented by the logistical and financial support of the *Mutualité Sociale Agricole* (MSA), the farmers' social welfare organization, to create an *ad hoc* research team; thus we would not step on the Parisians' toes on the "big" serious subjects and we would fill an abysmal void in terms of research in the French landscape. The enthusiastic support of other faculty members, medical doctors and pharmacists, an oncologist, an epidemiologist, a public health specialist and a biostatistician, and the addition of their own research groups to the structure defined in its objective, "health and rural environment", did the rest. In 1992, the SERF team was born and baptized at the ministerial baptismal font, a subtle play on words recalling the rural people of the

1 When I say "we", I am referring, depending on the passages in this chapter, to the "Health and Rural Environment-Franche-Comté University" research team, to the group dedicated to a particular study, the PASTURE study, described in the chapter, or, in a broader sense, to all the actors, academic, institutional, associative and professional, involved in this common action.

2 The report was quickly consigned to oblivion by the successors of those who had commissioned it: the fate of thousands of reports in the bureaucracy of many countries. The issue was apparently no longer relevant. Thirty years later, one wonders how far research on this topic would have progressed in France if this report had been followed up.

3 We may suppose the argument was: "why do badly in the countryside what we do very well in the city", especially Paris, the city of light, of course, with some provincial exceptions!

Middle Ages ("serf", in English as in French).[4] "The SERF" was a great adventure, not to say a great escape,[5] out of the usual "administrative mold" – and still as original as ever long after its creation.[6] We had opted for a "rotating" team leadership, in order to highlight each of the senior researchers of the team who desired it and to avoid long-lasting power grabs. The strength of several of the SERF members was to use their skills as clinicians or epidemiologists in synergy with the biology laboratories, mainly in mycology and immunology, and also in toxicology and bacteriology, and to acquire an international reputation in their respective research fields. It was through the international detour that we convinced the national authorities of our legitimacy.[7] For several of its members, the SERF team was also an apprenticeship in "research-action", with the MSA's transmission belt. The MSA[8] helped us to implement our research projects "in the field" and to apply many of our research results, first on "rural" infectious diseases, such as alveolar echinococcosis, or occupational agricultural diseases, such as farmer's lung. Developing quality research that has a real impact on society has gradually become the common denominator of the SERF, where research-supported actions and research tools have been developed: detection of echinococcosis, in the region, and also as far as China and elsewhere worldwide, and the creation of a national registry with a European vocation, screening and management of farmer's lung and also its prevention through changes in professional practices, setting up of one of the very first French cancer registers, which would allow reliable data to establish and evaluate health policies, the first serious (and immediately controversial) studies on hematological malignancies in the vicinity of nuclear reprocessing plants, or the role of incineration plants in the production of dioxins. The irruption of allergology within SERF only came later, as we will see further on in this chapter.

---

4 I don't know why in French it has always been masculinized, to the point that it became a wild animal ("cerf", the "deer"!) for some.

5 Who, indeed, among the ex-members of the SERF does not remember the green Citroën Evasion ("Escape" in English!) petrol (not diesel!) engine, to be ecological? Minibus, van, field lab, international transport, it did it all, the Evasion, after we were able to convince the university, not without difficulty, of its usefulness, in spite of the many administrative pitfalls we were confronted with. Moreover, we practiced excelling at this university version of martial arts that is the fight against administrative pitfalls, but it must be said that the exercise has now become increasingly difficult…

6 The theme of "health and the rural environment" is still not very popular…

7 Going through Beijing, London or New York to establish one's scientific credibility and to "pass on" multidisciplinarity sometimes seems to be a shorter route than the Province-Paris high-speed train… "No one is a prophet in his own land", said a countryside preacher with an international and lasting destiny.

8 But above all, the men and women of the "preventive-occupational medicine" department of the Doubs MSA.

3.2. Atopic allergic diseases: multifactorial, multidisciplinary and, paradoxically, not very rural

As early as the 19th century, two English physicians, John Bostock, physician and geologist, in 1828 (Bostock 1828), and Charles Harrison Blackley (1873), country doctor, in 1873, had described "hay fever" and rightly noted its greater prevalence in noble families or among the wealthy bourgeoisie and the inexplicable apparent protection of peasants, despite their contact with hay, identified at the time, and rightly so, as the main trigger of the clinical manifestations (Liu and Leung 2006). What was later called "allergy" and which two American physicians of the early 20th century, Coca and Cooke, described as "atopic" (i.e. "placeless", "unclassifiable" and "strange") was thus from the outset a disease of the "castle", the "rich", the "distinguished", as Bourdieu would have said, separated from the peasant masses by a way of life and an environment where, to use the well-known distinction of the English language dating back to William the Conqueror, *pork, beef and mutton* were on the tables of the well-to-do while *pigs, cows and sheep* were in the fields but often also shared the shelter of the poor – a linguistic premonition.

The very broad and clinical definition of atopy by Arthur Fernandez Coca (1875–1959) and Robert Anderson Cooke (1880–1960), two New York physicians and academics, in 1921, made it possible in a way (before the immunological foundations were revealed and, in particular, the role of IgE class antibodies) to "classify the unclassifiable": "individuals who, considered as a group, have the peculiar ability to become sensitive to certain proteins to which their environment and lifestyle expose them" (Coca and Cooke 1923).

As early as the Roaring Twenties, the big words were out: "environment and lifestyle" and, even then, the interference with a genetic predisposition was clearly seen (one could identify "families of atopic allergics" and the often hereditary character of "predisposition to allergy") (Edfors-Lubs 1971), and also the variety of clinical presentations of the "allergic" diseases affecting these families: hay fever and allergic rhinoconjunctivitis in general, but also asthma and atopic dermatitis, cutaneous and mucosal angioedema (laryngeal, intestinal, etc.), urticaria and anaphylactic shock, defying the convenient structuring of medicine into "organ" disciplines and immediately placing allergic diseases in the field of a "multidisciplinary" approach.[9]

9 The administrative translation of these observations took about a hundred years in France, since allergology was only officially recognized by a diploma validating the practice of a fully fledged speciality in 2017, after having been subjected to attacks from various disciplinary

The word "atopy" already implied in the last century, and continues to do so, genetic factors, environmental factors and the unavoidable intervention of the immune system. In order to understand allergy in the immunological sense of the term, which we will henceforth use to refer specifically to atopic allergy, it is therefore necessary to understand the immune system, which is the real interface between our genetic characteristics and our environment (Vuitton 2008). However, it should not be forgotten that this immune system, whose orientations are defined by the genetic characteristics of the species and of each individual, is likely to be permanently, but especially in particularly sensitive periods of life, "modulated" by the environment, within which the microbial ecosystem (viruses, bacteria, fungi and parasites) plays a major role. The learning of the immune system, from fetal life and in the first years of life, is thus associated with genetic predispositions to determine the aptitude of an individual to develop certain types of responses. When encountering the various components of the environment – microorganisms, and also inert molecules of a diverse biochemical nature (xenobiotics) – the immune system has roughly two options: to destroy the intruder, which is commonly called an "effector" response (with antibodies that designate the targets, and/or toxic cells or enzyme systems), or to tolerate the intruder, which is called a "tolerogenic" response. It took most of the 20th century to understand, on the one hand, that the so-called effector response required the simultaneous presence of antigenic determinants ("epitopes") of the intruder and of danger signals, identified by the so-called "innate" immunity, and, on the other hand, that the tolerogenic response was in fact much more frequent than the effector response and that it involved complex cellular and molecular phenomena, as active as those involved in the so-called "effector" response (Vuitton 2008). From an ecosystem perspective, a balance is most often desirable. Microorganisms have no interest in killing their host and the host has an interest in not depleting its resources to fight against proteins that feed it or non-pathogenic microorganisms that are otherwise indispensable in its microbiota (the community of tens of billions of microorganisms that cohabit with it, mainly but not exclusively, in its gut[10]) (Vuitton and Dalphin 2017; Hillon 2020). The mucous membranes (especially digestive ones) are the site of first contact: they are the most important immunological site of the organism and the "tolerogenic" arsenal is the most developed there: regulatory T-cells of several types, secreting cytokines such as TGF-β and IL-10, and IgA-class antibodies, secretory or not

corporatisms for over 40 years. Allergology has been a recognized medical specialty for several decades in other European countries and the United States.

10 In the course of evolution, integrated viruses have even become part of the masters of cellular function and metabolism in "higher" organisms.

(Vuitton 2008; Weiner *et al.* 2011).[11] The medical vision of the immune system has been disrupted since the very beginning of its study at the end of the 19th century by the use of immune markers for diagnosis and by technical aspects which mean that, on the one hand, the approach to the functioning of the system will generally be based on a blood test, and, on the other hand, only interactions with pathogenic microorganisms and with proteins will be retained when they result in a pathological reaction; the other interactions, with non-pathogenic microorganisms and with proteins apart from allergic reactions, are generally totally neglected, despite being much more numerous.

Since the 1970s, we have known the essential role of gut microbiota in the development of the immune system: the immune system does not develop without microbes; axenic mice (without contact with the microbial environment, thus with a sterile gut) remain at an embryonic level of immune development that makes them extremely vulnerable to any microbial attack when it occurs (Vuitton 2008; Vuitton and Dalphin 2017; Hillon 2020). However, it was not until the 21st century that this fundamental role of gut microbiota and exposure to the microbial environment was taken seriously, and research findings on the interference between genetic predispositions and the environment in fetal life (by the mother) and in early childhood led to the idea that this pre- and neo-natal environment not only influenced the ability to fight pathogens, but also the ability to tolerate the multiple components of the environment, inhaled or ingested, not only microbial but also protein or non-microbial lipopolysaccharide. In this context was conducted the research on the influence of the environment of a livestock farm in the protection against the occurrence of allergic diseases during the life of an individual, which will be discussed in this chapter (Vuitton *et al.* 2019).

3.3. The increasing prevalence of atopic allergic diseases in the population of developed countries: an enigma of the second half of the 20th century

It was the observation of an unprecedented increase in allergic diseases in the European, American and Japanese populations after World War II that precipitated the implementation of research in these areas. Each decade saw a doubling in the number of patients with the major clinical manifestations of atopy, asthma, allergic rhinitis and atopic dermatitis (Strachan *et al.* 1997; Von Mutius 1998; Williams *et al.* 1999); this increase in allergy prevalence, from less than 5% to more than 30%

11 Those antibodies that are never measured because they pose technical difficulties for their collection and are less informative – and for good reason – in the diagnosis of infectious diseases associated with pathogenic microorganisms.

of the general population in less than 50 years, seems, however, to have reached a plateau at the end of the 20th century, at least in the most developed countries in Europe, particularly for asthma (Aubier *et al.* 2005). Food allergy, which was extremely rare in the 1950s, regardless of its clinical manifestation (cutaneous, digestive, systemic leading to anaphylactic shock, and even respiratory), has also increased considerably over the same period (Poirot *et al.* 2020). Allergens rarely involved before were recognized, such as peanuts, latex and exotic fruits; there was also an increase in the frequency of so-called "cross-reacting" allergies, where allergic manifestations could be in the same "sensitized" patients triggered by food allergens (fruits, seeds, crustaceans, fish, etc.) or inhaled allergens (pollen, dust mites, etc.) (Faber *et al.* 2018). The observed increase in allergic diseases was initially attributed to better diagnostic recognition, as well as to respiratory or digestive contact with substances now present in the environment that were rarely encountered before the mid-20th century (e.g. pollen from imported trees and plants, foods of exotic origin[12]) or whose modes of encounter were different (e.g. cats and dogs as pets). The low prevalence of allergy in Africa has long been attributed to a lack of diagnosis and the absence of consideration of minor diseases in countries where major infectious problems and/or infant mortality were prevalent. However, in countries with equivalent medical care, in Europe and the United States, it was found that the prevalence of allergic diseases was higher in more industrially developed countries and in cities, and that immigrants from poor countries with low levels of allergy "catch up" to their host country in terms of prevalence of allergy in the next generation born in that country (Leung 1996; Grüber *et al.* 2002; Ventura *et al.* 2004). To simplify matters somewhat, it was easy to oppose rich and poor countries (or areas), with the level of "development" (in the sense given to this word at the end of the 20th century) being a determining factor: allergy in the population became, paradoxically, a marker of "development" (Williams *et al.* 1994, 1999; Strachan *et al.* 1997; da Costa Lima *et al.* 2003). Industrial pollution was then blamed for the increase in allergic diseases, as the most serious and visible manifestation of allergy was asthma, and it was quite natural to acknowledge a role for different types of air pollution in the origin and maintenance of respiratory diseases. It is now well accepted that pollution – industrial but especially motor vehicle pollution – is an important factor in the aggravation of allergy symptoms and especially asthma, and that it is not responsible for the increase in the prevalence of

12 These are the many decorative plants in cities, which were not used in the past (but which, in the end, are not as allergenic as all that), but above all birches (in the north) and cypresses (in the south), which were planted much more in France from the second half of the 20th century, and ragweed, whose allergenic potential is very important, which came to us from the United States in the 19th century but really took off along the Rhône valley due to the major motorway works undertaken after the World War II.

allergic diseases in the second half of the 20th century (Von Mutius 1998)[13]. The conceptual change initiated by the English allergist David P. Strachan, who proposed the so-called "hygiene hypothesis" in 1989 (Strachan 1989), has left many allergy researchers (mainly pulmonologists, who are very involved in urban pollution research) rather doubtful. The idea that "old-fashioned" life, with its large families and infant mortality, serious infections, poverty, malnutrition and dirtiness, could be a "positive" environment, which protected against frequent and potentially serious diseases, was shocking. It must have contributed to the fact that research in this field was not very popular and that the results of the cross-sectional epidemiological studies carried out in the 1990s, which supported Strachan's general hypothesis and refined its contours, were accused of suffering from irreducible biases that definitely made their conclusions untrustworthy. Pasteur and almost 100 years of hygienism had accustomed doctors to consider this "progress", hygiene, only as a factor promoting health. A reversal of the paradigm offended these convictions and still does.

It is in this context that the PASTURE (Protection Against Allergy: Study in Rural Environments) child cohort and its associated research projects were set up in 2001, in order to cut through the methodological criticisms inherent in population-based cross-sectional studies and to have a "longitudinal" follow-up, in real time, of children who had or had not been exposed from birth to the traditional environment of a cattle and dairy farm. This follow-up would concern prospectively, simultaneously and in real time the clinical events, environmental factors, lifestyle, and main immunological parameters supposed to be explanatory (Von Mutius *et al.* 2006). This "cohort study", which thus included an integrated approach to immunity in its microbial ecosystem, a new way of "thinking about immunity in terms of the environment", aimed to answer the many questions that arose in the late 1990s from cross-sectional studies (Figure 3.1): is the farm environment really involved in protecting children against atopic allergic diseases? If so, what are the independent elements determining protection (animals, microbes or raw milk, all of which had been identified by the cross-sectional studies, or other non-identified factors?) and what are the crucial periods of their intervention? What are the mechanisms (immune, in particular) of the environmental and dietary protective effect? Is the environment in the fetal period immunomodulatory before the child's exposure to its environment? Is there a "window of opportunity" for environmental influence in the postnatal period? What is the role of the microbiota? How does the microbiological composition of the environment and food intervene? Are non-microbial elements

13 The intervention of pollution in the epidemiology of allergy remains the most studied aspect in France (and in many other countries) and the one that mobilizes the most funds, both public and private, for research.

(from the "breathed" environment? from food?) also at work in the modulation of immunity by the environment? The children recruited are now between 17 and 19 years old; the study has given rise to some 60 scientific publications; the results confirm most of the initial hypotheses. It is this long-term, international and multidisciplinary process that the remainder of this chapter aims to describe, as well as the original, sometimes unexpected, results that were obtained and, in particular, the emphasis that they finally placed on the concept of "environmental diversity for an optimal balance of the immune system", biodiversity being at work from the pre-natal period.

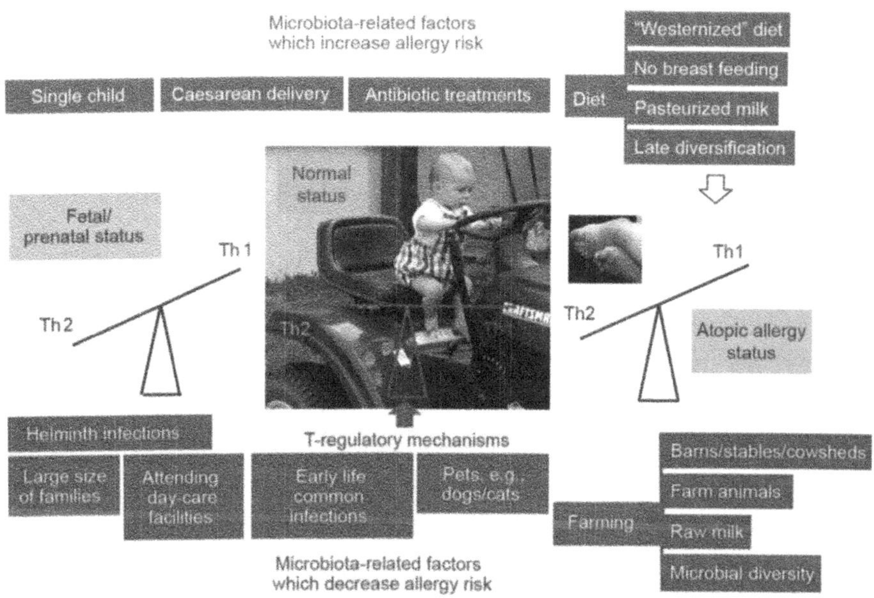

Figure 3.1. *Working hypotheses from cross-sectional epidemiological studies in the late 20th century, linking environmental factors, immune system development and protection against clinical manifestations of atopic allergy. For a color version of this figure, see www.iste.co.uk/giraudoux/socioecosystems.zip*

COMMENT ON FIGURE 3.1.– *In green: factors that promote the emergence of allergic diseases. In red: factors that promote protection against the emergence of these diseases. The fetal cytokine immune profile promotes a state of immune tolerance that shares many of its characteristics with the "atopic allergic" profile. The "environmental" learning of the immune system promotes a return to equilibrium, adapted to the fight against pathogens; these corrections are, among others, brought by the activation of regulatory lymphocytes.*

3.4. The role of the farm environment in protecting children from atopic allergic diseases

3.4.1. *Observations from cross-sectional epidemiological studies*

Since the 1990s, a large number of cross-sectional epidemiological studies have supplemented the initial observations of the 1980s. The most important studies have taken advantage of the existence of geographically contiguous regions with similar allergenic exposure but which, for reasons of political and socio-economic organization, were in profoundly different situations in terms of environment and lifestyle (Von Mutius *et al.* 1992; Von Mutius 1998; Vartiainen *et al.* 2002; Von Hertzen *et al.* 2006a, 2006b). The results of the various studies can be summarized as follows: atopic allergy was indeed more frequent in industrially developed countries and urban areas than in poor countries and rural areas, and more common in one-child families than in large families (but day care in a collective nursery, like the large family, was also "protective"), and in children or young adults who had had fewer infectious diseases.

The combination of these epidemiological results gave rise, as we have seen, to the explanatory hypothesis of "hygiene", capable of accounting for the observations by resituating them in the immunological concepts of the time (the period 1980–1995). These immunological concepts were based on the dichotomy of the orientation of the adaptive immune system (governed by T-cells), determined by the type of cytokines produced by the T *helper* cells (Th-cells). The system was then conceived as a two-sided Janus (or the two pans of a scale), the Th1 side[14] responsible for fighting viral and bacterial pathogens, and the Th2 side responsible for fighting helminthic pathogens (worms, intestinal and otherwise) and associated with allergy (Vuitton 2008), as summarized in Figures 3.1 and 3.2. Protection from IgE-dependent allergy through contact with microbial pathogens was initially central to the so-called "hygiene hypothesis". In this view, the protection against allergic diseases progressively reduced by the increase in the standard of living would have been caused by a series of associated factors, such as the decrease in family size (protecting children against common infectious diseases), the development of

14 The Th1 type T lymphocytes mainly secrete the cytokine Interferon-(IFN-) gamma (γ); in this way, they influence B lymphocytes to produce IgM and IgG type antibodies, and other lymphocytes that are "cytotoxic" by direct contact with the infected cells; the Th2 type T lymphocytes mainly secrete the cytokines Interleukin (IL)-4, -5 and -13; In this way, they influence B lymphocytes to produce IgE antibodies, which in turn are responsible for triggering vascular inflammatory reactions by mediators contained in basophilic and eosinophilic polynuclear cells and mast cells in the skin and mucous membranes; these "end-of-pipe" reactions are the cause of the clinical manifestations of atopic allergy (Figure 3.2).

hygiene, vaccination[15] and antibiotic therapy (reducing exposure to pathogens in early childhood), better water and waste management (avoiding food poisoning, hepatitis A in the barracks or high school, etc.), and "ready-to-eat" food, in plastic, with an expiry date (thus free of microbial contamination), and as a corollary to a certain weakening of Th1 type immune responses (Strachan 1996).

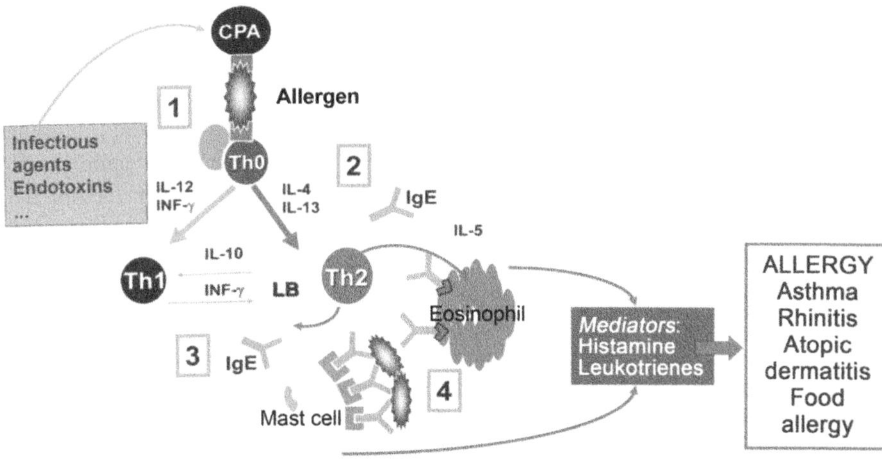

Figure 3.2. *Main steps in the immune response leading to the clinical manifestations of atopic allergy. For a color version of this figure, see www.iste.co.uk/giraudoux/socioecosystems.zip*

COMMENT ON FIGURE 3.2.– *1) Encounter of the allergen (an environmental antigen) with relatively undifferentiated Th0 T-cells, after presentation by an "antigen-presenting cell" (APC); this APC is influenced (in its maturation and cytokine production) by the prior or concurrent encounter with microbial agents or their products (e.g. endotoxins). 2) Cytokines produced by the APC-Th0 couple direct the response toward the Th1 profile (which promotes the response to microbial agents, by IgM and IgG class antibodies, and the activation of cytotoxic cells – not represented in the figure). The absence or paucity of these stimuli may, on the other hand, increase the capacity to respond with a Th2 orientation, which promotes the*

15 It has been shown that vaccination *per se* is not the cause; it can be assumed that the immune stimulation it generates is less "violent" than that generated by the infectious disease itself; but the studies on the influence of vaccination are the most discordant among those that have been carried out on this subject. Moreover, the benefits of vaccination against severe and frequent diseases are out of all proportion to the part played by vaccination in the relative "protection" against allergic diseases, as shown by the subsequent studies elaborated in this chapter (Smits *et al.* 2016).

stimulation of B cells for the production of IgE-type antibodies. 3) These IgE antibodies then bind to inflammatory mediator-producing cells, mast cells and eosinophils; when they encounter the allergen specific to them, they release chemical mediators (ready to use in granules or rapidly synthesized) which are the cause of the symptoms of allergic diseases (dilation of the small vessels, edema and, at the maximum, "anaphylactic shock", and/or contraction of the bronchi or intestine).

However, this unequivocal explanation was quickly challenged by observations showing that, like viral and bacterial infections, parasitic helminth infections also protected against allergy, which could seem paradoxical, given the "immune kinship" between parasitic infections and allergies, which share the same profile (Th2) (Yazdanbakhsh *et al.* 2001, 2002). On the other hand, other factors that suggested the positive action of non-pathogenic microbial agents in relation to changes in diet or contact with animals gave rise to other alternative explanations in which hygiene, as such, was not the most important determinant.

Among these, life on the farm represented a set of protective factors that several converging studies very well highlighted. The expression "hygiene hypothesis" has endured and has become established to designate all the factors of "modern life" likely to facilitate the onset of allergic diseases, even if hygiene is not ultimately the main driver of allergic susceptibility, and given that its "inventor" himself, David P. Strachan, helped to emphasize farm life and the role of non-pathogenic microbial ecosystems as a major protective element (Perkin and Strachan 2006; Genuneit *et al.* 2013).

3.4.2. *Farm life: a protective factor against the occurrence of allergic diseases?*

3.4.2.1. *Observations from dedicated cross-sectional epidemiological studies*

Observations from general epidemiological studies thus indicated a possible role of the rural environment and, more specifically, of life on a farm. The first specific studies, dedicated to the elucidation of the "city-country" paradox, pointed to the responsibility of birth and life on a farm in a certain level of protection against the subsequent occurrence of allergic diseases (Von Mutius and Vercelli 2010; Von Mutius 2012).[16] An analysis of a randomized sample of 6,251 adults aged 20–44 years included in the European Community Respiratory Health Survey confirmed these observations (Svanes *et al.* 1999). This somewhat anecdotal

16 These studies were conducted concurrently in Switzerland, Germany (Bavaria), Austria (Tyrol), Finland and Canada (Quebec).

and "ill-considered" observation of "protection by the farm", at the margin of the hygiene hypothesis, then became a key element in reflections on the increase in allergic diseases in the second half of the 20th century: moreover, did the increase in the prevalence of allergic diseases not also coincide with an unprecedented decrease in the agricultural population in developed countries? With an abandonment of the style of life, family structure, environment, food and contact with animals that had dominated for centuries? Not since the transition from the Paleolithic and its hunter-gatherers to the Neolithic and the introduction of agriculture 10,000 years ago have such profound societal upheavals occurred in the lifestyles of a large part of humanity.

While confirming the general findings on the protection of the farm environment against the occurrence of atopic allergic events, the European ALEX[17] study showed that exposure to the farm environment and the consumption of unpasteurized milk were independently and synergistically associated with a significantly lower prevalence of asthma, hay fever and atopic sensitization than those observed in unexposed children (Riedler *et al.* 2001)[18]. Since 2001, several studies have attempted to determine whether the protection conferred by farm life is a universal phenomenon (Braun-Fahrländer 2002). With some variations, represented mainly by the type of atopic manifestation most associated and/or the mode of expression of allergenic sensitization or atopic terrain, almost all studies conducted between 2002 and 2006[19] have confirmed the protection exerted by the agricultural environment, while at the same time providing details on the factors potentially involved (Vuitton and Dalphin 2006). The consumption of raw, unpasteurized milk soon emerged as an unexpected but major factor, independent and almost constantly found (Waser *et al.* 2007; Von Mutius 2012).[20] This type of consumption could represent an integration of microbial, antigenic and nutritional factors, capable, moreover, of intervening already during the gestation period through the mother's consumption. The PARSIFAL study, which studied in parallel the possible protective factors in children of farmers and children attending Steiner schools,[21] compared to control

17 This study was carried out among Swiss, German and Austrian families in a very similar environment (the mid-Alpine mountains).

18 The observed prevalences are, for asthma: 0.8% versus 11.8%, for hay fever: 0.8% versus 16%, and for atopic sensitization: 8.2% versus 32.9%; these prevalences are very significantly lower than those observed in unexposed children.

19 Mainly in Europe – Austria, Germany, Sweden, Finland, Norway, France and Denmark – but also in the United States and Canada.

20 Following initial observations in alpine regions (Riedler *et al.* 2001), it was also found in Great Britain, where its influence was confirmed to be independent of farm life and shown to be associated with a lower frequency of allergenic sensitization (Perkin and Strachan 2006).

21 These schools use pedagogical principles based on anthroposophy, an esoteric doctrine, often accused of sectarianism, at least in some countries; this pedagogy is based on the development

children, also fully confirmed the protection against atopy of children living on the farm, and the part played by raw milk consumption in this protection, also for children who do not live on a farm but attend Steiner schools (Alfvén *et al.* 2006).

Apart from completely opposite results from a study conducted in Vietnam (Chai *et al.* 2004), the less clear results obtained in other studies have sometimes been used as an argument against the reality of a protective effect of life on the farm.[22] These mixed results then raised the question of the type of farm concerned by the "protective" effect, and the traditional dairy farm, especially breeding cattle, in mid-mountain regions of Europe appeared to be the most protective (Von Mutius and Vercelli 2010). The main factors evoked to explain the effects of this environment on the prevention of allergy emergence were then of two types: microorganisms, mostly non-pathogenic, present in the stable, hay, straw, silage and the substances they produce, first and foremost the bacterial endotoxins; animals, mainly livestock, especially cattle, through their intestinal flora, their feces and manure or slurry from them, but also their secretions and their milk more or less contaminated by microorganisms. Finally, an additive role of guard animals and/or pets, dogs and cats, traditionally present on these farms, could not be excluded, a role strongly suggested by the results of the ALEX study (Waser *et al.* 2005).

3.4.2.2. *Immunological correlations from dedicated cross-sectional studies*

Endotoxins were the first substances of microbial origin mentioned as one of the elements of protection conferred by this agricultural environment of dairy production (Von Mutius *et al.* 2000; Braun-Fahrländer 2003). Lipo-polysaccharide substances derived from the walls of many bacteria, including enterobacteria, block the production of specific IgE in experimental animals and, in children at high risk

---

of the child, the personalization of teaching, and the early learning of the arts and languages. Families who adhere to the philosophy developed by Rudolf Steiner, an Austrian spiritualist and occultist philosopher from the end of the 19th century, refuse, among other things, vaccination and the use of medicines, and are in favor of a traditional and organic diet (including raw milk) and a life close to nature.

22 In 2001, a study in Crete failed to show a significant difference between children living on farms and children in the capital, Heraklion; however, there was a negative association between allergenic sensitization and contact with animals, as well as with the consumption of raw farm milk, an association that was particularly found in urban children (Barnes *et al.* 2001). A study of New Zealand children found a high frequency of asthma-like respiratory events in farm children; however, allergenic sensitization was low and there was a negative association between weekly yogurt consumption and hay fever, unpasteurized milk consumption and atopic dermatitis, the presence of cats and/or dogs and both diseases, and between the presence of pigs on the farm and sensitization to common atopic allergens (Wickens *et al.* 2002).

of asthma, are likely to induce a Th1 response associated with a decreased frequency of sensitization to house dust allergens (Gereda *et al.* 2000). Cattle are considered to be the major contributors of endotoxins to the farm environment, but cats and dogs also disperse high amounts of endotoxins in their feces (Heinrich *et al.* 2001); at the same time, studies in mainly urban environments have shown that the presence of pets in children is protective when these animals are present from birth or even before birth, during the mother's pregnancy (Hesselmar *et al.* 1999; Roost *et al.* 1999). Cats and dogs, which are common in dairy farming, could therefore participate in the protective effect of the farms via their droppings, which disseminate endotoxins. Endotoxins were thus "ideal candidates" for linking cats, dogs, cows and allergy. The results of the ALEX study showed that endotoxin concentrations were particularly high in dairy farming habitats and that their concentration in the domestic environment was inversely correlated with the allergic risk in school children[23] (Braun-Fahrländer *et al.* 2002) and with atopic asthma in adult farmers (Eduard *et al.* 2004).

The demonstration in farm children of a correlation between protection and the level of expression or polymorphism of certain receptors of innate immunity[24] (Lauener *et al.* 2002; Eder *et al.* 2004, 2006) (the TOLL receptors, or TLRs, discovered in the same period at the end of the 20th century) contributed to support the hypothesis of a particular role of endotoxins, ligands of these receptors, and also opened up reflection on the intervention of other compounds of microbial origin that

23 The production of Th1-type cytokines (but also of IL-10, a "regulatory" cytokine) by the lymphocytes of the children in this study was inversely correlated with the levels of endotoxins measured in the mattresses of these children, suggesting a modulation of their overall immune responses (Braun-Fahrländer *et al.* 2002). The effect of the farm environment seemed to be as strong, or even stronger, in children of atopic parents than in those without atopic heredity, a key observation suggesting interactions between genetic predisposition and environmental influences (Remes *et al.* 2002).

24 Innate immunity is a "first line" system that recognizes, on the one hand, molecular groupings typical of different types of microbes (molecular "motifs" or PAMPs, Pathogen-Associated Molecular Patterns) and, on the other hand, "danger signals". It is the combination of the two that will orientate the following immune response toward an "effector" direction (to get rid of the intruder) or "tolerant" (to support the presence of the intruder while preventing it from becoming too much invasive). The endotoxin-TLR4/CD14 couple is an example of the recognition of constituent elements of certain bacteria (without prejudging their species) by innate immunity. The discovery of this "innate" recognition system was the reason for the most recent award of a Nobel Prize in Medicine to a Frenchman, Dr Jules Hoffmann (Institute of Molecular and Cellular Biology, Strasbourg), in 2010.

are not necessarily pathogenic (Beutler 2002; Moresco *et al.* 2011)[25]. Despite their conceptual interest, the clinico-microbio-immunological correlations found at school age did not automatically prejudge a cause-and-effect relationship for phenomena that had likely occurred from the earliest years of life, or even only during the perinatal period, which is crucial for immune system development. Other studies on older children in other countries, for example, have provided conflicting results (Raby *et al.* 2002; Nakashima *et al.* 2006).

3.5. Setting up the PASTURE cohort in Germany, Austria, Finland, Switzerland and Franche-Comté

The reflection was at the stage where cross-sectional studies were reaching their limits when several European research teams noted the convergence of their intentions: to establish a cohort (or cohorts) of children observed from before birth, making it possible to confirm with an indisputable methodology the hypothesis of a particular relationship between the farm environment and protection against atopic allergy, and to explain the factors and mechanisms. Until then, the pulmonologists of the SERF research team had focused on the specific respiratory manifestations of work on livestock farms, such as farmer's lung and the various types of agricultural bronchopathies, occupational diseases related to antigens specifically present in the farm environment (Reboux *et al.* 2001). They had noted in passing that sensitization to allergens common in the atopic population and/or a history of atopy and/or the frequency of allergic asthma were rather less represented in this "farming" population (Dalphin *et al.* 1998). This group of the SERF team, with the support of the Doubs MSA, had become in about 10 years an essential reference point in the field of agricultural respiratory pathologies, in France and worldwide, and the mycology laboratory of the Regional University Hospital of Besançon had acquired recognized expertise in the study of the microbial content of professional environments, especially dairy farms (Dalphin *et al.* 1991; Reboux *et al.* 2001; Roussel *et al.* 2005). The presence of a clinical immunologist-allergist in the same research team provided the opportunity for joint projects in the field of "rural environment and allergies".

After initially considering setting up a French cohort of children recruited in the main traditional livestock farming regions (Jura, Alps, Auvergne) with the advice of a team of epidemiologists from Inserm (Inserm U 472), discussions with Erika von Mutius, who had just brought together the Austrian and Swiss groups already

25 Indeed, children living on the farm and protected by their lifestyle had not only high expression of CD14, a TLR-4 co-receptor for endotoxins, but also high expression of other TLRs, such as TLR-2, which recognizes components of Gram-negative bacteria and mold peptidoglycans (Lauener *et al.* 2002).

involved in the ALEX study and a Finnish team, convinced us that it was more operational to join this emerging consortium by setting up a Franche-Comté cohort as soon as possible in line with the "case-control" methodology (in this case "exposure-non-exposure") adopted for the European PASTURE study (Von Mutius *et al.* 2006). The possible addition of 200 French children to the initially planned cohort of 800 children strengthened the statistical power of the study, and the areas of expertise specific to the SERF research team were complementary to those already present in the rest of the consortium of the PASTURE project, which had just been accepted for funding by the 5th Framework Programme for Research and technological Development (FPRD) of the European Commission's Directorate General for Research and Innovation. Recruitment of the families enrolled in the study had just begun. The SERF team was therefore faced with three imperatives: to find start-up funding, to find the necessary staff to manage the questionnaires and samples and, above all, to recruit 200 pregnant women in their third trimester, if possible within a year.

Since the period 1985–1990, the close collaboration with the MSA of the group dedicated to respiratory diseases in rural areas and the group dedicated to the study of alveolar echinococcosis (AE) (see Chapter 4) was one of the bases for the creation of the SERF research team. The MSA had been a major player with the SERF team in the creation in 2000 of the PRISMAL project (Perception and representation of health risks in rural areas), a "regional public health research network", in response to an Inserm call for tenders. This network allowed the joint recruitment of a research mediator, in charge of the link between the SERF team and the MSA, and of the communication on the research-action activities in public health carried out by the two entities. The positive nature of the "allergy protection" issue for the dairy farmers of the Franche-Comté region, at the end of a decade that had seen profound upheavals in the agricultural sector owing to the outbreak of mad cow disease, seemed to the MSA occupational medicine-prevention team to come at a crucial time and to represent an opportunity that had to be seized. The prospect of a slightly more optimistic view of the profession and of the way it is viewed by the general population could both motivate young farming families and other MSA beneficiaries in the same geographical areas to take part in an ambitious scientific study, and convince the administrative bodies responsible for funding MSA prevention activities to subsidize them.[26]

The setting up of the cohort and the associated studies had to involve professionals and researchers from various disciplines (epidemiologists, pulmonologists, dermatologists, allergists, gynecologists, pediatricians, biologists,

26 Without the support of the MSA, the legal, ethical and financial obstacles could never have been removed and the French branch of PASTURE never allowed to see the light of day.

immunologists, microbiologists – including bacteriologists, mycologists, parasitologists – geneticists and physiologists), both for the development of the questionnaires and the standard operating procedures essential to the quality of a study taking place in five countries and for more than 15 years, and for the collection of medical data, the technical management of samples and measurements in the laboratory, and their analysis. In addition, the "environmental" part of the study involved sampling in homes, carried out by specialized personnel at the beginning of the study (visiting families at the age of 2 months for children), or for sub-studies dedicated to a sample of farms, for example, as well as calling for the active cooperation of families throughout the study (sending sampling material and instructions for use, collecting samples taken by the families and then sending them to the various laboratories in charge of the different types of measurements and analyses). One of the most sensitive aspects, or rather the "weakest link" in a study of this type, was the logistics of transporting the samples to the dedicated laboratories all over Europe.[27]

However, the cohesion of the PASTURE cohort steering group and, in particular, the high professionalism and exceptional human qualities of Erika von Mutius, the study coordinator, made it possible to overcome all obstacles (with the exception of Austria, where, when the coordinator of the Austrian cohort retired, it was not possible to find a successor ready to take over), to maintain the cohort until 2021 with children now on the verge of adulthood and to continue to envisage new studies. The adherence of families to the project, in all countries, and particularly in Franche-Comté, was a major element of the study's success: it benefited greatly from the help of the research mediators.[28]

27 In particular, samples taken for the detailed study of the functioning of the immune system, which require either extemporaneous samples that must be processed within 24 hours, or samples frozen in liquid nitrogen that must be managed under very strict conditions for their transport; trivial problems of transport and thawing without explanation by the sending or receiving laboratories (and even less by the carriers or airlines!) which may have compromised this or that part of the analyses in this ambitious project.

28 Nadia Guillou and then Cécile Travers, first co-financed by the SERF team and the MSA, then by the MSA and the *Association pour la santé et la prévention sur les territoires* (ASEPT), who successively ensured warm communication with the families whenever the slightest problem was encountered, managed the reimbursement of travel expenses to the families, regularly reported on the progress of the study in annual newsletters sent with the New Year's greetings, selected small gifts for the children during the visits, organized events bringing together the "PASTURE families" and the researchers and research staff involved on several occasions in a festive atmosphere, and provided the interface with the media, in particular when filming for television programs.

As much as it was a scientific adventure, the PASTURE study was also a human adventure where multidisciplinarity and multinationalism, with unavoidable but enriching cultural differences, not only required effort from everyone but also brought great satisfaction. The adventure ended earlier than expected, in 2019, for Jean-Charles Dalphin who had initiated it. This chapter is dedicated to his memory. He had succeeded in bringing on board a young intern in pediatrics who wished to specialize in allergology,[29] and in sharing with her our enthusiasm for this study; it is now up to her to lead the PASTURE cohort toward new developments.

The methodological outline, based on a longitudinal "exposed-not exposed" cohort design, the recruitment regions and the conduct of the PASTURE study are presented in Figure 3.3.

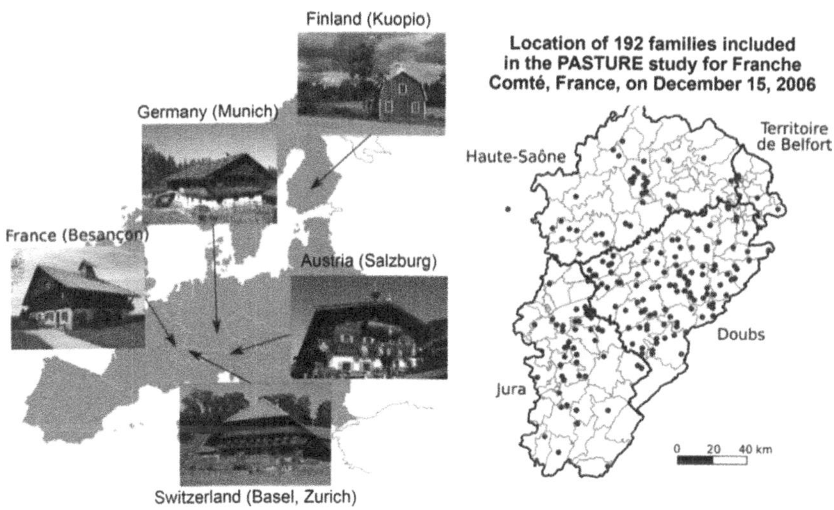

Figure 3.3. *PASTURE cohort study sites. For a color version of this figure, see www.iste.co.uk/giraudoux/socioecosystems.zip*

COMMENT ON FIGURE 3.3.– *The study involved five rural areas (settlements with fewer than 5,000 inhabitants) in European regions with traditional dairy farming, as shown on the map of Europe. In each country, 100 women in their third trimester of pregnancy living on a farm and 100 from the same rural areas not living on a farm were recruited. On the map of Franche-Comté, each circle represents a family followed at least until the children were four and a half years old (192 families).*

29 Amandine Divaret-Chauveau, now a university hospital practitioner at Nancy University Hospital, co-author of this chapter.

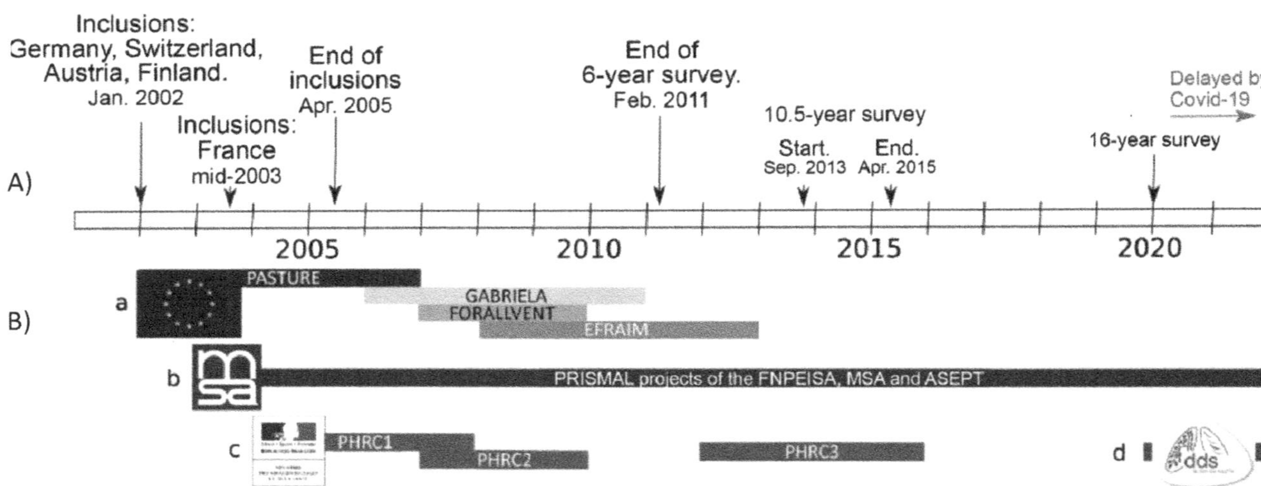

Figure 3.4. *Timeline of the study. The timeline shows the main stages of the study (A) and its funding arrangements (B)*

COMMENT ON FIGURE 3.4.– *a) European Commission Framework Programmes for Research and Technical Development (FPRD). b) French National Fund for Agricultural Prevention, Education and Health Information (*Fonds national de prévention, d'éducation et d'information sanitaire agricole, *FNPEISA) of the* Mutualité Sociale Agricole *(MSA) and Association for Health, Education, and Prevention in Territories (*Association Santé, Education et Prévention sur les Territoires, *ASEPT). c) Hospital clinical research programs of the Ministry of Health (through the Regional University Hospital of Besançon Clinical Research Department). d) TSaR-PATURE project of the "Gift of Breath" Association (Association* Don du Souffle).

Funding for the short-term maintenance of the cohort and associated studies was provided for the five countries by two European contracts, FORALLVENT, from the 6th FPRD, the objective of which was to place "farm life" in the broader context of exposure to microbial agents (including parasites), and "allergies" in the broader context of the increasing prevalence of inflammatory and autoimmune diseases, and EFRAIM, from the 7th FPRD, more focused on the impact of protective factors on immunity and particularly the acquisition of immune tolerance, and including ancillary studies on milk and studies in experimental models of allergy (Figure 3.4). The SERF team was also involved in other projects of the same consortium, which involved the fine analysis of the farm environment and the microbiological composition of raw milk, such as the GABRIELA study (Loss *et al.* 2011). The long-term maintenance of the cohort could only be ensured by the implementation of "local" funding, for each country, and in particular for France by the French Ministry of Health via the regional and interregional hospital clinical research projects (*Projets Hospitaliers de Recherche Clinique*, PHRC), and the national prevention fund of the MSA.

3.5.1. *Protection against allergic diseases by the farm environment: a confirmation*

The results of the PASTURE study have fully confirmed the protective role of birth and farm life against the occurrence of various allergic diseases, as suggested by cross-sectional studies (Riedler *et al.* 2001). This role is consistently found after eliminating confusing factors, such as parental allergic predisposition, maternal smoking exposure or breastfeeding. The duration of stay in the stable and/or barn of the mother during pregnancy and of the child during the first years of life, as well as contact with animals, is significantly associated with a lower prevalence of allergy manifestations in children and its associated immunological parameters (Von Mutius and Vercelli 2010).

3.5.2. Exposure to animals and the farm microbial ecosystem

Contact with animals, especially cattle, is thus associated with protection; but it is interesting to note that the diversity of contact with animals likely to be present on a farm, such as pigs and barnyard animals, cats and dogs, enhances the effect of contact with cattle (Ege *et al.* 2008; Pfefferle *et al.* 2010; Roduit *et al.* 2011; Bergroth *et al.* 2012; Loss *et al.* 2012; Michel *et al.* 2013). Furthermore, the protective effect is greater the earlier in life the exposure to the farm environment (Loss *et al.* 2012). This contact can be indirectly measured by the level of specific antibodies that are likely to develop in individuals exposed to animals[30]. Used as markers in children in the PARSIFAL study and the PASTURE cohort, they are significantly correlated with protection against asthma (Frei *et al.* 2018). Dust, rich in microorganisms and plant particles, breathed in the barn comes into play: on traditional farms, where the family dwelling is in the farm building, this dust and that of the barns are transported to the family dwelling, where it is found in the mattresses of mother and child (Normand *et al.* 2011; Karvonen *et al.* 2012). Microbial elements are predominant in this dust: endotoxin-producing Gram-positive and Gram-negative bacteria, and many species of molds (Karvonen *et al.* 2012).

However, the predominant role of a specific microbial agent, or even a genus or family of microorganisms, could not be shown; and more than the quantity, it is the diversity of microbial agents present on farms that is significantly associated with protection (Heederik and Von Mutius 2012; Karvonen *et al.* 2014). The most traditional farms (Roussel *et al.* 2011) are the most "protective" against asthma and allergenic sensitization, and microbial diversity is much higher on these farms[31] than on more "modernized" farms. The protective effect is even more pronounced when these farms are grouped together in hamlets or villages (Müller-Rompa *et al.* 2018); the proximity of the families' dwellings to a manure facility (regardless of the type of living space, farm or other) is also a "protective" factor.

30 These are antibodies to N-glycolylneuraminic acid, a compound found in non-human mammals; as expected, they are significantly correlated with life on the farm; the presence of these antibodies, which are not associated with any pathological manifestation, shows that environmental contacts induce immune responses outside of any pathogenicity, even if in this case, we can only deduce the role of exposure to animals on protection against asthma, without being able to show that there is a cause and effect relationship between the presence of these antibodies and the observed protection.

31 Actinomycetes and fungi, such as *Absidia spp.*, *Eurotium spp.*, *Cladosporium spp.* and *Penicillium spp.*, predominate.

Other dust components, such as grass pollens, which can be inhaled to a massive extent when children stay in the barn during manual feeding or play in the hay barns, could also be involved in protection[32] (Sudre *et al.* 2009) (Figure 3.5); this could be one of the mechanisms that explain the particularly high protection of farm children against hay fever (seasonal pollen rhinoconjunctivitis).

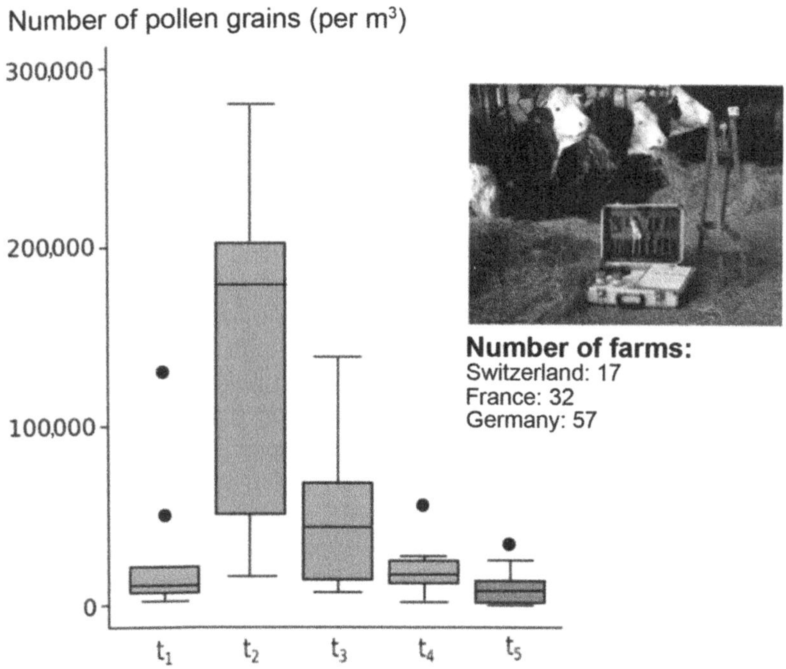

Figure 3.5. *Exposure to Poaceae pollen in barns at different stages of cattle feeding; t_1 before feeding, t_2 5 minutes after the start of feeding, t_3 5 minutes after the end, t_4 30 minutes after the end, t_5 55 minutes after the end. Based on a study carried out by the UMR Chrono-environment group on a sample of farms from the PASTURE study, according to Sudre* et al. *(2009)*

32 The "immune paralysis" (tolerance to specific antigens) by exposure to large doses of antigens in a limited period of time has been known to immunologists since the early 1900s (Vuitton 2008). The additional effect of non-microbial contacts is also suggested by other studies: in a sample of children from the GABRIELA study, contact with cows and straw (associated with raw milk consumption), for asthma, and exposure to hay stored in barns and manure, for atopic dermatitis, explained the entire "protection by farm environment" effect; the different factors included in the questionnaires did not, however, explain the whole of the overall "farm" effect on hay fever and allergenic sensitization, which makes it possible to evoke still unknown factors, even if they must be relatively minor (Illi *et al.* 2012).

As a side note to the PASTURE study, the study of American Amish and Hutterite communities[33] showed that differences in farming practices and, most importantly, in the presence or absence of mothers and young children on the farm premises explained major differences in children's protection against allergy, while the genetic basis and lifestyles were otherwise very similar (Gozdz *et al.* 2016). This is a great example of a situation where exposure to a complex ecosystem, such as farm animals and their microbial corollary, is combined with societal practices inherited from the past but modified over time, all against a common genetic background.

The correlations between exposure to the farm environment and/or animals and immunological parameters, observed from birth, in the cord blood, thus before any "personal" exposure of the child to this environment, are a strong point of the PASTURE cohort study, which combined epidemiological aspects collected by questionnaires with environmental measurements and immunological and immunogenetic analyses. We were able to observe an inverse correlation between allergy antibodies (IgE antibodies) in the child at birth and the exposure of the mothers to animals in the barn (Ege *et al.* 2008), as well as, with reference to the Th1/Th2 balance discussed above, a correlation between the production of Th1 cytokines by the child's lymphocytes and the exposure of the mothers to the farm environment (Pfefferle *et al.* 2010); the intervention of maternal environmental factors on TLR gene expression (Loss *et al.* 2012) and on gene methylation (Michel *et al.* 2013) could also be measured: epigenetic phenomena on innate immunity are thus at work from fetal life.[34] This effect of the farm environment on the gestating

33 The prevalences of asthma and allergenic sensitization among Amish children are lower than those of Swiss "farmer" children (5.2% vs. 6.8%), while those of Hutterite children are notably higher (21.3%) (Gozdz *et al.* 2016). These communities come from common European Anabaptist ancestors (essentially Swiss, then Franc-Comtois or German from the Rhine Valley, then Alsatians) through successive migrations, a certain level of endogamy having prevailed quite widely, for reasons of religious community and exclusion by Catholic or Protestant communities from the same geographical areas, since the 16th century. In the 21st century, they still share the same "ancestral" way of life in terms of food and family and religious behavior. However, their agricultural practices are very different, still "ancestral" for the Amish and highly mechanized and modern for the Hutterites, and the direct exposure of children and their mothers to the farm environment is important for the Amish, but almost non-existent for the Hutterites, where mechanization makes it unnecessary for women to participate in farm work.

34 The decrease in the occurrence of atopic dermatitis in nearly 50% of the children studied at the age of 2 years, when their mothers had been in contact with more than three different species of animals, was significantly associated with elevated expression of genes encoding TLR5 and TLR9 in cord blood; the effect of maternal contact with cats was, moreover, associated with the TLR2 polymorphism (Roduit *et al.* 2011; Loss *et al.* 2012).

child may seem surprising at first glance. The "familial" aspect of the manifestations of allergy characterizes atopy (Coca and Cooke 1923). Genetic determinants of allergic sensitization and the distinct determinants of asthma have been demonstrated (Anderson and Cookson 1999). However, there is a greater risk with maternal rather than paternal atopy; this long-observed "parent-of-origin" phenomenon may be caused by selection bias, true genetic imprinting or immunological interactions between mother and child in utero and/or postpartum, primarily through the transfer of vaginal microbiota from mother to child and through breastfeeding (Anderson and Cookson 1999). Observations made on the PASTURE cohort provide solid arguments for this last hypothesis and the influence of the mother's environment during pregnancy. The maternal influence is indeed also manifested after birth through breastfeeding: IgA concentrations in breast milk are inversely associated with the occurrence of atopic dermatitis in the first two years of the child's life; these IgA concentrations are also inversely associated with the occurrence of atopic dermatitis in the second year; these IgA concentrations, which vary from country to country, are significantly correlated with animal exposure (Orivuori *et al.* 2014). A recent analysis of data from the PASTURE study suggests that both breastfeeding and exposure to the farm environment explain the epigenetic interactions observed at variants of 17q21, a locus associated with asthma susceptibility (Gorlanova *et al.* 2018).

The German and French branches of the cohort were used to study the regulatory T-cell system[35] as a function of environmental exposure on the farm: an association between regulatory T-cell generation and "on-farm" exposure is observed in children at the age of four and a half years; there is an inverse association between regulatory T-cell generation, asthma and allergenic sensitization in the same children (Lluis *et al.* 2014). This "regulatory T atmosphere" is no longer observed at the age of 6 years, which confirms the existence of a "window of opportunity", already suggested by cross-sectional studies (Schröder *et al.* 2017). The genetic studies, on the other hand, found the genetic characteristics of the predisposition to asthma and atopy, and their parentage; however, the most original point of the contributions of the PASTURE study was to show that these same genes, which are susceptibility factors toward asthma, IgE-dependent immunological responses or the generation of regulatory T-systems, are also the ones influenced by the farm environment. The farm environment would thus turn a genetic point of vulnerability

35 More than the Th1/Th2 imbalance, it is the intervention of regulatory lymphocytes (among others the Treg, which express the FoxP3 factor – forkhead box P3 – sometimes called the "rheostat of the immune response") that modulates the immune response in a "tolerogenic" direction; this explains, at least partially, that childhood parasitic infections contribute to protection against allergies, even though they share the same Th2 profile.

into an asset, which only becomes a vulnerability when the environment changes (city life, loss of contact with animals and other factors that will be discussed below)[36].

3.5.3. *Consumption of raw milk and dairy products and exposure to the microbial ecosystem of raw milk*

As in the population-based studies, it was the observation of protection by raw milk consumption that was also the most consistent finding in the analyses from the PASTURE cohort (Loss *et al.* 2012; Lluis *et al.* 2014; Martikainen *et al.* 2015; Schröder *et al.* 2017). Raw milk consumption was even associated with protection against early childhood upper airway infections[37]; a similar effect, but to a lesser degree, was found for raw milk boiled in the family (Loss *et al.* 2015). The complementary studies of the GABRIELA project, on more than 8,000 children, confirmed the major influence of raw milk consumption, inversely correlated with clinical parameters of allergic manifestations,[38] and this independently of other farm exposures. The protective effect of a farm in the vicinity (within 100 m of the child's home) of non-farmers is totally explained by the consumption of untreated farm milk.

Not only clinical effects, but also immunological parameters were associated[39] with the consumption of raw milk, sometimes reinforced by the consumption of yogurts or cheeses, or even butter, "from the farm" (Pfefferle *et al.* 2010).

Furthermore, raw milk consumption was significantly associated with increased generation of activated T-cells and regulatory T-cells after lymphocyte stimulation

---

36 During the first years of life (from the first year especially, but still notable at four and a half years), there is an influence of farm life on the expression levels of TLR genes and the generation of different types of myeloid dendritic cells and regulatory T-cells, which show an orientation toward a "tolerance" type response (Martikainen *et al.* 2015).

37 With average relative risks of 0.70 for rhinitis, respiratory infections and fever, and even 0.14 for ear infections (Loss *et al.* 2015).

38 With observed relative risks of 0.50 for hay fever, 0.59 for asthma and 0.74 for allergic sensitization.

39 For example, maternal consumption of unskimmed milk and butter made from raw farm milk was associated with the greatest effect on Th1-type cytokine secretion by cord blood mononuclear cells (Pfefferle *et al.* 2010), suggesting a role for milk lipids in the observed clinical effects.

in German and French children of the PASTURE cohort studied at the age of four and a half years (Lluis *et al.* 2014)[40].

The endotoxin of Gram-negative bacteria, as we have seen above, was at the top of the list of protective factors suggested by the results of previous cross-sectional studies (Braun-Fahrländer 2003). The first surprise brought by the results of the PASTURE study was the impossibility of demonstrating a difference in the endotoxin content of raw milks, taken from "farmers'" families, compared to pasteurized/sterilized milk, opened and stored in the family refrigerator, whether taken from "non-farmers" or "farmers" (Gehring *et al.* 2008). Additional results from the associated GABRIELA study also failed to show a significant association between the total number of viable bacteria in milk and the occurrence of asthma or allergenic sensitization (Loss *et al.* 2011). These observations are paradoxical, as they do not agree with the initial hypothesis, especially since a complementary study in 16 farms in the Franche-Comté region shows that, on the farms, there is indeed a microbial transfer to the milk of most of the fungal species and environmental bacteria found in the cowshed or the milking parlor (Vacheyrou *et al.* 2011) (Figure 3.6). These observations can be seen as the result of drastic measures to control the microbiological quality of milk to avoid pathogens among farmers in the regions studied (Montel *et al.* 2014), measures which have not, however, eliminated the protective effect of raw milk.

This can also be seen as an additional element in favor of the importance of biodiversity: even though the total number of microorganisms has decreased considerably in raw milks over the last 20 years, the diversity of living species of microorganisms found in raw milk remains the major difference with heat-treated milk, where the number of identifiable species of microorganisms is low after the heat treatment (Vacheyrou *et al.* 2011).

It soon became clear, however, that the microbiology of raw milk did not explain the entire protective effect of raw milk against atopic allergy and that non-microbial components, which may be either more present in milk harvested in the relevant farming areas or destroyed by the technical sterilization processes, could explain

---

40 As well as to the demethylation of FoxP3, characteristic as we have seen above (see note no. 30) of "natural" regulatory T lymphocytes. The generation of regulatory T-cells is correlated with a lower incidence of asthma and allergenic sensitization with relative risks close to 0.25 in these children, and the protective effect of raw milk consumption against the occurrence of asthma and allergenic sensitization can be partially explained by its effect on regulatory T-cells (Lluis *et al.* 2014). The effect observed is even more pronounced when children have little contact with the barn and when raw milk consumption has involved both the mother during pregnancy and the child up to the age of 4 years.

part of the immunological properties of raw milk (Braun-Fahrländer and Von Mutius 2011; Von Mutius 2012). Heat treatment modifies some of the milk proteins; homogenization, on the other hand, by producing smaller fat globules, modifies their membrane, increases the total surface area of these globules, resulting in an adsorption of proteins normally present in whey and a greater and spatially different exposure of casein and lactoglobulins, the main milk allergens, to their surface; these physical modifications could be involved in the modulation of immune responses, directly or via modifications of the gut microbiome (Von Mutius and Vercelli 2010). The GABRIELA study confirmed that milk concentrations of bovine serum albumin, α-lactalbumin, and β-lactoglobulin were inversely associated with the occurrence of asthma, but not allergenic sensitization (Loss *et al.* 2011).

Regarding the lipid fraction of unpasteurized milk, there are differences in composition depending on the type and location of rearing; in particular, high-altitude milks, from grazing cows, are richer in ω-3 polyunsaturated fatty acids; indeed, higher concentrations of ω-3 polyunsaturated fatty acids in raw milk are related to a reduction in the occurrence of asthma (Brick *et al.* 2016). The richness of alpine milk in ω-3 polyunsaturated fatty acids is largely due to the biodiversity of the plant diet of cattle (and other dairy species), also involved in the taste of derived products when using raw milk (especially cheeses) (Von Mutius 2012; Montel *et al.* 2014); the contribution of summer grazing and floral biodiversity – and the lipid composition of milk – also intervenes on the vitamin composition, particularly vitamin D, which the PASTURE study also showed to be among the protective factors against allergy (Rochat *et al.* 2010).[41] Finally, recent results show that the content of short-chain fatty acids, butyrate and propionate, in the feces of children in the PASTURE cohort at the age of 1 year is inversely correlated with the onset of allergenic sensitization and asthma between 3 and 6 years of age. Yogurt consumption, but not raw milk consumption, is significantly associated with high butyrate levels; children with the highest butyrate concentrations also develop the least food allergy and allergic rhinitis (Roduit *et al.* 2018). In this study, however, the consumption of cheeses, especially hard cheeses made from raw milk, whose ripening flora produces propionates (Montel *et al.* 2014), was not specifically investigated.

41 In the French and Finnish children in the cohort, vitamin D supplementation increased the expression of immunoglobulin-like transcripts (ILT) 3 and ILT4 genes, which promote "tolerogenic" presentation by antigen-presenting cells; this increase in ILT3 and ILT4 expression is also observed in the "farm" children, independently of vitamin D intake (Rochat *et al.* 2010)

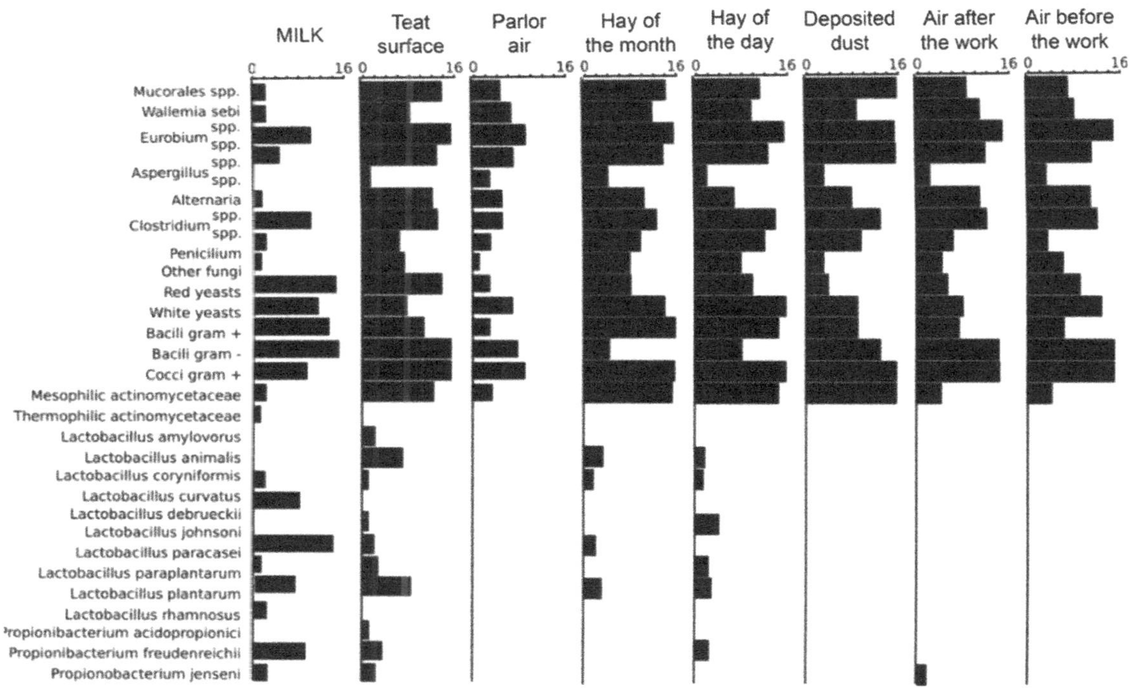

Figure 3.6. *Composition of the microbial flora of milk analyzed in the framework of the GABRIELA study by the UMR Chrono-environnement group, according to Vacheyrou et al. (2011). Comparative composition of the microbial ecosystem of raw milk and the microbial ecosystem of the different components of the farm where it was collected*

3.5.4. *Dietary diversification in the first year of life and cheese consumption*

The role of diversity and early exposure is also apparent in other dietary habits than in raw milk consumption. A common attitude in most so-called "developed" countries, in order to avoid early contact with allergens, has been for decades to postpone as much as possible the introduction of "allergenic" foods, an attitude that has not been followed by the expected effects (West *et al.* 2010). Analysis of the diets of children in the PASTURE cohort showed that diversity in the introduction of foods other than milk in the first year of life was associated with a reduced risk of developing atopic dermatitis between birth and 4 years of age.[42] The introduction of yogurt during the first year of life also significantly reduced the risk of atopic dermatitis by almost half (Roduit *et al.* 2012). A greater diversity of foods introduced during the first year of life is also significantly associated with a dose-response relationship between the number of new types of food introduced and the reduction in the onset of allergic diseases, with protection against the onset of asthma, food allergy and allergenic sensitization; the introduction of yogurt, butter or other dairy products before the age of 1 year is associated with a reduced risk of asthma and food allergy at 6 years of age (Roduit *et al.* 2014). As for the farm environment in general and contact with animals, the relationship between dietary diversification and immune parameters has been shown[43] (Roduit *et al.* 2014).

More specifically, we studied the consumption of cheese by the children in the PASTURE cohort during their first 18 months of life.[44] Cheeses were classified into six types according to their manufacturing technique[45]; again, it was the diversity of cheeses consumed during the first 18 months of life that could be related to protection against atopic dermatitis (halving the risk of developing the disease before 6 years of age) and food allergy (2/3 reduction in this case) (Nicklaus *et al.*

42 Even when the reverse causality was taken into account (avoidance or delay of diversification in families at risk of allergy).

43 The consequences of dietary diversification in terms of protection against the occurrence of asthma, food allergy and allergenic sensitization at the age of 6 years were associated with higher expression of the FoxP3 gene, a marker for regulatory T-cells, and lower expression of the gene coding for the IgE heavy chain, the antibodies involved in atopic allergy, measured in children at the age of 6 years (Roduit *et al.* 2014).

44 In collaboration with Sophie Nicklaus, INRAE research director, in charge of the team "Determinants of food behavior over the course of a lifetime, relations with health" at the *Centre du goût et de l'alimentation*, in Dijon.

45 However, it was not possible to differentiate between cheeses made with unpasteurized or pasteurized milk, as the PASTURE steering group considered this distinction to be too difficult to obtain from parents.

2019). The role of cheese diversification was independent of the general food diversification mentioned above. Cheese consumption alone was also associated with protection, as children who had no or very low cheese consumption during the first 18 months of life had an increased risk of developing allergic rhinitis or food allergy by the age of 6 years. In addition to the results meeting its main objectives, this analysis of cheese consumption provided some sociological information on cheese consumption by young children: contrary to what might have been expected, there was no difference between "farm" and "non-farm" children; quantitatively, cheese consumption was more frequent in Finnish families, but the diversity in terms of types of cheese was greater among the French and in families where the mothers had attained a higher level of education (Nicklaus *et al.* 2019).

3.6. At the crossroads: assessment and avenues still to be explored by the PASTURE cohort

The Covid-19 pandemic has delayed, at least for France which was 1 year "late" in the inclusion of children, the medical check-up of children known as "at 16 years old", which was supposed to take place in spring 2020. Despite the cessation of the Austrian cohort, the four other sites, including Franche-Comté, have managed to maintain their regional cohorts in activity "with bits and pieces" of funding. The search for funding always comes up against the same obstacles: without a specific and "sexy" study plan (i.e. implementing "fashionable" research avenues or technologically "high-end" analyses or, at least, considered as such), the simple maintenance of a cohort in activity, even if it is at the origin of totally innovative observations and more than 60 international publications, with the objective of completing epidemiological and immunological analyses over the long term, is not considered worthy of interest by the funding bodies. The refusal of funding not only by the inter-regional PHRC in 2019, but also by ANSES, the *Fondation de France*, and others, is quite characteristic of this misunderstanding. The permanence of the interest of the families – and the teams involved in the study – is more pleasing. During the visit of the 10- to 12-year old children, more than 60% of the families remained faithful to the study on all the sites, 76% in Franche-Comté, which is an exceptional result for a study of this type; the "rurality" of the population, the relative rarity of removals, the desire of farming families to defend a "positive" effect, against the general and media opinion of their profession, may be the reasons. This attachment to the study would certainly merit further sociological analysis.

The participation in this "long-term" study has also allowed us to experiment "in real time and place" with multidisciplinarity combined with multinationality.[46] That the PASTURE publications council continues to meet annually, including in the year 2020 by videoconference, and that, against the Covid-19 pandemic, the so-called "16-year visit" is being achieved show the solidity of the links that have been established. The unity and originality of the research subject, the necessary contribution of many disciplinary fields to elucidate the ins and outs of the effects of a complex environment on health, and the internationally recognized competencies of the research teams involved, each in its own field of disciplinary expertise, are probably among the assets of the success.

3.6.1. *The scientific achievements of nearly 20 years of the PASTURE adventure*

The main scientific achievements have largely benefited from this "crossing of disciplines" exploited throughout the PASTURE study; they have allowed: (1) confirmation of the protective effect of life on the farm, contact with animals and consumption of raw milk in protecting against the onset of atopic allergic diseases; in addition to the demonstration of the independence of these three protective factors, it is indeed a lifestyle that is highlighted, which probably combines and associates other aspects whose independence could not be demonstrated; (2) the role of the mother's exposure during gestation on the protection acquired by a proportion of children ranging from 30% to 60%, depending on the type of allergic manifestation considered and the level of exposure of the mothers; (3) the demonstration of the link between environmental exposures and parameters that reflect the establishment of immune tolerance, within the limits of our current knowledge on this subject, including the exposure of mothers during pregnancy; (4) the demonstration that epigenetic environmental factors have as targets genes known to be associated with predisposition to allergic diseases, including those

46 Among others: German, Finnish and Swiss epidemiologists and biostatisticians, clinicians (pulmonologists, dermatologists, allergists, infectious diseases specialists, gynecologists, pediatricians, etc.), immunologists and microbiologists from the five countries, German and French geneticists and nutritionists, French farm environment specialists and Dutch domestic environment specialists, and Austrian and Irish specialists of milk production and products. The list is not exhaustive, but it gives an idea of the complex network established during 20 years of collaboration. To have been able to overcome the obstacles linked to language, working habits, prejudices about the faults and qualities of the nationalities involved, logistical blunders that involved the reputation of the countries or, at least, of their laboratories or their transporters, and also the death of two major players and the health problems of the two historical coordinators, is an achievement in itself.

specific to asthma, independent of those involved in the functioning of the immune system; and, finally, (5) more globally, the association between "diversity of exposures" (including microbiological biodiversity) and the "establishment of a balanced immune system", protecting against both infectious agents and allergic slippage; this is certainly the common thread of the PASTURE study's findings. The analyses carried out on the basis of this cohort attempted, as is the rule in scientific studies, to "unwind the threads" of the "farm environment" ball to try to extract the individual factors and the mechanisms brought into play by these factors. The end of the quest could have been the discovery of a silver bullet, that is the "microbial agent" responsible for the "farm effect" and the "raw milk effect"... and the development of a profitable start-up. However, at the end of 20 years of study, it is indeed a "lifestyle" and its complex ecosystem (including microbial, but not only) more than a "microbe" that emerges from these analyses. Moreover, this lifestyle is not limited to the "quality of life" (in the health sense of the term) of the rural environment, which is once again attracting many city dwellers in large cities to rural areas. As the GABRIEL study has shown, the quality of life of rural children is better than that of city dwellers; however, this parameter alone is not significantly associated with the "protective" and immunoregulatory effect of life on the farm (Stöcklin *et al.* 2012).

In addition to answering questions about allergy protection and its mechanisms, the follow-up of the PASTURE cohort has also allowed technical developments (Bomert *et al.* 2011; Leppänen *et al.* 2014, 2015), environmental field studies (Karvonen *et al.* 2009), a review of the clinical phenotypes of allergic diseases, especially asthma (Depner *et al.* 2014; Hose *et al.* 2017), and the study of other risk factors for the diseases studied, such as the role of vitamins (Rochat *et al.* 2010; Nwaru *et al.* 2014) and, above all, the role of antibiotics, already suggested by population studies (Metzler *et al.* 2019). As the Franche-Comté cohort had added the results of skin testing, usually considered more relevant for allergy diagnosis, among the collected parameters, we were also able to conduct a longitudinal and comparative evaluation of allergy diagnostic tests (Chauveau *et al.* 2016, 2017). Finally, the PASTURE project, in general, has prompted several experimental studies, themselves involving research teams from various disciplines, from agriculture to veterinary medicine, microbiology and pharmacology.

3.6.2. *Intestinal microbiota: the missing link? First results from the PASTURE study*

The study of the children's intestinal microbiota had been planned from the start of the cohort, in particular by collecting the babies' stools at the age of 2 months, during the visit to the families' homes. For reasons related to the financing of this specific study (which was not included in the first European contracts, because an improvement in the study techniques had been anticipated and it had been considered preferable to postpone the analyses to a later date), to the logistics of conservation and routing of the samples and to difficulties in the genomic and bioinformatics analysis of the results, the composition of the children's microbiota was not analyzed until very recently. However, the composition of the microbiota is the result of the environment both breathed and ingested by children (and, before them, by their mothers, who pass on their vaginal flora), and it is decisive for the establishment and orientation of the immune system. The only in-depth analysis of children's stools was until recently that of fecal fatty acids published by Roduit *et al.* (2018). The results of the study of the children's microbiome between 2 and 12 months of age were just published in November 2020 in *Nature Medicine* (Depner *et al.* 2020). This first study of the microbiome focused on its degree of maturation. This maturation, assessed by the "estimated age of the microbiome", in children at the age of 1 year, is indeed significantly associated with farm life and reduced risk of asthma at school age. From the results obtained, we can consider that the effect of the farm environment on the maturation of the microbiome can explain 19% of the overall "farm effect". In agreement with the inverse correlation already observed between allergic risk and fecal butyrate concentrations (Roduit *et al.* 2018), this microbiome study confirms this association, as well as its relationship with the presence of butyrate-producing bacterial taxa and the relative abundance of butyryl-coenzyme A (CoA)-acetate-CoA-transferase, an important enzyme in butyrate metabolism (Depner *et al.* 2020). We can only hope for new sources of funding to complement the studies already initiated. In addition, studies of the respiratory microbiota are underway based on samples taken at the age of 10 and a half years, and those taken at the "16 years old" visit.

3.6.3. *Debates on the "benefit-risk" ratio and the possible contribution of the human and social sciences*

The results of the PASTURE study generated debates and controversies concerning early contact with animals and the risks it could entail, but especially the consumption of raw milk and its by-products (yogurt, butter and traditional "raw milk" cheeses). The first and main debate concerned the microbiological (pathogenic) risks of this double exposure. As soon as the results of the population studies were available, the scientific community, in particular that of infectious diseases specialists, gynecologists, and pediatricians, was somewhat shocked by the observation of a "protective" effect with regard to allergies, linked to the consumption of unpasteurized milk by the child during the first year of life and by its mother during pregnancy (Von Mutius 2012). The dairy and cheese industry (which depends on pasteurization for its production processes and wished to obtain a pasteurization authorization for a prestigious cheese with a Protection Designation of Origin, PDO) had already reacted violently[47].

Just as hygiene is, by definition, associated with progress and the reduction of diseases, the pasteurization of milk is associated with modernity and the reduction of the risk of infection, an association that is particularly accurate for tuberculosis or brucellosis, which were likely to be transmitted by the consumption of unpasteurized milk until the end of the 1950s in France. This notion seems to be deeply anchored in the collective subconscious, including that of scientific experts who invariably conclude their reports on collective food poisoning with the danger of unpasteurized milk, even if they have stressed throughout the text that infectious accidents associated with the consumption of unpasteurized milk or unpasteurized milk products were exceedingly rare in countries with a high level of hygiene, such as European countries, and that products from pasteurized milk and other foods were much more frequently incriminated in epidemiological observations or systematic studies.[48,49] The "beneficial effect of raw milk" against allergies was therefore an observation that many thought would not hold up when it came to longitudinal studies with impeccable methodology. The results of the PASTURE study not only confirmed the protective effect, but also provided solid arguments for the mechanisms that could explain it. In the late 2000s, the risk of fatal cerebral-meningeal listeriosis in the immuno-deficient elderly and the risk of listeriosis-related abortion in pregnant women were prominent in the debate (Centers for

---

47 It had engaged in a kind of blackmail against a public television channel that had planned to broadcast a documentary on traditional cheeses, "*Ces fromages qu'on assassine*", at the end of December 2008. It had even threatened to sue those, including researchers, who simply echoed the scientific information on the benefits of unpasteurized milk in this film. For an overview of this debate and the role played by the food industry, see: https://www.lemonde.fr/vous/article/2008/03/13/ces-fromages-qu-on-assassine_992195_3238.html.

48 For example, ready-to-eat smoked salmon and dairy products made from pasteurized milk for listeriosis (Centers for Disease Control and Prevention (CDC) 2013); Lopez-Valladares *et al.* 2018) or ground beef for Shiga-toxin-producing *Escherichia coli* (STEC) (Bruyand *et al.* 2019); in a recent meta-analysis, dairy products (in general) did not emerge significantly among the foodstuffs at risk of sporadic STEC infection in America and Europe (Devleesschauwer *et al.* 2019).

49 This is evident from the Norwegian report in the context of the approval of unpasteurized milk and unpasteurized cheeses: https://vkm.no/english/riskassessments/allpublications/aqualitativeassessmentoftherisksoftransmissionofmicroorganismstohumansresultingfromtheconsumptionofrawmilkandrawcreaminnorway.4.72c3261615e09f2472f4b05c.html, and various reports from ANSES: https://www.anses.fr/fr/content/point-sur-la-bact%C3%A9rie-escherichia-coli-productrice-de-shigatoxines-stec.

Disease Control and Prevention 2013). Curiously, in the late 2010s, it shifted to that of hemolytic uremic syndrome (HUS) by *Escherichia coli* STEC strains in young children (Bruyand *et al.* 2019; Jones *et al.* 2019). An episode of collective food poisoning from cheese[50] offered on trays in supermarkets by an artisanal dairy (Jones *et al.* 2019) led to a ban on serving raw milk cheeses in nursery and kindergarten canteens to children below 5 years of age,[51] a paradoxical injunction if we take into account the results of the PASTURE study and the "window of opportunity" described above, and also the recommendations to make local products available to children in school canteens and to favor short circuits. One could certainly discuss the reasons for the hasty drafting and publication of this "technical instruction", which provides for precise and heavy sanctions for offenders. In any case, it seems that the action of the agri-food industry (once again in the forefront during this period for the authorization of the heat treatment of a PDO cheese) was more discreet and subtle than in 2008. Before and after the publication of this new regulatory text, emotions were high this time among raw milk distributors and traditional cheese producers at all stages of the production chain. At their request, as the results of the PASTURE study are considered to be a reliable source of scientific information on the "positive health effects" of consuming raw milk and raw milk cheeses, in 2019 and 2020, we multiplied the number of lectures and talks not only at conferences, meetings and workshops over a large part of the French territory, but also at an international congress in Valencia (Spain) and for a professional association in Norway.

These professionals, gathered in "collectives" at both the French and European levels, are calling for an evaluation of the benefit-risk ratio of the consumption of raw milk cheeses by young children, which would weigh the very rare occurrence[52] of an effectively serious infection, in a specific, probably genetic, but still poorly determined, terrain, against the possible prevention of frequent allergic diseases, also potentially serious (even fatal, such as asthma and food anaphylaxis), but occurring at a distance from this consumption. Working groups have been set up to undertake this assessment. The potential prevention of early childhood respiratory

50 Cheeses made from unpasteurized milk and pasteurized milk, incriminated by tracing from family purchases, without it being possible to demonstrate the identity of the bacterial strain between those present in the cheeses made from unpasteurized milk and the bacteria isolated from the affected patients.

51 Technical Instruction: DGAL/DGSSA/2019-365 of 02/05/2019, https://info.agriculture.gouv.fr/gedei/site/bo-agri/instruction-2019-365.

52 Approximately one fatal case of HUS every 3 years in this age group, according to a calculation based on official French surveillance of this condition by the HUS network (Bruyand *et al.* 2019).

infections, shown in the PASTURE study[53] (Loss *et al.* 2015), should also be estimated. It can be observed that, like the regulatory bodies, the medical community is clearly more sensitive to risks than to benefits when it comes to raw milk. Indeed, following our scientific article on the association between consumption of a greater diversity of cheeses and protection against certain manifestations of allergy, discussed above (Nicklaus *et al.* 2019), the journal *Allergy* received a letter to the editor from French nephro-pediatricians, based on recent cases of HUS in France and the ministerial instruction just mentioned, urging us to state publicly in the journal that consumption of raw milk cheeses posed unbearable risks to children below 5 years of age. This request was all the more surprising since our article did not even mention the consumption of raw milk cheeses[54] and nothing in the article resembled any kind of apology for the consumption of raw milk cheese. Our reply, which mentioned this surprise and pleaded for a real benefit-risk study on the consumption of unpasteurized milk and cheese, was rejected by a reviewer who declared this proposal inadmissible ("the risks – the death of a child – were obvious while the benefits remained hypothetical"), a very unscientific position in the spirit.[55]

The debates that have arisen in society from this work go beyond the risk of infection but are now relevant to the considerations outlined in Chapter 2 of this book. Until recently, the results of the PASTURE study were used by supporters of "traditional", extensive, pasture-based breeding, as practiced in the areas where the PASTURE study families were recruited. They put forward the biodiversity of dairy breeds and floral biodiversity, guaranteeing the quality of raw milk, its richness in ω-3 fatty acids, a virtuous circle for animal welfare and consumer health, against the intensive breeding in permanent stalling of a single breed of cattle (as well as goats), strong producers of mediocre quality milk, highly transformed into pasteurized and standardized products. The pollution of the rivers of the Doubs and the possible responsibility of the cattle breeders in the production area of the PDO Comté has reshuffled the cards and brought in new actors, with renewed arguments.[56]

It is clear that a better understanding of the forces at work and the obvious or hidden motives of the social groups concerned would benefit greatly from the contribution of a socio-anthropological analysis. The historical analysis of the

53 It is also legitimate to ask whether the inhomogeneous distribution of Covid-19, both in the world and in a single country, could not be explained by the environmental and dietary history of individuals and populations.

54 Data that we were unable to collect, as described above.

55 Finally, neither the nephro-pediatricians' letter nor our response was published by the journal *Allergy*.

56 See Chapter 2 for a detailed socio-anthropological analysis.

perception and representation of health risks associated with unpasteurized milk would perhaps also allow us to understand the extreme reluctance of scientists (infectious diseases specialists, pediatricians, health economists) to implement a benefit-risk analysis as it is done in many other fields of public health. It is remarkable to see that the risk of HUS by contact with animals (during a visit to a farm, for example, or by human-to-human contact after a first contact with farm animals) is higher than the risk of HUS by consumption of unpasteurized milk cheeses; yet, of the two protective factors found in the PASTURE study (contact with farm animals and consumption of raw milk), it is only the second, never the first, that is put forward by pediatricians or by the media. In any case, the return "for all" to the dairy farmers' lifestyle, the world "before", is illusory in our countries and the search for a "farm pill" has not yielded the expected results (Vuitton and Dalphin 2017)! In terms of prevention of allergic diseases, the consumption of probiotics has proven to be notoriously insufficient, even if it may have generated some hopes (Vuitton and Dalphin 2017). A calm – and scientifically documented – debate on the possible interest of raw milk produced in optimal conditions of personal and environmental safety (Berge and Baars 2020) would perhaps make it possible to compensate for this impossible return to the world "before".

KEY POINTS.–

Birth and life on a livestock farm protect against the onset of atopic allergic diseases. This protection is all the more important when the number of animal species encountered on a daily basis is high and the microbial diversity encountered is significant, a biodiversity which is itself associated with the "traditional" character of farms.

The consumption of raw milk by the mother during pregnancy and/or by the child during the first years of life protects against allergic diseases and acute respiratory infections. The microbial biodiversity of raw milk and also its composition in fatty acids and proteins (which is altered by the industrial processes of heating and homogenization of milk) are involved in the protective effect of raw milk.

Early dietary diversification of infants and, independently, early and diversified consumption of cheese are also protective against the onset of atopic allergic diseases.

The environmental factors of this protection are associated with an early orientation of immune responses toward a "regulatory" cellular and cytokine profile of both innate and adaptive immunity; the mother's environment during pregnancy influences the child's immune profile even before the child is exposed to these environmental factors.

Interferences between genetic and environmental factors are multiple: epigenetic modifications of genes predisposing to different manifestations of allergy and to the immunological mechanisms associated with them are observable from birth when the mother is exposed to protective environments, and then in young children; the intermediary role of the different microbiota (intestinal, respiratory, cutaneous) remains, however, still to be fully explained.

The diversity of exposures, whether environmental or dietary, appears to be an essential element of an immune balance that allows both proper defenses against pathogens (fight against infections) and the absence of abnormal and excessive reactions to the environment (allergic manifestations); this state of balance is not ensured by a single type of exposure, which could be reproduced at will from a preventive or therapeutic perspective, or by the quantitative importance of microbial exposure.

The PASTURE case-control cohort study was able to provide answers as to the "independent" factors involved in protection thanks to its multi- and transdisciplinarity (and multinationality), but its results also highlight the interest of a lifestyle; some of its results also raise a societal debate: should we accept an extremely low but existing risk in the short term (infection of a few vulnerable individuals by a pathogen) for a benefit affecting more than a third of a population in the long term (protection against allergy)? A "benefit-risk" analysis, and also a sociological analysis of the forces at work in institutional decisions, should be conducted to answer this question.

3.7. Acknowledgments

The authors wish to thank the multiple actors who made the PASTURE study possible and those who are taking up the torch after 18 years of the cohort's existence. Jean-Charles Dalphin was the soul of the French arm of the study until his untimely passing in October 2019, which left us all distraught but determined to continue his work. This chapter, which owes everything to his enthusiasm, his tenacity and his sense of human relationships, is dedicated to him. In a study of this type, each person, in his or her own way, is indispensable, and it is difficult to name individually – from master's and thesis students to university professors, from clinical research and biology technicians to managers and logisticians – the members of the shadow army who make such an adventure possible. I would like to mention as representatives of their respective teams Professors Josef Riedler in Austria, Juha Pekkanen in Finland, Charlotte Braun-Fahrlander and Roger Lauener in Switzerland, and *last but not least*, Erika von Mutius in Germany, who was the first to put her trust in us and whose friendship is precious. We have been closer to some of our European collaborators on specific sub-projects and, in this respect, we would like to thank especially Caroline Roduit in Switzerland and Bianca Schaub in Germany. For the French branch, Vincent Kaulek has been in charge of the practical organization, logistics and contact with families for several years; Bertrand Sudre, a pure product of the DEA "environment, health, society", has made his first steps in public health research thanks to PASTURE, in France and in Switzerland; the microbiological support for the specific studies on the farm environment has benefited from the expertise of Gabriel Reboux and his team; Sophie Nicklaus allowed us to approach original aspects of child nutrition and demonstrated that the Besançon-Dijon cooperation was not an empty word. Finally, the understanding of the pediatric department of Besançon University Hospital and the investment of Marie-Laure Dalphin were decisive in the smooth running of the hospital visits of the children, just as the smiling efficiency of Nadia Guillou and Cécile Travers and their colleagues, at the MSA and then the ASEPT, brought a note of conviviality and exchanges with the "subjects included in the study", which is often lacking in scientific research. We thank them all for their unfailing commitment. Our last and warmest thanks are addressed to the families and especially to the parents who remain mobilized almost 20 years after the inclusion of their baby in a scientific study "which did not bring them anything", except for a little consideration, which is not to be overlooked in modern-day rural France. May they be assured of our immense esteem.

3.8. References

Alfvén, T., Braun-Fahrländer, C., Brunekreef, B., Von Mutius, E., Riedler, J., Scheynius, A., van Hage, M., Wickman, M., Benz, M.R., Budde, J., Michels, K.B., Schram, D., Ublagger, E., Waser, M., Pershagen, G., PARSIFAL Study Group (2006). Allergic diseases and atopic sensitization in children related to farming and anthroposophic lifestyle – The PARSIFAL study. *Allergy*, 61(4), 414–421.

Anderson, G.G. and Cookson, W.O. (1999). Recent advances in the genetics of allergy and asthma. *Molecular Medicine Today*, 5(6), 264–273.

Aubier, M., Neukirch, F., Annesi-Maesano, I. (2005). Épidémiologie de l'asthme et des allergies. La fréquence des allergies augmente partout dans le monde, l'asthme a atteint sa prévalence maximale en Europe : quelles explications ? *Bulletin de l'Académie nationale de médecine*, 189(7), discussion 1434, 1419–1434.

Barnes, M., Cullinan, P., Athanasaki, P., MacNeill, S., Hole, A.M., Harris, J., Kalogeraki, S., Chatzinikolaou, M., Drakonakis, N., Bibaki-Liakou, V., Newman Taylor, A.J., Bibakis, I. (2001). Crete: Does farming explain urban and rural differences in atopy? *Clinical and Experimental Allergy*, 31(12), 1822–1828.

Berge, A.-C. and Baars, T. (2020). Raw milk producers with high levels of hygiene and safety. *Epidemiology and Infection*, 148, e14.

Bergroth, E., Remes, S., Pekkanen, J., Kauppila, T., Büchele, G., Keski-Nisula, L. (2012). Respiratory tract illnesses during the first year of life: Effect of dog and cat contacts. *Pediatrics*, 130(2), 211–220.

Beutler, B. (2002). Toll-like receptors: How they work and what they do. Current Opinion. *Hematology*, 9(1), 2–10.

Blackley, C. (1873). *Experimental Researches on the Causes and Nature of Catarrhus Æstivus (Hay-fever or Hay-asthma)*. Tindall & Cox, London.

Bomert, M., Köllisch, G., Roponen, M., Lauener, R., Renz, H., Pfefferle, P.I., Al-Fakhri, N. (2011). Analytical performance of a multiplexed, bead-based cytokine detection system in small volume samples. *Clinical Chemistry and Laboratory Medicine*, 49(10), 1691–1693.

Bostock, J. (1828). On the Catarrhus aestivus or summer catarrh. *Medico-Chirurgical Transactions*, 14(2), 437–446.

Braun-Fahrländer, C. (2002). Do only European cattle protect from allergies? *Allergy*, 57(12), 1094–1096.

Braun-Fahrländer, C. (2003). Environmental exposure to endotoxin and other microbial products and the decreased risk of childhood atopy: Evaluating developments since April 2002. *Current Opinion in Allergy and Clinical Immunology*, 3(5), 325–329.

Braun-Fahrländer, C. and Von Mutius, E. (2011). Can farm milk consumption prevent allergic diseases? *Clinical and Experimental Allergy*, 41(1), 29–35.

Braun-Fahrländer, C., Riedler, J., Herz, U., Eder, W., Waser, M., Grize, L., Maisch, S., Carr, D., Gerlach, F., Bufe, A., Lauener, R.P., Schierl, R., Renz, H., Nowak, D., Von Mutius, E., Allergy and Endotoxin Study Group (2002). Environmental exposure to endotoxin and its relation to asthma in school-age children. *The New England Journal of Medicine*, 347(12), 869–877.

Brick, T., Schober, Y., Böcking, C., Pekkanen, J., Genuneit, J., Loss, G., Dalphin, J.-C., Riedler, J., Lauener, R., Nockher, W.A., Renz, H., Vaarala, O., Braun-Fahrländer, C., Von Mutius, E., Ege, M.J., Pfefferle, P.I., PASTURE Study Group (2016). ω-3 fatty acids contribute to the asthma-protective effect of unprocessed cow's milk. *The Journal of Allergy and Clinical Immunology*, 137(6), 1699–1706.

Bruyand, M., Mariani-Kurkdjian, P., Le Hello, S., King, L.-A., Van Cauteren, D., Lefevre, S., Gouali, M., Jourdan-da Silva, N., Mailles, A., Donguy, M.-P., Loukiadis, E., Sergentet-Thevenot, D., Loirat, C., Bonacorsi, S., Weill, F.-X., De Valk, H., Réseau français hospitalier de surveillance du Shu pédiatrique (2019). Paediatric haemolytic uraemic syndrome related to Shiga toxin-producing *Escherichia coli*, an overview of 10 years of surveillance in France, 2007 to 2016. *Euro Surveillance*, 24(8).

Centers for Disease Control and Prevention, CDC (2013). Vital signs: Listeria illnesses, deaths, and outbreaks–United States, 2009–2011. *MMWR. Morbidity and Mortality Weekly Report*, 62(22), 448–452.

Chai, S.K., Nga, N.N., Checkoway, H., Takaro, T.K., Redding, G.J., Keifer, M.C., Trung, L.V., Barnhart, S. (2004). Comparison of local risk factors for children's atopic symptoms in Hanoi, Vietnam. *Allergy*, 59(6), 637–644.

Chauveau, A., Dalphin, M.-L., Kaulek, V., Roduit, C., Pugin, A., Von Mutius, E., Vuitton, D.A., Dalphin, J.-C. (2016). Disagreement between skin prick tests and specific IgE in early childhood. *International Archives of Allergy and Immunology*, 170(2), 69–74.

Chauveau, A., Dalphin, M.-L., Mauny, F., Kaulek, V., Schmausser-Hechfellner, E., Renz, H., Riedler, J., Pekkanen, J., Karvonen, A.M., Lauener, R., Roduit, C., Vuitton, D.A., Von Mutius, E., Dalphin, J.-C., PASTURE Study Group (2017). Skin prick tests and specific IgE in 10-year-old children: Agreement and association with allergic diseases. *Allergy*, 72(9), 1365–1373.

Coca, A. and Cooke, R. (1923). On the classification of the phenomena of hypersensitiveness. *Journal of Immunology*, 8, 163–171.

da Costa Lima, R., Victora, C.G., Menezes, A.M.B., Barros, F.C. (2003). Do risk factors for childhood infections and malnutrition protect against asthma? A study of Brazilian male adolescents. *American Journal of Public Health*, 93(11), 1858–1864.

Dalphin, J.-C., Pernet, D., Reboux, G., Martinez, J., Dubiez, A., Barale, T., Depierre, A. (1991). Influence of mode of storage and drying of fodder on thermophilic actinomycete aerocontamination in dairy farms of the Doubs region of France. *Thorax*, 46(9), 619–623.

Dalphin, J.-C., Dubiez, A., Monnet, E., Gora, D., Westeel, V., Pernet, D., Polio, J.-C., Gibey, R., Laplante, J.-J., Depierre, A. (1998). Prevalence of asthma and respiratory symptoms in dairy farmers in the French province of the Doubs. *American Journal of Respiratory and Critical Care Medicine*, 158(5 Pt 1), 1493–1498.

Depner, M., Fuchs, O., Genuneit, J. *et al.* (2014). Clinical and epidemiologic phenotypes of childhood asthma. *American Journal of Respiratory and Critical Care Medicine*, 189(2), 129–138.

Depner, M., Taft, D.H., Kirjavainen, P.V. *et al.* (2020). Maturation of the gut microbiome during the first year of life contributes to the protective farm effect on childhood asthma. *Nat. Med.*, 26, 1766–1775.

Devleesschauwer, B., Pires, S.M., Young, I., Gill, A., Majowicz, S.E., Study Group (2019). Associating sporadic, foodborne illness caused by Shiga toxin-producing *Escherichia coli* with specific foods: A systematic review and meta-analysis of case-control studies. *Epidemiology and Infection*, 147, e235.

Eder, W., Klimecki, W., Yu, L., Von Mutius, E., Riedler, J., Braun-Fahrländer, C., Nowak, D., Martinez, F.D., ALEX Study Group (2004). Toll-like receptor 2 as a major gene for asthma in children of European farmers. *The Journal of Allergy and Clinical Immunology*, 113(3), 482–488.

Eder, W., Klimecki, W., Yu, L., Von Mutius, E., Riedler, J., Braun-Fahrländer, C., Nowak, D., Holst, O., Martinez, F.D., ALEX Study Group (2006). Association between exposure to farming, allergies and genetic variation in CARD4/NOD1. *Allergy*, 61(9), 1117–1124.

Edfors-Lubs, M.L. (1971). Allergy in 7000 twin pairs. *Acta Allergologica*, 26(4), 249–285.

Eduard, W., Douwes, J., Omenaas, E., Heederik, D. (2004). Do farming exposures cause or prevent asthma? Results from a study of adult Norwegian farmers. *Thorax*, 59(5), 381–386.

Ege, M.J., Herzum, I., Büchele, G. *et al.* (2008). Prenatal exposure to a farm environment modifies atopic sensitization at birth. *The Journal of Allergy and Clinical Immunology*, 122(2), 407–412.

Faber, M.A., Van Gasse, A.L., Decuyper, I.I., Sabato, V., Hagendorens, M.M., Mertens, C., Bridts, C.H., De Clerck, L.S., Ebo, D.G. (2018). Cross-reactive aeroallergens: Which need to cross our mind in food allergy diagnosis? *The Journal of Allergy and Clinical Immunology: In Practice*, 6(6), 1813–1823.

Frei, R., Ferstl, R., Roduit, C. *et al.* (2018). Exposure to nonmicrobial N-glycolylneuraminic acid protects farmers' children against airway inflammation and colitis. *The Journal of Allergy and Clinical Immunology*, 141(1), 382-390.e7.

Gehring, U., Spithoven, J., Schmid, S., Bitter, S., Braun-Fahrländer, C., Dalphin, J.-C., Hyvärinen, A., Pekkanen, J., Riedler, J., Weiland, S.K., Büchele, G., Von Mutius, E., Vuitton, D.A., Brunekreef, B., PASTURE Study Group (2008). Endotoxin levels in cow's milk samples from farming and non-farming families – The PASTURE study. *Environment International*, 34(8), 1132–1136.

Genuneit, J., Strachan, D.P., Büchele, G., Weber, J., Loss, G., Sozanska, B., Boznanski, A., Horak, E., Heederik, D., Braun-Fahrländer, C., Von Mutius, E., GABRIELA Study Group (2013). The combined effects of family size and farm exposure on childhood hay fever and atopy. *Pediatric Allergy and Immunology*, 24(3), 293–298.

Gereda, J.E., Leung, D.Y., Thatayatikom, A., Streib, J.E., Price, M.R., Klinnert, M.D., Liu, A.H. (2000). Relation between house-dust endotoxin exposure, type 1 T-cell development, and allergen sensitisation in infants at high risk of asthma. *Lancet*, 355(9216), 1680–1683.

Gorlanova, O., Illi, S., Toncheva, A.A., Usemann, J., Latzin, P., Kabesch, M., Dalphin, J.-C., Lauener, R., Pekkanen, J.R., Von Mutius, E., Riedler, J., Kuehni, C.E., Röösli, M., Frey, U., BILD and PASTURE Study Groups (2018). Protective effects of breastfeeding on respiratory symptoms in infants with 17q21 asthma risk variants. *Allergy*, 73(12), 2388–2392.

Gozdz, J., Holbreich, M., Metwali, N., Thorne, P.S., Sperling, A.I., Martinez, F.D., Ober, C., Von Mutius, E., Vercelli, D. (2016). Amish and Hutterite environmental farm products have opposite effects on experimental models of asthma. *Annals of the American Thoracic Society*, 13, Suppl. 1, S99.

Grüber, C., Illi, S., Plieth, A., Sommerfeld, C., Wahn, U. (2002). Cultural adaptation is associated with atopy and wheezing among children of Turkish origin living in Germany. Clinical and experimental allergy. *Journal of the British Society for Allergy and Clinical Immunology*, 32(4), 526–531.

Heederik, D. and Von Mutius, E. (2012). Does diversity of environmental microbial exposure matter for the occurrence of allergy and asthma? *The Journal of Allergy and Clinical Immunology*, 130(1), 44–50.

Heinrich, J., Gehring, U., Douwes, J., Koch, A., Fahlbusch, B., Bischof, W., Wichmann, H.E., INGA Study Group (2001). Pets and vermin are associated with high endotoxin levels in house dust. *Clinical and Experimental Allergy*, 31(12), 1839–1845.

Hesselmar, B., Aberg, N., Aberg, B., Eriksson, B., Björkstén, B. (1999). Does early exposure to cat or dog protect against later allergy development? *Clinical and Experimental Allergy*, 29(5), 611–617.

Hillon, P. (2020). *Le Microbiote : un ami qui vous protège ?* Eyrolles, Paris.

Hose, A.J., Depner, M., Illi, S. *et al.* (2017). Latent class analysis reveals clinically relevant atopy phenotypes in 2 birth cohorts. *The Journal of Allergy and Clinical Immunology*, 139(6), 1935–1945.

Illi, S., Depner, M., Genuneit, J., Horak, E., Loss, G., Strunz-Lehner, C., Büchele, G., Boznanski, A., Danielewicz, H., Cullinan, P., Heederik, D., Braun-Fahrländer, C., Von Mutius, E., GABRIELA Study Group (2012). Protection from childhood asthma and allergy in Alpine farm environments – The GABRIEL Advanced Studies. *The Journal of Allergy and Clinical Immunology*, 129(6), 1470–1477.

Jones, G., Lefèvre, S., Donguy, M.-P., Nisavanh, A., Terpant, G., Fougère, E., Vaissière, E., Guinard, A., Mailles, A., de Valk, H., Fila, M., Tanné, C., Le Borgne, C., Weill, F.-X., Bonacorsi, S., Jourdan-Da Silva, N., Mariani-Kurkdjian, P. (2019). Outbreak of Shiga toxin-producing *Escherichia coli* (STEC) O26 paediatric haemolytic uraemic syndrome (HUS) cases associated with the consumption of soft raw cow's milk cheeses, France, March to May 2019. *Euro Surveillance*, 24(22).

Karvonen, A.M., Hyvärinen, A., Roponen, M., Hoffmann, M., Korppi, M., Remes, S., Von Mutius, E., Nevalainen, A., Pekkanen, J. (2009). Confirmed moisture damage at home, respiratory symptoms and atopy in early life: A birth-cohort study. *Pediatrics*, 124(2), e329–338.

Karvonen, A.M., Hyvärinen, A., Gehring, U. *et al.* (2012). Exposure to microbial agents in house dust and wheezing, atopic dermatitis and atopic sensitization in early childhood: A birth cohort study in rural areas. *Clinical and Experimental Allergy*, 42(8), 1246–1256.

Karvonen, A.M., Hyvärinen, A., Rintala, H. *et al.* (2014). Quantity and diversity of environmental microbial exposure and development of asthma: A birth cohort study. *Allergy*, 69(8), 1092–1101.

Lauener, R.P., Birchler, T., Adamski, J., Braun-Fahrländer, C., Bufe, A., Herz, U., Von Mutius, E., Nowak, D., Riedler, J., Waser, M., Sennhauser, F.H., ALEX Study Group (2002). Expression of CD14 and Toll-like receptor 2 in farmers' and non-farmers' children. *Lancet*, 360(9331), 465–466.

Leppänen, H.K., Nevalainen, A., Vepsäläinen, A., Roponen, M., Täubel, M., Laine, O., Rantakokko, P., Von Mutius, E., Pekkanen, J., Hyvärinen, A. (2014). Determinants, reproducibility, and seasonal variation of ergosterol levels in house dust. *Indoor Air*, 24(3), 248–259.

Leppänen, H.K., Täubel, M., Roponen, M., Vepsäläinen, A., Rantakokko, P., Pekkanen, J., Nevalainen, A., Von Mutius, E., Hyvärinen, A. (2015). Determinants, reproducibility, and seasonal variation of bacterial cell wall components and viable counts in house dust. *Indoor Air*, 25(3), 260–272.

Leung, R. (1996). Asthma and migration. *Respirology*, 1(2), 123–126.

Liu, A.H. and Leung, D.Y.M. (2006). Renaissance of the hygiene hypothesis. *The Journal of Allergy and Clinical Immunology*, 117(5), 1063–1066.

Lluis, A., Depner, M., Gaugler, B. *et al.* (2014). Increased regulatory T-cell numbers are associated with farm milk exposure and lower atopic sensitization and asthma in childhood. *The Journal of Allergy and Clinical Immunology*, 133(2), 551–559.

Lopez-Valladares, G., Danielsson-Tham, M.-L., Tham, W. (2018). Implicated food products for listeriosis and changes in serovars of listeria monocytogenes affecting humans in recent decades. *Foodborne Pathogens and Disease*, 15(7), 387–397.

Loss, G., Apprich, S., Waser, M., Kneifel, W., Genuneit, J., Büchele, G., Weber, J., Sozanska, B., Danielewicz, H., Horak, E., van Neerven, R.J.J., Heederik, D., Lorenzen, P.C., Von Mutius, E., Braun-Fahrländer, C., GABRIELA Study Group (2011). The protective effect of farm milk consumption on childhood asthma and atopy: The GABRIELA study. *The Journal of Allergy and Clinical Immunology*, 128(4), 766–773.

Loss, G., Bitter, S., Wohlgensinger, J. *et al.* (2012). Prenatal and early-life exposures alter expression of innate immunity genes: The PASTURE cohort study. *The Journal of Allergy and Clinical Immunology*, 130(2), 523–530.

Loss, G., Depner, M., Ulfman, L.H. *et al.* (2015). Consumption of unprocessed cow's milk protects infants from common respiratory infections. *The Journal of Allergy and Clinical Immunology*, 135(1), 56–62.

Martikainen, M.-V., Kääriö, H., Karvonen, A., Schröder, P.C., Renz, H., Kaulek, V., Dalphin, J.-C., Von Mutius, E., Schaub, B., Pekkanen, J., Hirvonen, M.-R., Roponen, M. (2015). Farm exposures are associated with lower percentage of circulating myeloid dendritic cell subtype 2 at age 6. *Allergy*, 70(10), 1278–1287.

Metzler, S., Frei, R., Schmaußer-Hechfellner, E., Von Mutius, E., Pekkanen, J., Karvonen, A.M., Kirjavainen, P.V., Dalphin, J.-C., Divaret-Chauveau, A., Riedler, J., Lauener, R., Roduit, C., PASTURE-EFRAIM Study Group (2019). Association between antibiotic treatment during pregnancy & infancy and the development of allergic diseases. *Pediatric Allergy and Immunology*, 30, 423–433.

Michel, S., Busato, F., Genuneit, J., Pekkanen, J., Dalphin, J.-C., Riedler, J., Mazaleyrat, N., Weber, J., Karvonen, A.M., Hirvonen, M.-R., Braun-Fahrländer, C., Lauener, R., Von Mutius, E., Kabesch, M., Tost, J., PASTURE Study Group (2013). Farm exposure and time trends in early childhood may influence DNA methylation in genes related to asthma and allergy. *Allergy*, 68(3), 355–364.

Montel, M.-C., Buchin, S., Mallet, A., Delbes-Paus, C., Vuitton, D.A., Desmasures, N., Berthier, F. (2014). Traditional cheeses: Rich and diverse microbiota with associated benefits. *International Journal of Food Microbiology*, 177, 136–154.

Moresco, E.M.Y., LaVine, D., Beutler, B. (2011). Toll-like receptors. *Current Biology: CB*, 21(13), R488–493.

Müller-Rompa, S.E.K., Markevych, I., Hose, A.J., Loss, G., Wouters, I.M., Genuneit, J., Braun-Fahrländer, C., Horak, E., Boznanski, A., Heederik, D., Von Mutius, E., Heinrich, J., Ege, M.J., GABRIELA Study Group (2018). An approach to the asthma-protective farm effect by geocoding: Good farms and better farms. *Pediatric Allergy and Immunology*, 29(3), 275–282.

Nakashima, K., Hirota, T., Obara, K., Shimizu, M., Jodo, A., Kameda, M., Doi, S., Fujita, K., Shirakawa, T., Enomoto, T., Kishi, F., Yoshihara, S., Matsumoto, K., Saito, H., Suzuki, Y., Nakamura, Y., Tamari, M. (2006). An association study of asthma and related phenotypes with polymorphisms in negative regulator molecules of the TLR signaling pathway. *Journal of Human Genetics*, 51(4), 284–291.

Nicklaus, S., Divaret-Chauveau, A., Chardon, M.-L., Roduit, C., Kaulek, V., Ksiazek, E., Dalphin, M.-L., Karvonen, A.M., Kirjavainen, P., Pekkanen, J., Lauener, R., Schmausser-Hechfellner, E., Renz, H., Braun-Fahrländer, C., Riedler, J., Vuitton, D.A., Mutius, E.V., Dalphin, J.-C., PASTURE Study Group (2019). The protective effect of cheese consumption at 18 months on allergic diseases in the first 6 years. *Allergy*, 74, 788–798.

Normand, A.-C., Sudre, B., Vacheyrou, M., Depner, M., Wouters, I.M., Noss, I., Heederik, D., Hyvärinen, A., Genuneit, J., Braun-Fahrländer, C., Von Mutius, E., Piarroux, R., GABRIELA Study Group (2011). Airborne cultivable microflora and microbial transfer in farm buildings and rural dwellings. *Occupational and Environmental Medicine*, 68(11), 849–855.

Nwaru, B.I., Virtanen, S.M., Alfthan, G., Karvonen, A.M., Genuneit, J., Lauener, R.P., Dalphin, J.-C., Hyvärinen, A., Pfefferle, P., Riedler, J., Weber, J., Roduit, C., Kaulek, V., Braun-Fahrländer, C., Von Mutius, E., Pekkanen, J., PASTURE Study Group (2014). Serum vitamin E concentrations at 1 year and risk of atopy, atopic dermatitis, wheezing, and asthma in childhood: The PASTURE study. *Allergy*, 69(1), 87–94.

Orivuori, L., Mustonen, K., Roduit, C., Braun-Fahrländer, C., Dalphin, J.-C., Genuneit, J., Lauener, R., Pfefferle, P., Riedler, J., Weber, J., Von Mutius, E., Pekkanen, J., Vaarala, O., PASTURE Study Group (2014). Immunoglobulin A and immunoglobulin G antibodies against β-lactoglobulin and gliadin at age 1 associate with immunoglobulin E sensitization at age 6. *Pediatric Allergy and Immunology*, 25(4), 329–337.

Perkin, M.R. and Strachan, D.P. (2006). Which aspects of the farming lifestyle explain the inverse association with childhood allergy? *The Journal of Allergy and Clinical Immunology*, 117(6), 1374–1381.

Pfefferle, P.I., Büchele, G., Blümer, N. *et al.* (2010). Cord blood cytokines are modulated by maternal farming activities and consumption of farm dairy products during pregnancy: The PASTURE Study. *The Journal of Allergy and Clinical Immunology*, 125(1), 108–115.

Poirot, E., He, F., Gould, L.H., Hadler, J.L. (2020). Deaths, hospitalizations, and emergency department visits from food-related anaphylaxis, New York City, 2000–2014: Implications for fatality prevention. *Journal of Public Health Management and Practice*, 26(6), 548–556.

Raby, B.A., Klimecki, W.T., Laprise, C., Renaud, Y., Faith, J., Lemire, M., Greenwood, C., Weiland, K.M., Lange, C., Palmer, L.J., Lazarus, R., Vercelli, D., Kwiatkowski, D.J., Silverman, E.K., Martinez, F.D., Hudson, T.J., Weiss, S.T. (2002). Polymorphisms in toll-like receptor 4 are not associated with asthma or atopy-related phenotypes. *American Journal of Respiratory and Critical Care Medicine*, 166(11), 1449–1456.

Reboux, G., Piarroux, R., Mauny, F., Madroszyk, A., Millon, L., Bardonnet, K., Dalphin, J.-C. (2001). Role of molds in farmer's lung disease in Eastern France. *American Journal of Respiratory and Critical Care Medicine*, 163(7), 1534–1539.

Remes, S.T., Pekkanen, J., Soininen, L., Kajosaari, M., Husman, T., Koivikko, A. (2002). Does heredity modify the association between farming and allergy in children? *Acta Paediatrica*, 91(11), 1163–1169.

Riedler, J., Braun-Fahrländer, C., Eder, W., Schreuer, M., Waser, M., Maisch, S., Carr, D., Schierl, R., Nowak, D., Von Mutius, E., ALEX Study Group (2001). Exposure to farming in early life and development of asthma and allergy: A cross-sectional survey. *Lancet*, 358(9288), 1129–1133.

Rochat, M.K., Ege, M.J., Plabst, D., Steinle, J., Bitter, S., Braun-Fahrländer, C., Dalphin, J.-C., Riedler, J., Roponen, M., Hirvonen, M.-R., Büchele, G., Renz, H., Lauener, R., Krauss-Etschmann, S., Von Mutius, E., PASTURE Study Group (2010). Maternal vitamin D intake during pregnancy increases gene expression of ILT3 and ILT4 in cord blood. *Clinical and Experimental Allergy*, 40(5), 786–794.

Roduit, C., Wohlgensinger, J., Frei, R. *et al.* (2011). Prenatal animal contact and gene expression of innate immunity receptors at birth are associated with atopic dermatitis. *The Journal of Allergy and Clinical Immunology*, 127(1), 179–185.

Roduit, C., Frei, R., Loss, G. *et al.* (2012). Development of atopic dermatitis according to age of onset and association with early-life exposures. *The Journal of Allergy and Clinical Immunology*, 130(1), 130–136.

Roduit, C., Frei, R., Depner, M. *et al.* (2014). Increased food diversity in the first year of life is inversely associated with allergic diseases. *The Journal of Allergy and Clinical Immunology*, 133(4), 1056–1064.

Roduit, C., Frei, R., Ferstl, R. *et al.* (2018). High levels of butyrate and propionate in early life are associated with protection against atopy. *Allergy*, 74(4), 799–809.

Roost, H.P., Künzli, N., Schindler, C., Jarvis, D., Chinn, S., Perruchoud, A.P., Ackermann-Liebrich, U., Burney, P., Wüthrich, B. (1999). Role of current and childhood exposure to cat and atopic sensitization. European Community Respiratory Health Survey. *The Journal of Allergy and Clinical Immunology*, 104(5), 941–947.

Roussel, S., Reboux, G., Dalphin, J.-C., Pernet, D., Laplante, J.-J., Millon, L., Piarroux, R. (2005). Farmer's lung disease and microbiological composition of hay: A case-control study. *Mycopathologia*, 160(4), 273–279.

Roussel, S., Sudre, B., Reboux, G., Waser, M., Buchele, G., Vacheyrou, M., Dalphin, J.-C., Millon, L., Braun-Fahrländer, C., Von Mutius, E., Piarroux, R. (2011). Exposure to moulds and actinomycetes in Alpine farms: A nested environmental study of the PASTURE cohort. *Environmental Research*, 111(6), 744–750.

Schröder, P.C., Illi, S., Casaca, V.I. *et al.* (2017). A switch in regulatory T cells through farm exposure during immune maturation in childhood. *Allergy*, 72(4), 604–615.

Smits, H.H., Hiemstra, P.S., Prazeres da Costa, C. *et al.* (2016). Microbes and asthma: Opportunities for intervention. *The Journal of Allergy and Clinical Immunology*, 137(3), 690–697.

Stöcklin, L., Loss, G., Von Mutius, E., Genuneit, J., Horak, E., Braun-Fahrländer, C., GABRIELA Study Group (2012). Health-related quality of life does not explain the protective effect of farming on allergies. *Pediatric Allergy and Immunology*, 23(6), 519–521.

Strachan, D.P. (1989). Hay fever, hygiene, and household size. *British Medical Journal*, 299(6710), 1259–1260.

Strachan, D.P. (1996). Socioeconomic factors and the development of allergy. *Toxicology Letters*, 86(2/3), 199–203.

Strachan, D.P., Sibbald, B., Weiland, S. *et al.* (1997). World-wide variations in prevalence of symptoms of allergic rhinoconjunctivitis in children: The International Study of Asthma and Allergies in Childhood (ISAAC). *Pediatric Allergy and Immunology*, 8(4), 161–176.

Sudre, B., Vacheyrou, M., Braun-Fahrländer, C., Normand, A.-C., Waser, M., Reboux, G., Ruffaldi, P., Von Mutius, E., Piarroux, R., PASTURE Study Group (2009). High levels of grass pollen inside European dairy farms: A role for the allergy-protective effects of environment? *Allergy*, 64(7), 1068–1073.

Svanes, C., Jarvis, D., Chinn, S., Burney, P. (1999). Childhood environment and adult atopy: Results from the European Community Respiratory Health Survey. *The Journal of Allergy and Clinical Immunology*, 103(3 Pt 1), 415–420.

Vacheyrou, M., Normand, A.-C., Guyot, P., Cassagne, C., Piarroux, R., Bouton, Y. (2011). Cultivable microbial communities in raw cow milk and potential transfers from stables of sixteen French farms. *International Journal of Food Microbiology*, 146(3), 253–262.

Vartiainen, E., Petäys, T., Haahtela, T., Jousilahti, P., Pekkanen, J. (2002). Allergic diseases, skin prick test responses, and IgE levels in North Karelia, Finland, and the Republic of Karelia, Russia. *The Journal of Allergy and Clinical Immunology*, 109(4), 643–648.

Ventura, M.T., Munno, G., Giannoccaro, F., Accettura, F., Chironna, M., Lama, R., Hoxha, M., Panetta, V., Ferrigno, L., Rosmini, F., Matricardi, P.M., Barbuti, S., Priftanji, A., Bonini, S., Tursi, A. (2004). Allergy, asthma and markers of infections among Albanian migrants to Southern Italy. *Allergy*, 59(6), 632–636.

Von Hertzen, L., Laatikainen, T., Mäkelä, M.J., Jousilahti, P., Kosunen, T.U., Petays, T., Pussinen, P.J., Haahtela, T., Vartiainen, E. (2006a). Infectious burden as a determinant of atopy: A comparison between adults in Finnish and Russian Karelia. *International Archives of Allergy and Immunology*, 140(2), 89–95.

Von Hertzen, L., Mäkelä, M.J., Petäys, T., Jousilahti, P., Kosunen, T.U., Laatikainen, T., Vartiainen, E., Haahtela, T. (2006b). Growing disparities in atopy between the Finns and the Russians: A comparison of 2 generations. *The Journal of Allergy and Clinical Immunology*, 117(1), 151–157.

Von Mutius, E. (1998). The rising trends in asthma and allergic disease. *Clinical and Experimental Allergy*, 28(Suppl 5), discussion 50–51, 45–49.

Von Mutius, E. (2012). Maternal farm exposure/ingestion of unpasteurized cow's milk and allergic disease. *Current Opinion in Gastroenterology*, 28(6), 570–576.

Von Mutius, E. and Vercelli, D. (2010). Farm living: Effects on childhood asthma and allergy. Nature Reviews. *Immunology*, 10(12), 861–868.

Von Mutius, E., Fritzsch, C., Weiland, S.K., Röll, G., Magnussen, H. (1992). Prevalence of asthma and allergic disorders among children in united Germany: A descriptive comparison. *British Medical Journal*, 305(6866), 1395–1399.

Von Mutius, E., Braun-Fahrländer, C., Schierl, R., Riedler, J., Ehlermann, S., Maisch, S., Waser, M., Nowak, D. (2000). Exposure to endotoxin or other bacterial components might protect against the development of atopy. *Clinical and Experimental Allergy*, 30(9), 1230–1234.

Von Mutius, E., Schmid, S., PASTURE Study Group (2006). The PASTURE project: EU support for the improvement of knowledge about risk factors and preventive factors for atopy in Europe. *Allergy*, 61(4), 407–413.

Vuitton, D.A. (2008). *Immunologie*. Pradel, Paris.

Vuitton, D.A. and Dalphin, J.-C. (2006). Hygiene and allergy: Are farm microorganisms protective? *Journal de mycologie médicale*, 16, 220–38.

Vuitton, D.A. and Dalphin, J.-C. (2017). From farming to engineering: The microbiota and allergic diseases. *Engineering*, 3, 98–109.

Vuitton, D.A., Divaret-Chauveau, A., Dalphin, M.-L., Laplante, J.-J., Von Mutius, E., Dalphin, J.-C. (2019). Protection contre l'allergie par l'environnement de la ferme : en 15 ans, qu'avons-nous appris de la cohorte européenne PASTURE ? *Bulletin de l'Academie nationale de médecine*, 203, 618–630.

Waser, M., Von Mutius, E., Riedler, J., Nowak, D., Maisch, S., Carr, D., Eder, W., Tebow, G., Schierl, R., Schreuer, M., Braun-Fahrländer, C., ALEX Study Group (2005). Exposure to pets, and the association with hay fever, asthma, and atopic sensitization in rural children. *Allergy*, 60(2), 177–184.

Waser, M., Michels, K.B., Bieli, C., Flöistrup, H., Pershagen, G., Von Mutius, E., Ege, M., Riedler, J., Schram-Bijkerk, D., Brunekreef, B., Van Hage, M., Lauener, R., Braun-Fahrländer, C., PARSIFAL Study Group (2007). Inverse association of farm milk consumption with asthma and allergy in rural and suburban populations across Europe. *Clinical and Experimental Allergy*, 37(5), 661–670.

Weiner, H.L., da Cunha, A.P., Quintana, F., Wu, H. (2011). Oral tolerance. *Immunological Reviews*, 241(1), 241–259.

West, C.E., Videky, D.J., Prescott, S.L. (2010). Role of diet in the development of immune tolerance in the context of allergic disease. *Current Opinion in Pediatrics*, 22(5), 635–641.

Wickens, K., Lane, J.M., Fitzharris, P., Siebers, R., Riley, G., Douwes, J., Smith, T., Crane, J. (2002). Farm residence and exposures and the risk of allergic diseases in New Zealand children. *Allergy*, 57(12), 1171–1179.

Williams, H.C., Strachan, D.P., Hay, R.J. (1994). Childhood eczema: Disease of the advantaged? *British Medical Journal*, 308(6937), 1132–1135.

Williams, H.C., Robertson, C., Stewart, A. *et al.* (1999). Worldwide variations in the prevalence of symptoms of atopic eczema in the International Study of Asthma and Allergies in Childhood. *The Journal of Allergy and Clinical Immunology*, 103(1 Pt 1), 125–138.

Yazdanbakhsh, M., van den Biggelaar, A., Maizels, R.M. (2001). Th2 responses without atopy: Immunoregulation in chronic helminth infections and reduced allergic disease. *Trends in Immunology*, 22(7), 372–377.

Yazdanbakhsh, M., Kremsner, P.G., van Ree, R. (2002). Allergy, parasites, and the hygiene hypothesis. *Science*, 296(5567), 490–494.

4

Ecology of *Echinococcus multilocularis* Transmission

Patrick GIRAUDOUX[1,4], Dominique Angèle VUITTON[2] and Philip Simon CRAIG[3]

[1] *Chrono-environnement UFC-CNRS, Université Bourgogne Franche-Comté, Besançon, France*

[2] *Université Bourgogne Franche-Comté, Besançon, France*

[3] *School of Environment and Life Sciences, University of Salford, Greater Manchester, UK*

[4] *Yunnan University of Finance and Economics, Kunming, China*

GUIDING QUESTIONS.–

How does the landscape and its evolution determine the transmission of a zoonotic parasite lethal to humans, *Echinococcus multilocularis*, in different socioecosystems worldwide?

How are the dynamics of intermediate host populations, rather than the species richness of the community, a crucial factor in enhancing transmission?

How does the typically aggregative distribution of the parasite, at all scales, remain an enigma and complicate the understanding of transmission processes?

What are the contributions of a typology of socioecosystems to the understanding of transmission modes and to health management?

What are the advantages of an eco-epidemiological approach and spatial modeling in such socioecosystems?

4.1. Introduction

Echinococcus multilocularis is a cestode parasite that circulates in a lifecycle alternating larval and adult forms (Figure 4.1). The larval form is harbored in the liver of small mammals, called intermediate hosts, such as voles (rodents) and pikas (lagomorphs). Its proliferation by asexual budding generates micro-cysts, or "alveoli", producing larvae called "protoscoleces" which are, in a way, the "invaginated" form of the adult worm. The adult worm will then live in carnivore hosts, called "definitive hosts", where it will reproduce sexually. If these protoscoleces are ingested by a suitable carnivore, such as a fox or dog definitive host, then they become scoleces, capable of attaching to the mucous membrane of the intestine where they develop into adult worms. These then produce eggs in three to four weeks which are dispersed with the carnivore's droppings in the environment. If humans accidentally ingest these eggs (by eating soiled plants or stroking the soiled coat of a definitive host, for example), they can hatch and migrate into the liver. The parasite is eliminated by the human immune system in 95% of cases (Bresson-Hadni *et al.* 1994). Otherwise, the parasitic larva proliferates in the manner of a very slowly developing cancer (sometimes over several decades). The pseudotumor thus developed is made up of a mixture of parasitic and inflammatory tissue. It can even, in some cases, but less often than in animal intermediate hosts, produce protoscoleces.

The human disease is named alveolar echinococcosis, because of the "alveolar" structure of the parasitic pseudotumor caused by the development of the *E. multilocularis* larva. It is a rare disease, with a few hundred new cases each year worldwide (Deplazes *et al.* 2017), but its prevalence in the population may be as high as 1 in 1,000 or 15 in 100 in highly endemic areas. It is fatal in the absence of medical management. Removal of the affected liver lobe(s), when possible, leads to a cure. To date, there is no effective parasiticide drug to eliminate the parasite. Only a high-dose parasitostatic drug, albendazole, allows control of parasite development when removal of the parasitic pseudotumor is not possible (Vuitton *et al.* 2017).

Although alveolar echinococcosis was described by Buhl in 1852 in Germany, then recognized as a parasitic disease by the great German scientist Virchow a few years later and identified in France as early as 1890, the infectious agent that caused it, *E. multilocularis*, was not identified as a distinct species from other *Echinococcus* species until the 1950s, following the work of Rausch and Schiller in Alaska and Vogel in Germany, who demonstrated the nature of the cycle through experimental infections (Vuitton *et al.* 2011). Other than in the Arctic tundra of St. Lawrence Island in the Bering Strait, where it had been masterfully studied in Inuit communities by Robert Rausch (Rausch and Schiller 1956; Rausch 1967), its ecology was virtually unknown in the early 1980s in other ecosystems where the parasite circulated. In France, René Houin, Martine Liance and their team identified the parasite for the first time in an intermediate host, the montane water vole, in the Jura Mountains (Houin *et al.* 1980). The importance of the climate and the habitat of the species that serve as intermediate hosts were then pointed out in the distribution of the vulpine infestation, by Gilot *et al.* (1988) in the Massif Central and by Pesson and Carbiener (1989) in Alsace.

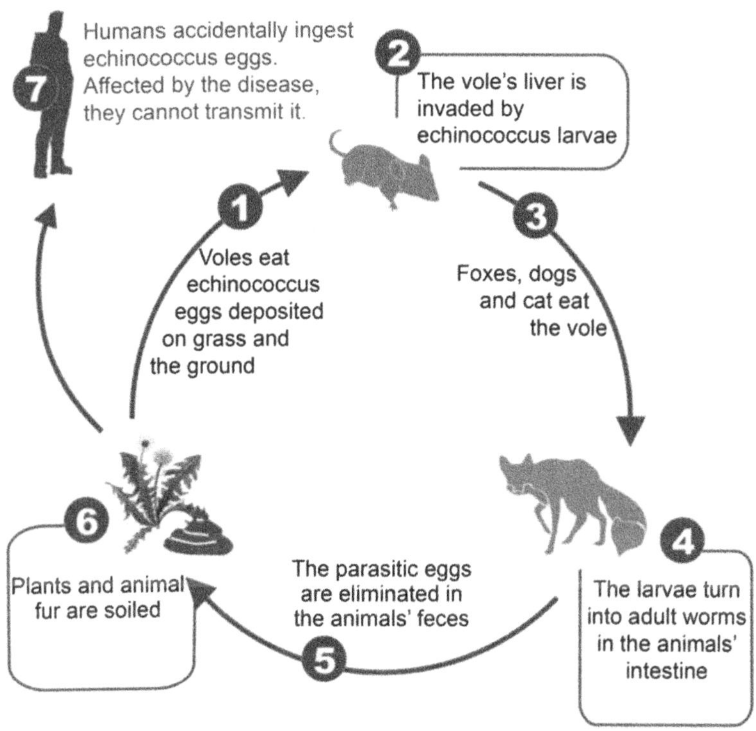

Figure 4.1. *Lifecycle of the Echinococcus multilocularis (infography CHRU Jean Minjoz)*

In the Jura Massif, under the direction of Pierre Delattre, INRA researchers sampled local small mammal populations from 1979 to 1996 in order to understand the causation of grassland vole outbreaks and their decline phases (see Chapter 1). The parasite was found in several rodent species over that period (Delattre *et al.* 1990). A medical team at the University of Franche-Comté, led by Jean-Philippe Miguet and later by Dominique Angèle Vuitton, had been interested in the human component of the cycle since the 1970s following what appeared to be a resurgence of cases (Miguet 1973; Vuitton *et al.* 2011). The joining of this team with that of Pierre Delattre, through work with the team of René Houin and Martine Liance, led, from the mid-1980s onward, not only to the collaboration of ecologists, agronomists and physicians in a common program, but also to the involvement of students in science and veterinary medicine. The veterinary laboratories of the Doubs and Jura departments were also involved in the work on the vulpine component, with the support of the guards of the French National Office for Hunting and Wildlife (*Office national de la chasse et de la faune sauvage*), who carried out night shootings to collect specimens. A comprehensive eco-epidemiological approach could thus be achieved by Patrick Giraudoux's thesis, directed by Pierre Delattre, in collaboration with all of the parties (Giraudoux 1991). The aim of the joint studies was to ensure the description of the transmission system and to answer the main questions that health officials had about the particular epidemiology of this infection. These questions were: is there a relationship between the distribution of human cases, the infection of the fox definitive host and the risk factors for grassland rodent population outbreaks? If so, at what spatial and temporal scales? Since the intermediate host of the parasite is a possible prey of its definitive host, can these observations be explained by a landscape composition that may modulate the predator-prey relationship? Can interventions to control rodent outbreaks (justified by reasons unrelated to alveolar echinococcosis, but by agricultural imperatives) interfere with the maintenance of the parasite reservoir (intermediate hosts) and/or definitive host populations and, *ultimately*, with the transmission of the parasite to humans? And more broadly: can human interventions motivated by economic or even political reasons, whether local or global, unexpectedly or inadvertently modify the epidemiology of a rare zoonotic disease, such as alveolar echinococcosis?

We propose here to present the results obtained over decades in the Jura socioecosystem using this multidisciplinary approach, and to show how this research has been successfully applied to other transmission systems in Asia, mainly in western China and also in southern Kyrgyzstan.

4.2. The Jura transmission system

4.2.1. *Seasonality and microfoci in rodents*

Echinococcus multilocularis is usually a rare parasite in its intermediate hosts. Its prevalence on a landscape scale rarely exceeds one per thousand in a vole population, even in highly endemic areas (Giraudoux *et al.* 2002). Detecting the natural infection of individuals under such conditions is therefore akin to finding a needle in a haystack. Long-term observational studies (1979–1996) of rodent and shrew populations to understand their dynamics involved the autopsy of 22,851 small mammals in the Septfontaines-Le Souillot study area (see Chapter 1). Incidentally, they confirmed the capture of 27 animals parasitized by *E. multilocularis*. Detailed study of their spatial and temporal distribution, and comparison with other cestode species with a similar cycle, such as the cat tapeworm (*Taenia taeniaeformis*), allowed an understanding of the seasonal variations of infections and their dependence on the age structure of the host population. Figure 4.2 shows, for example, the relationship between prevalence and age structure of a common vole population: all animals above the black dotted curve were born before winter, those below were born in the current year. The young animals are only slightly infected. The abnormally high number of parasitized individuals on the edge of ploughed fields, even though these areas represent only a small fraction of the sampled area, also indicates the existence of microfoci where prevalences are thousands of times higher than elsewhere in the landscape. These are places frequented by rodents, where fox droppings (which contain the parasite's eggs) are present at densities 10–100 times higher than elsewhere (territorial marking zones), and where the humidity conditions (favored by the burying of droppings, for example, by trampling by cattle or ploughing) allow a longer survival of the parasite's eggs (Delattre *et al.* 1988, 1990; LePesteur *et al.* 1992; Giraudoux *et al.* 2002).

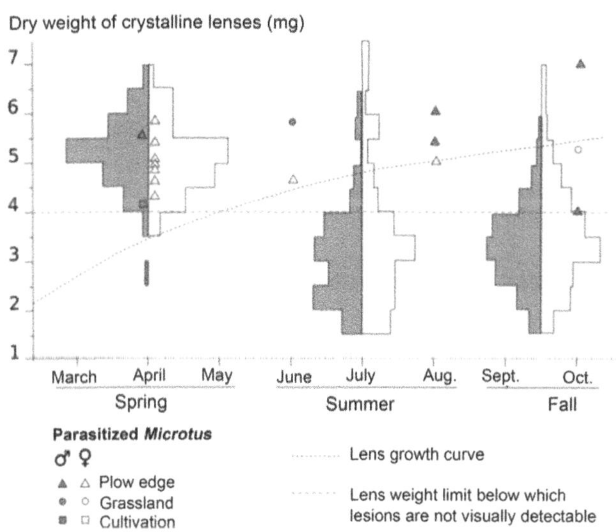

Figure 4.2. *Age structure of a common vole (Microtus arvalis) population and distribution of Echinococcus multilocularis, from Delattre* et al. *(1988). The mass of the crystalline lenses increases with age: it is an indirect measure of the relative age of individuals. For a color version of this figure, see www.iste.co.uk/giraudoux/socioecosystems.zip*

These studies also confirmed the very low specificity of *E. multilocularis* with respect to its intermediate hosts: practically all rodent species present in each of the environments of the study area were susceptible hosts that can harbor the parasite (Delattre *et al.* 1990) (Figure 4.3), but the very low prevalences of the parasite in these intermediate hosts, combined with the fact that the age structures of the populations must be known to make the figures comparable (Figure 4.2), limit the comparisons that could be made between species.

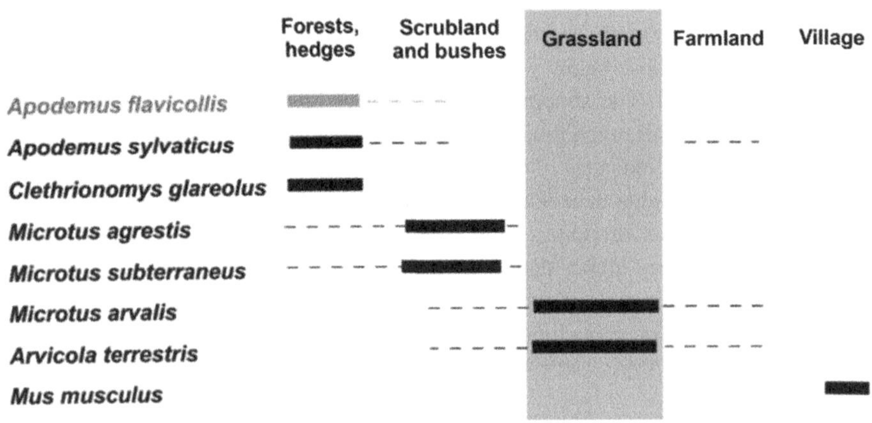

Figure 4.3. *Distribution of rodent species according to their habitat in the study area, from Giraudoux* et al. *(1994, 2003). Those in black are those that have been shown to be naturally infected with E. multilocularis. In this landscape, only two species, both grassland voles, are likely to reach population densities exceeding several hundred individuals per hectare over large areas*

4.2.2. *Vole outbreaks and distribution of human cases*

The epidemiological situation of human alveolar echinococcosis evolved during the 1990s in Europe from a disease strictly limited to "traditional" endemic areas, such as eastern France, southern Germany, Switzerland and Austria, to a disease covering almost all of non-Mediterranean Europe (Thompson *et al.* 2017). As early as the late 1980s, spatial aggregation of human cases was detected at several scales in France, based on clinical epidemiology data (Vuitton *et al.* 1990). Most cases, even today, are located in the eastern part of the country, particularly in the Jura Mountains and Savoy, and in the Ardennes-Lorraine, in the form of nested clusters whose location may vary over decades (Kern *et al.* 2003; Piarroux *et al.* 2013; Said-Ali *et al.* 2013). The question of what factors preside over this strange distribution in "clusters" is still very open, as the disease is not transmissible between humans. On a European scale, the location of cases in temperate or boreal climatic zones, and their absence in the Mediterranean zone, suggests that relatively cold climatic conditions are necessary for the completion of the parasite cycle, probably because of the known sensitivity of the parasite eggs to heat and desiccation (Veit *et al.* 1995). But what about at more local scales? Eco-epidemiological studies conducted in the Doubs department in the 1990s showed the superposition of areas where outbreaks of grassland vole were observed and those where a prevalence of human cases was observed that was nearly four times higher than in the surrounding area (Viel *et al.* 1999), confirming an association that had been considered as early as the 1970s (Houin *et al.* 1980). Thus, it appeared that although potential intermediate rodent host species were present in all habitats in the Doubs landscape, a higher prevalence of the disease in humans was correlated only with habitats where some of these rodent species regularly reached densities of several hundred individuals per hectare.

4.2.3. *Infection in definitive hosts*

Raoul *et al.* (2001), in a study conducted from 1996 to 1999, confirmed the uneven distribution of parasite burdens in the red fox population (*Vulpes vulpes*): nearly 76% of the parasite load was carried by less than 12% of the foxes. The youngest foxes, born in the year, are the most contaminated and successive reinfections lead to a certain degree of immunity, a phenomenon confirmed by laboratory studies. Infection of a definitive host is accompanied by the shedding of parasitic material (cells, eggs, DNA, etc.) with the feces. Parasitic material can be detected by different techniques. The ELISA technique consists of producing labeled antibodies, capable of revealing the presence of parasitic antigens (copro-antigens). Pleydell *et al.* (2004) demonstrated, in the Jura Mountains, the strong spatial autocorrelation of the level of copro-antigens detected by this method in fox feces collected on a local scale, and the correlation of this contamination with the ratio of grassland to total area derived from the CORINE Land Cover (CLC) biophysical land use database, on a regional scale. This observation corroborates the results obtained from human epidemiology: there are "at risk" landscapes where the transmission of the parasite is more intense, characterized by a high proportion of permanent grassland. Raoul *et al.* (2010) confirmed the strong food preference of foxes for grassland voles: montane water voles and common voles dominate the fox diet as soon as they are present in the ecosystem (see also Chapter 1, section 1.2.6 and Figure 1.7). The feeding response and parasite contamination are furthermore nonlinear, rapidly asymptotic (Figure 4.4) and complicated by the differences in fluctuations of these two species (Raoul *et al.* 2010; Baudrot *et al.* 2016) (see also Chapter 1, section 1.2.5 and Figure 1.6).

In addition, fox infection is subject to significant inter-annual variation. Winter prevalences of *E. multilocularis* increased from 27% to 46% in 1989–1990 to 53% to 65% in 1996–1999 in the highly endemic areas of the Jura Mountains, in parallel with the increase in populations of the red fox (Giraudoux *et al.* 2001). These prevalences also increased during the 2000s in traditionally less favorable peripheral areas, sometimes reaching local values comparable to those of high endemic areas (Combes *et al.* 2012). Genetic analysis of parasite samples from foxes across Europe indicates that parasite emergence has occurred from the historic endemic foci of the Jurassic and Alpine arc (Knapp *et al.* 2009).

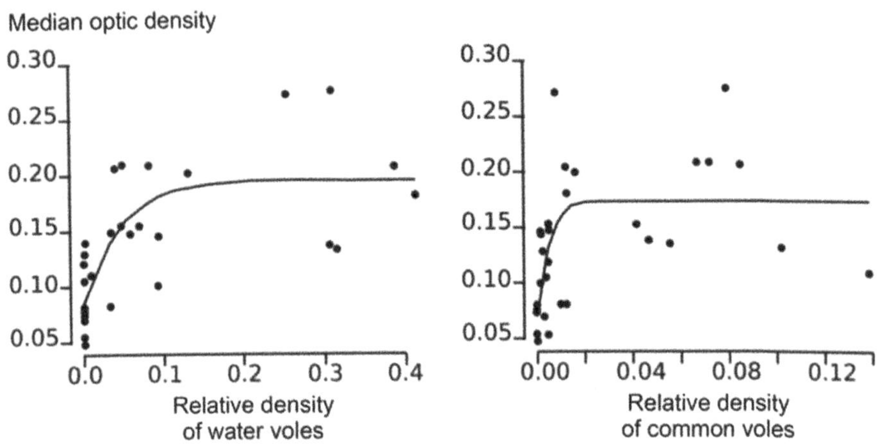

Figure 4.4. *Contamination of fox feces by E. multilocularis as a function of the relative densities of the water vole and the common vole, from Raoul* et al. *(2010). Relative densities are expressed as percentages of intervals occupied by indices of the species, counted when transects were conducted in appropriate environments (see Chapter 1, section 1.2.3); optical density is proportional to the amount of labeled E. multilocularis antigen detected by ELISA*

On a local scale, Knapp *et al.* (2018) confirmed the aggregative spatial distribution of fox feces contaminated with eggs of the parasite. Furthermore, a massive decrease in the level of contamination of feces by *E. multilocularis* could be observed where the massive use of rodenticide anticoagulants for the control of water vole populations led, through secondary poisoning, to a very strong decline in fox populations (Raoul *et al.* 2003). However, experiments to control parasite transmission on a large scale by lowering the fox population through shooting and trapping alone did not succeed in lowering the fox population sufficiently, and may even have had the opposite effects to those expected, by increasing the parasite prevalence (Comte *et al.* 2017). This type of control therefore poses logistical, ethical, ecological and efficiency problems that make it inappropriate.

4.2.4. *Conclusion*

All the results obtained in eastern France thus indicate the aggregative distribution of the parasite at all scales and in all of its hosts, and show that the intensity of its circulation depends on the density of host populations and the interactions between them. For intermediate hosts, this density is determined by the composition and structure of the landscape (see Chapter 1). The sensitivity of the parasite eggs to moderate heat and drought, the virtual disappearance of intermediate hosts for several seasons during a phase of decline (see Chapter 1) and the significant drop in density of the definitive host can each or all explain a temporary halt in the lifecycle, and then its subsequent re-emergence. The variations of those factors are probably the origin of the shifting clusters that were observed from the earliest studies in which this fact was evoked as "sparks" on the periphery of more stable "foci" (Vuitton *et al.* 1990; Delattre *et al.* 1991). Further epidemiological observations, made possible by the establishment of an exhaustive register of human echinococcosis cases in France in 1997, going back to 1982 (Kern *et al.* 2003), have corroborated this hypothesis, in particular in Haute-Saône, a peripheral and low-intensity focus in the 1990s, (re)activated in the early 2000s (Said-Ali *et al.* 2013).

The knowledge gained in the French focus of alveolar echinococcosis with the highest prevalence has allowed the development of a conceptual model of transmission which, beyond the "classical" parasite cycle patterns (Figure 4.1), in addition highlights the ecological and population factors that may be responsible for the intensity of transmission (Figure 4.5). The existence of landscapes where transmission is more likely related to the population dynamics of certain small mammal intermediate hosts therefore offers an opportunity to identify "at risk" landscapes for public health and to optimize the organization of prevention.

The Far East, mainly western China, was then to provide a number of endemic study sites to test and complete this approach and model.

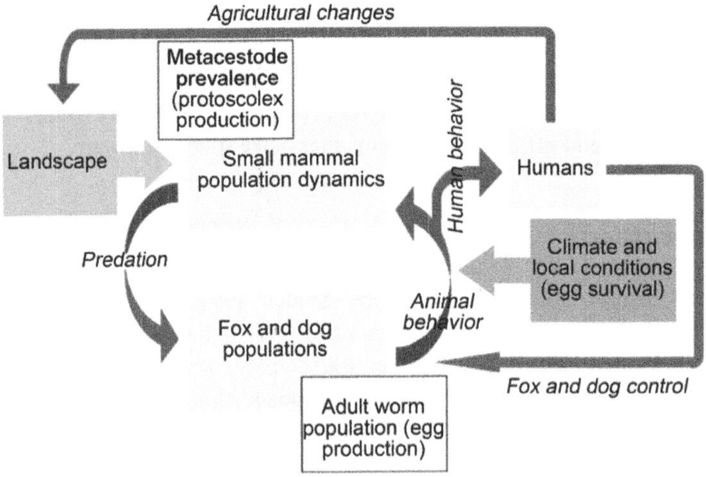

Figure 4.5. *Global variables and ecological model of Echinococcus multilocularis transmission, from Giraudoux* et al. *(2003)*

4.3. Ecology of Echinococcus multilocularis transmission in China and Kyrgyzstan

4.3.1. *The background*

In China, the first cases of human alveolar echinococcosis were recognized in 1956 in the Xinjiang Uyghur Autonomous Region, and were followed in the 1960s by the discovery of other cases in Qinghai and Gansu Provinces. Published at the time in Chinese, they remained unknown to the international public for a long time (Vuitton *et al.* 2011). The opening up of China in the late 1970s ushered in an era of cooperation between Chinese and Western doctors and researchers. Professors Philip Simon Craig, then from the Liverpool School of Tropical Medicine (UK), Liu Deshan, from the Gansu Medical College in Lanzhou, and Ding Zhaoxun, from the Xinjiang Medical University in Ürümqi, were the initiators in the field of echinococcosis and carried out the first joint exploratory missions in the late 1980s (Craig *et al.* 1991). This cooperation was extended in the early 1990s to the network of French researchers working on this disease and its ecology in the Jura Mountains. Initially, in 1991, the French researchers were approached by Phil Craig specifically for their experience in the clinical field and, in particular, therapeutic developments including the use of antiparasitic drugs of the benzimidazole family which they had helped to validate in studies under the aegis of the World Health Organization (WHO) (Davis *et al.* 1986, 1989), and the setting up of a liver transplant program for patients with the most severe clinical forms (Gillet *et al.* 1988). The training of surgeons from endemic areas in China, in particular Professor Wen Hao, who became one of the leading figures in echinococcosis research in China, was the pretext for this rapprochement, as well as the expertise from Besançon in the echography of alveolar echinococcosis, in view of an ambitious screening program in the Chinese countryside. However, the work carried out on the ecology of the hosts in the Franche-Comté region and the ability of local researchers and medical doctors to integrate the multiple dimensions of a transmission system through an eco-epidemiological approach quickly convinced the British and Chinese researchers of the relevance of adding this expertise and these dimensions to the research project they were setting up. The first study campaign, involving jointly British, Chinese and French researchers, supported by European funds, began in 1994 in Zhang County, north of the Min Mountains in southern Gansu (Figure 4.6), following the confirmation of a major outbreak of human alveolar echinococcosis discovered during exploratory missions in the late 1980s. The prevalence of the disease averaged 4%, and in some villages was as high as 15% (Craig *et al.* 1992, 2000). Other foci of similar importance were subsequently discovered, notably on the northeastern edge of the Tibetan plateau, in southern Ningxia and in the upper Ili valley (Tian Shan mountain range). They were explored with the support of the Fogarty International Center of the US National Institutes of Health (NIH) from 2001 to 2008, and later by the Wellcome Trust, the Chinese Center for Disease Control (CDC) and the French National Centre for Scientific Research (CNRS), among others. Later, in the 2010s, a major outbreak was studied by Kyrgyz doctors and veterinarians in the vicinity of Sary Mogul in the Alay Valley of southern Kyrgyzstan (Raimkulov *et al.* 2015), with support for ecological and veterinary studies from the same international network that was working in China[1]. These academic initiatives have been followed since 2005 by a national disease surveillance and management program in China, which involves 14 ministries and still remains active in 2022. Indeed, some of the work conducted showed, for example, that Tibetan pastoralist communities in northwestern Sichuan lost 0.81 DALYs (disability adjusted life years) owing to alveolar and cystic echinococcosis, compared to an average of 0.18 DALYs lost in the general Chinese population because of all communicable and non-communicable diseases combined (Budke *et al.* 2005b). This demonstrated that a disease that is apparently rare on a global and even a Chinese national scale can nevertheless have a considerable human and economic impact in certain regions. China's national control program is certainly the most ambitious state-funded project ever implemented to diagnose, treat and prevent alveolar echinococcosis in the world (Vuitton *et al.* 2011).

1 https://gdri-ehede.univ-fcomte.fr.

The approach described in the previous section was developed in each of the regions where an outbreak had been identified, with the necessary adaptations and optimizations, taking into account the logistical and collaborative possibilities that could be established locally. The eco-epidemiological approach here consisted of: (1) spatializing as precisely as possible the prevalences of human alveolar echinococcosis, on the basis of data from local hospital registers, and also of specific mass-screenings combining abdominal ultrasound and serology organized in the village communities by Chinese clinicians with the support of French doctors, with an epidemiological objective and also for training purposes; (2) describing as precisely as possible the distribution of small mammal communities and their habitats, including those likely to be intermediate hosts of the parasite; (3) collecting, when available, data on the interannual dynamics of their populations; (4) quantifying, from satellite data and field measurements, the habitats that make up the regional landscapes; and (5) looking for correlations between these variables. However, the protection status of the different fox species in China and the scarcity of the red fox in some regions did not allow us to explore in as much detail as in Europe the spatial distribution of the parasite in foxes. Moreover, it quickly became apparent that, contrary to the European situation, the dog also played a major role in the functioning of the parasite cycle and transmission to humans. Clinicians, parasitologists, veterinarians, taxonomists, ecologists, geographers and modelers have systematically collaborated on these ("One Health") studies for more than 30 years.

4.3.2. *Transmission and biodiversity on a continental scale*

Mainland China stretches from Siberia to northern Xinjiang, from Inner Mongolia and Heilongjiang to the Yunnan rainforest, and includes high-altitude areas such as the Tibetan Plateau, which has an average elevation close to 5,000 m, several deserts (some of them below sea level), and low-lying agricultural and coastal areas (Figure 4.6). A total of 25 biogeographic regions and 77 subregions are defined. Red foxes (or Tibetan foxes on the plateau) are present throughout, but transmission of *E. multilocularis* is sustained in diverse and dissimilar small mammal communities. While some of these communities have been studied, many remain unexplored. The distribution of the 274 known species of non-flying small mammals in mainland China shows that the distribution of human alveolar echinococcosis is not related to the diversity of these species as a whole, nor is it related to the diversity of known intermediate hosts, including lagomorphs (especially pikas, *Ochotona sp.*) that do not occur in European endemic areas (Figure 4.7).

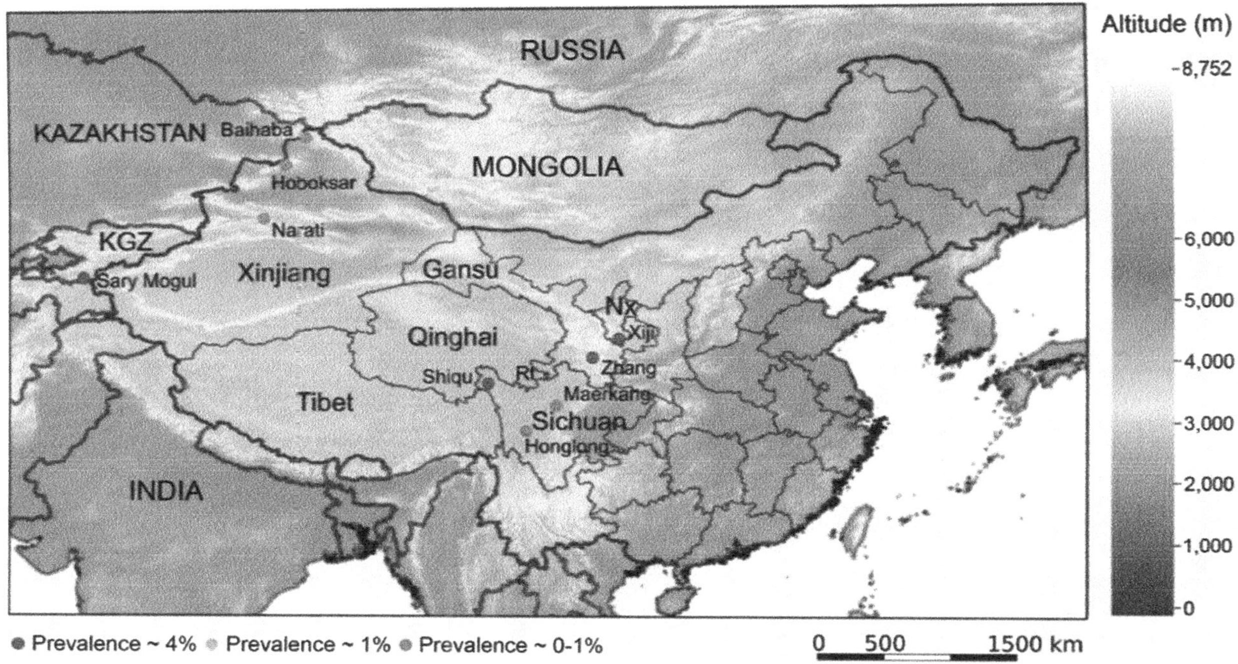

Figure 4.6. *Map of the Asian study areas. The background is a 1 x 1 km resolution digital terrain model (US National Oceanic and Atmospheric Administration). KGZ, Kyrgyzstan; Nx, Ningxia; Rt, Rangtang, from Giraudoux* et al. *(2013a). For a cclor version of this figure, see www.iste.co.uk/giraudoux/socioecosystems.zip*

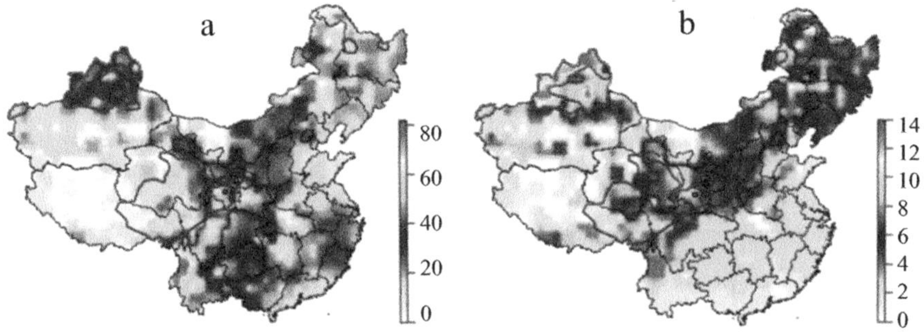

Figure 4.7. *Species richness of non-flying small mammals in mainland China. Alveolar echinococcosis endemic areas are surrounded by a red line. a) All species combined. b) Known intermediate hosts, from Giraudoux* et al. *(2013b). For a color version of this figure, see www.iste.co.uk/giraudoux/socioecosystems.zip*

4.3.3. *The Tibetan plateau*

From 2001 to 2008, 17,589 people were screened in 94 villages on the eastern edge of the Tibetan plateau[2]. The overall prevalence of alveolar echinococcosis was then 3.3%, but varied considerably from one village to another (Giraudoux *et al.* 2013b) confirming the aggregative spatial distribution of the parasite, already observed in Europe. On this basis, the distribution of prevalences was analyzed according to internal (age, gender, occupation, ethnicity, etc.) and external (altitude, rainfall, landscape composition, etc.) factors. The greatest difficulty with this type of analysis is that in such regional systems, the explanatory variables that may determine transmission, directly or indirectly, are not independent. For example, grass productivity and forest cover are correlated with precipitation and temperature, which are (locally) correlated with altitude. Moreover, the aggregative spatial distribution of cases considerably increases the phenomenon of spatial autocorrelation (close locations show identical prevalences), which poses specific statistical problems. Finally, the relationship between two variables may be nonlinear (exponential or bell-shaped response), which further complicates matters (Hanf *et al.* 2014). These problems have been circumvented by specific multivariate approaches (modeling the explanatory variables together and accounting for their covariations), as well as through the use of generalized additive mixed models to better account for possible nonlinear responses and spatial autocorrelation (Figure 4.8).

2 The compilation of such a database, combining several years of screening in several Chinese provinces, was made possible by late Professor Akira Ito, who brought together Chinese researchers Li Tiaoying and Han Xiuming and French eco-epidemiologist Patrick Giraudoux every year from 2008 to 2010, to the very wintry and warm isolation of his parasitology laboratory at Asahikawa Medical University, on the island of Hokkaidō, Japan.

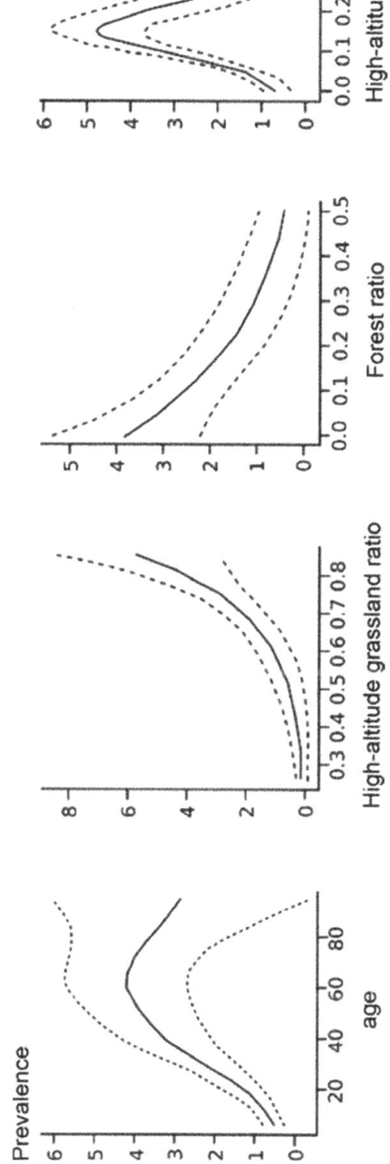

Figure 4.8. *Examples of relationships between human alveolar echinococcosis prevalence and age and environmental factors. Predicted values (solid line) and 95% confidence intervals (dashed line) were calculated using a generalized additive mixed model with a binomial link, a term smoothed by a cubic spline regression on age and a random effect on villages accounting for the aggregation of values (Giraudoux et al. 2013b)*

The main results are that the proportion of infected individuals increases with age and is higher among Tibetan herders and females than in the rest of the population. The proportion of infected individuals is also higher in landscapes with a high proportion of permanent grassland (and is, in contrast, lower in forested areas) (Giraudoux *et al.* 2013b).

The wild intermediate hosts are mainly the Tibetan fox (*Vulpes ferrilata*) and the red fox. Both species have been found to be naturally infected at similar prevalences to those observed in Europe in highly endemic areas. However, unlike in Europe, dogs are frequently infected at high levels. Observations of dogs feeding on small mammals in and around villages are common and frequent consumption of small mammals is evidenced by diet analysis (Wang *et al.* 2010). Budke *et al.* (2005c) showed that the consumption of one infected small mammal every 1–2 years is sufficient to explain the 10%–20% prevalence of *E. multilocularis* observed in dogs in Serxu County. In some areas, given the high number of dogs in villages and the much lower number of foxes in the vicinity, it is likely that most of the biomass of the *E. multilocularis* population is in the dog population (Craig *et al.* 2019). The large proportion of females among patients with alveolar echinococcosis may at least in part be explained by Tibetan females' closer contact with dogs, which they feed and sometimes may be the only ones allowed to approach. Furthermore, the spatial distribution of fox feces is aggregative, and that of dog feces doubly aggregative: concentrated in villages (vs. outside villages) and within villages, in some areas (Vaniscotte *et al.* 2011).

Many species of small mammals occur on the Tibetan plateau and virtually all of them, as in Europe, are susceptible to *E. multilocularis*. The number of species present in the plateau grasslands is, however, lower than that counted in nearby highland forests (Raoul *et al.* 2006; Vaniscotte *et al.* 2009; Craig *et al.* 2019). It is in these grasslands that outbreaks of plateau pikas (*Ochotona curzoniae*), a small colonial lagomorph, and various species of voles and dwarf hamsters can be observed, reaching several hundred to several thousand individuals per hectare (Figure 4.9). The very fact that the spatial distribution of human cases was related to the extent of the permanent grassland, and not to the number of species present or other types of environment (forest, etc.), indicates once again that in a favorable cold and wet climate, the abundance of one or more intermediate host species, favored by the extension of their optimal habitat, is a determining factor in the intensity of parasite transmission, rather than species richness itself. This factor may, moreover, be modulated locally by unexpected distal causes. For example, in the Serxu region, a change in the property law in the 1980s led to the privatization of some pastures, which were enclosed, thus reducing the area of communal pastures, which were then overused. The enclosures of winter pastures near human habitations then increased the risk of alveolar echinococcosis (Wang *et al.* 2004) by altering the prevalence of *E. multilocularis* in dogs, correlated with observed changes in the distribution of small mammals between enclosed and overgrazed areas (Wang *et al.* 2006, 2007, 2010).

Figure 4.9. *Typical grassland landscape of the eastern edge of the Tibetan plateau (Rizha hamlet, Serxu county, 4,200 m altitude). Inset, a plateau pika (photos P. Giraudoux)*

These observations and the resulting models, taking into account both landscape and climatic variables, allowed the confirmation and prediction of human alveolar echinococcosis foci over a 200,000 km^2 area of the Tibetan plateau (Figure 4.10). Notably, three foci and one low-prevalence area, which were not in the dataset used for model development, were correctly "predicted": on the eastern edge, the Zhang/Puma focus in Gansu (see section 4.4), a focus northwest of Naqu and yet another south of Dingqing in the Tibet Autonomous Region, as well as the low-prevalence area around Lake Qinghai (Giraudoux *et al.* 2013b). In addition, the results obtained were used to model the potential distribution of plateau pika (*O. curzoniae*) in Shíqú (Serxu) County, Sichuan, based on Landsat ETM+ and elevation data, explaining 71% of the variance in the distribution, with an accuracy of 90.1% (Marston *et al.* 2014).

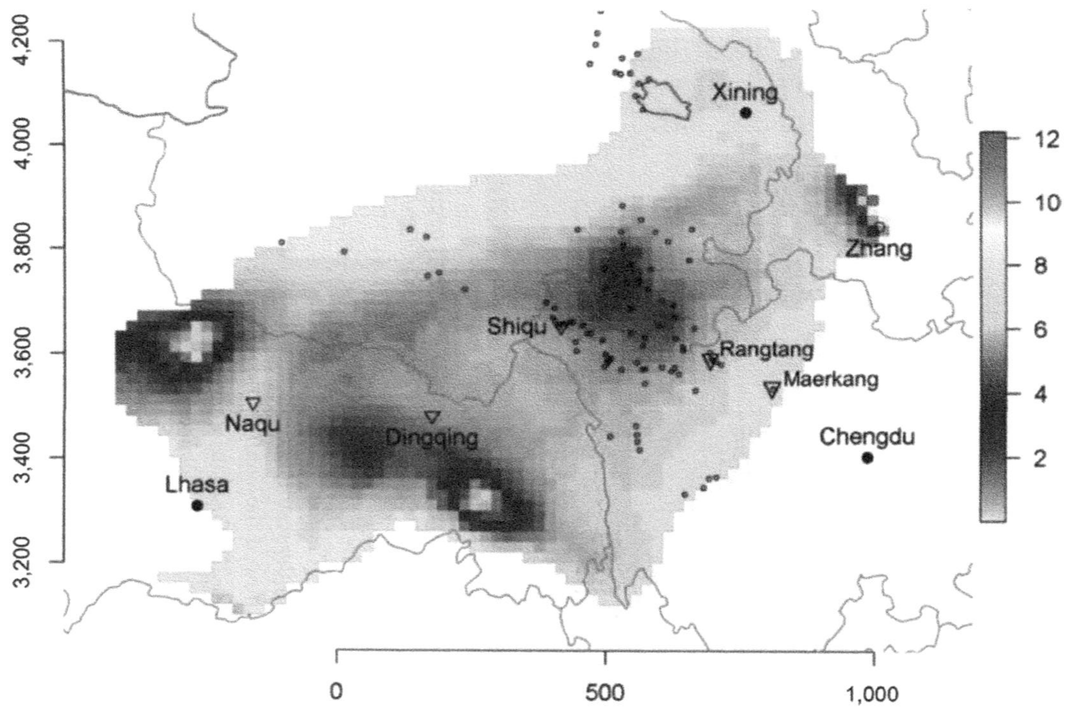

Figure 4.10. *Model for predicting the prevalence of human alveolar echinococcosis on the eastern Tibetan plateau, from Giraudoux* et al. *(2013b). The vertical gauge gives the prevalence in percentage. Empty circles locate villages, triangles locate major localities and solid circles locate major cities. The units of the axes are in kilometers (UTM 47). For a color version of this figure, see www.iste.co.uk/giraudoux/ socioecosystems.zip*

4.4. The mountains of southern Gansu and Ningxia

4.4.1. *Gansu*

In the late 1980s, a large focus of human alveolar echinococcosis was identified in the poor farming communities of the southern highlands of Gansu province, in Min and Zhang counties. Unlike the Tibetan plateau, where 95% of the inhabitants are yak and sheep farmers, the villages in these mountains (2,300–2,600 m above sea level) are inhabited by Han farmers, who grow soybeans, rapeseed, potatoes, flax and buckwheat, and who also keep small family herds of a few *pian niu*, a hybrid of female yak (*Bos grunniens*) and domestic bull (*Bos taurus)*, a few pigs and a few sheep (Figure 4.11).

In 1991, 5% of 1,312 individuals screened by abdominal ultrasound were diagnosed with liver lesions caused by alveolar echinococcosis. In addition, 6 of a sample of 58 dogs necropsied had the adult form of the parasite in their intestines, with parasitic loads ranging from 20 to 5,000 worms (75% of families had at least one dog). The parasite cycle was therefore active at the time, but the details of its ecology (nature of intermediate hosts, possible role of foxes, details of transmission mechanisms to the dog, etc.) were unknown (Craig *et al.* 1992). Since 1994, this focus was the subject of international multidisciplinary monitoring (China, France and the United Kingdom) until 2015.

Eco-epidemiological studies conducted from 1994 to 1997 mapped human cases and the distribution of small mammals, including potential intermediate hosts. At that time, dogs, which had been very numerous a few years earlier, had virtually disappeared, as had red foxes, which had been relatively abundant according to local hunters. The reasons for this disappearance remain unclear to this day. From information obtained from farmers, it is reasonable to assume that it is the consequence of their secondary poisoning during chemical treatment campaigns against rodents, conducted in the early 1990s. In the absence of definitive hosts, the parasite cycle was interrupted, although still high prevalences of human alveolar echinococcosis could be observed. Indeed, the development of the parasite in the liver can take up to several decades in humans, who are naturally resistant to the parasite; the patients identified in 1994–1997 had been infected in the 1980s or even earlier, or in the early 1990s, when the parasite lifecycle was still active. It was therefore essential for researchers and health officials to understand the conditions that had allowed the parasite cycle to intensify in the past in order to assess the risks of its re-emergence.

Figure 4.11. *Landscapes in May 2015 in Zhang County. Ploughing on steep slopes, still practiced until the 1990s, has been banned (abandoned terraces) and these areas are left to the natural recolonization of vegetation. Reforestation programs have also been implemented since the mid-1990s (photos P. Giraudoux). For a color version of this figure, see www.iste.co.uk/giraudoux/socioecosystems.zip*

From 1994 to 1997, 2,482 people in the villages of the endemic area had their livers examined by ultrasound. Combined with earlier studies from the late 1980s, the observed prevalences of human alveolar echinococcosis were found to be extremely different from one village to another, ranging from 0% to nearly 16% (as also described in the Tibetan plateau villages). These prevalences increased with age, and for the same age, they were also found to be higher in women and in dog owners during the previous decade. In addition, prevalences were more than three times higher in villages surrounded by grassland and scrubland and bushes than in those dominated by ploughed land (Figure 4.12).

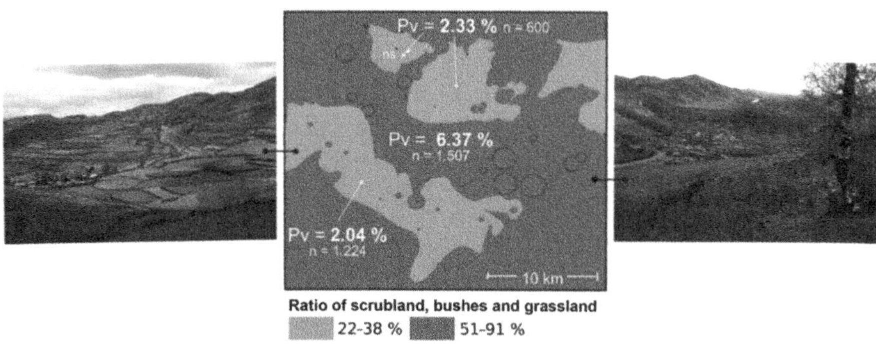

Figure 4.12. *Mapping of villages, from Giraudoux* et al. *(2003). Red circles are proportional to the prevalence of alveolar echinococcosis in the village (0%–16%). Pv means prevalence within landscape clusters defined by their ratio of scrubland and bushes to grassland; n, number of people screened. The photos illustrate the general physiognomy of these areas (photos P. Giraudoux). For a color version of this figure, see www.iste.co.uk/giraudoux/socioecosystems.zip*

The study of the distribution of small mammals, combining indexing and trapping methods (Giraudoux *et al.* 1998), initially came up against the difficulty of a species taxonomy that had yet to be defined. For example, the identification of the lacustrine vole (*Microtus limnophilus*), whose proven distribution area was, in principle, located more than a thousand kilometers to the north, took several years (Courant *et al.* 1999). In fact, each type of environment in the local landscapes was found to harbor one or more species that could serve as an intermediate host for the parasite (Figure 4.13). Again, rather than the mere presence of one or more species likely to serve as an intermediate host, the extension of habitats favorable to species likely to reach large population densities, such as lacustrine voles, was associated with more intense parasite transmission (Giraudoux *et al.* 2003). This observation confirmed that, as in Europe or on the Tibetan plateau, it is indeed the population dynamics of certain small mammal species that can reach high densities that is responsible for the intensity of transmission, rather than the richness of intermediate host species likely to be infected.

The diachronic study of Landsat satellite images collected from 1975 to 1997 indicated that, in the case of Zhang and Min counties, it was mainly the deforestation practiced until the early 1990s in order to gain agricultural land that was responsible for the extension of intermediate stages of clearing, as scrubland and grassland are favorable to vole outbreaks (Danson *et al.* 2004).

A "follow-up" screening was subsequently conducted in 2005–2006 on 2,500 individuals. This revealed a decrease in the prevalence of human alveolar echinococcosis, most marked in the younger age groups, indicating that, as expected from the 1994–1997 fieldwork, they had been less exposed to the infection (Figure 4.14). However, the discovery of a patient younger than 10 years of age indicated that the parasite was again circulating, albeit less intensely, in the late 1990s or early 2000s.

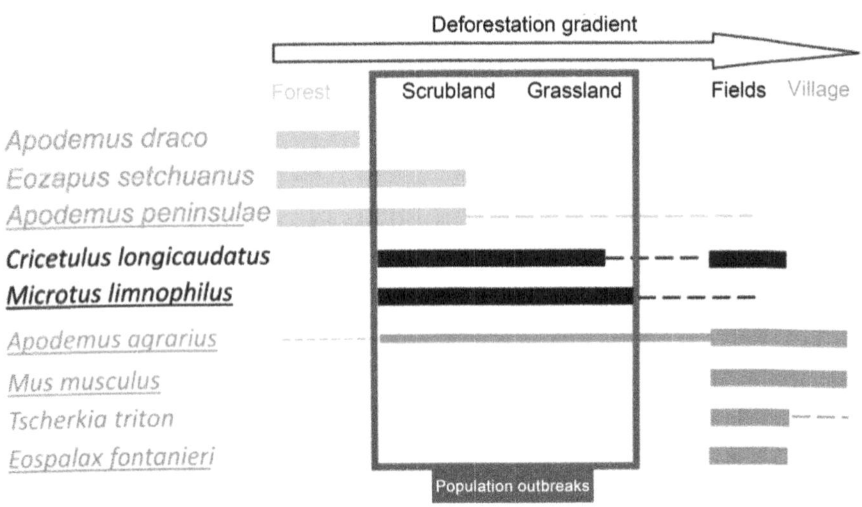

Figure 4.13. *Distribution of small mammal species by habitat in Min and Zhang counties, from Giraudoux* et al. *(1998). The underlined species are known to be possible intermediate hosts of the Echinococcus multilocularis. The expansion of scrubland, shrubs and grassland during deforestation operations in the 1980s explains why some of these species may have undergone outbreaks at the outskirts of villages and increased transmission. For a color version of this figure, see www.iste.co.uk/giraudoux/socioecosystems.zip*

In May 2015, in contrast to the period 1994–1997 when the disappearance of dogs from villages was a major observation, a large number of dogs were present in villages, most of them tethered. At least 50% of the dogs were treated with a deworming agent distributed by the government as part of the national echinococcosis control program implemented in China from 2005, but at a much lower frequency than recommended. Some of the dogs had been imported from Maqu County (Gannan Tibetan Autonomous Prefecture, in southern Gansu Province), itself a focus of alveolar echinococcosis. There is therefore a risk of possible reintroduction of the parasite from endemic areas far from the eastern Tibetan plateau (this fear extends to cities in eastern and southern China, where Tibetan dogs are highly prized and traded). However, PCR testing for the parasite conducted in 2015 on the feces of 256 dogs from study villages in Gansu failed to detect the presence of the parasite.

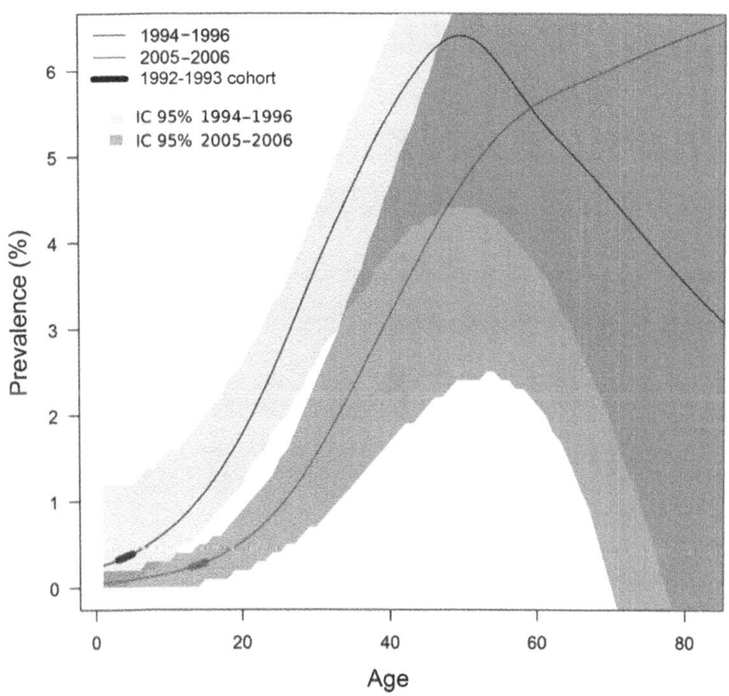

Figure 4.14. *Age-specific prevalence in the screened populations from 1994 to 1996 and 2005 to 2006. The thick line represents the fate of the 1992–1993 cohort born just after 1991 (when the dogs were still infected) and before 1994 (when the dog population was virtually extinct). CI, confidence interval. For a color version of this figure, see www.iste.co.uk/giraudoux/socioecosystems.zip*

In addition, since the late 1990s, soil protection and reforestation programs have greatly expanded areas of bushes and forest on slopes too steep for cultivation without deleterious erosion (Figure 4.11). This change in land use provides more potential habitat for rodents of the genus *Microtus* and *Cricetulus* and the red fox, the latter being more frequently observed by farmers in 2015 than in the 1990s, when the red fox was thought to be virtually extinct in the region, as reported by local farmers interviewed regularly. In 2015, a large number of small mammal records (species of the genera *Microtus*, *Cricetulus* and *Eospalax*) could be observed at similar frequencies to those observed in the 1990s.

The overall pattern of change in the transmission system can therefore be reconstructed as shown in Figure 4.15.

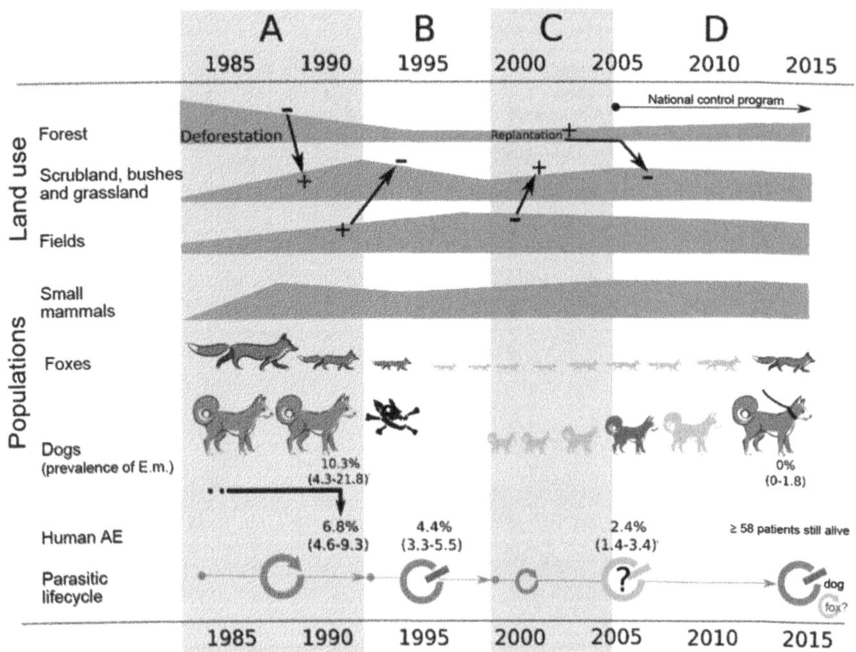

Figure 4.15. *Scenario of transmission system evolution in Min and Zhang counties, Gansu, from Giraudoux* et al. *(2019). Prevalences of human alveolar echinococcosis (AE) are standardized to an age of 35 years. E.m., Echinococcus multilocularis. See below for comments on periods A, B, C and D*

Four periods can be identified:

– *Period A. Deforestation and expansion of agriculture (1970s to early 1990s)*: logging, increase in scrubland, bushes and pasture, and later in ploughed areas and human and dog populations. Chinese zokors (*Eospalax fontanierii*) were trapped by farmers to protect crops and regularly fed to dogs, facilitating transmission of the parasite. Red foxes are relatively abundant and other iconic species of local highland forest wildlife such as leopard (*Panthera pardus*), Tibetan bear (*Ursus thibetanus*) and takin (*Budorcas taxicolor*) were still present. Scrubland and grassland were favorable for small mammal species that can reach high densities (*M. limnophilus*, etc.). These species are easily accessible to dogs and create conditions extremely favorable for intensive transmission. This lead to a high prevalence of *E. multilocularis* in dogs and a high local prevalence of human alveolar echinococcosis – perhaps the highest ever described in the world.

– *Period B. Crash of the dog and fox population (1992–1997)*: the crash occurred in 1992–1993 and the near extinction of the dog population could be observed from 1994 to 1997. It most probably interrupted the lifecycle of *E. multilocularis*. The cases of human alveolar echinococcosis identified in 1994–1996 were probably all infected during period A. However, a residual prevalence of human alveolar echinococcosis averaging about 4% was still observed. Furthermore, the prevalence of human alveolar echinococcosis was three times lower in areas where villages were surrounded by ploughland rather than grassland, scrubland and bushes. The emblematic species of highland forests mentioned above, which are incompatible with the expansion of agriculture and human penetration, have all disappeared from the ecosystem.

– *Period C. Ban on logging and reforestation (1998–2006)*: regulations imposed a ban of ploughing on steep slopes and increased replanting of trees, which lead to a decrease in agricultural land, an increase in the area of natural vegetation (grassland, bushes) and, finally, an increase in the area of scrubland and forest plots. The fox population is gradually recovering. At least one case of human alveolar echinococcosis whose infection occurred during this period was identified, indicating that transmission of *E. multilocularis* is probably active again in either foxes or dogs or both. However, the shift to a lower prevalence of human alveolar echinococcosis in the younger age cohorts of the 2005–2006 screening indicated that zoonotic transmission was much lower than in the past.

– *Period D. National echinococcosis control program (from 2007 to 2015)*: most dogs were tethered and a portion of the dog population was treated with a dewormer (praziquantel), but not as frequently as recommended. No evidence of *E. multilocularis* infection was found in dogs in 2015 and a very low prevalence (~0.4%) of *Taenia hydatigena* was found, contrasting with the 43% found in 1991, indicating either the effectiveness of deworming treatment or reduced access of dogs to livestock offal or an absence of predation in the village environment. However, trade in dogs during this period increased the risk of reinfection from other endemic areas (e.g. the Tibetan plateau). Nearly 60% of patients identified in the 1994–1996 were still alive in 2014 and there was no evidence of recent infection in the human population.

4.4.2. *Ningxia*

The Hui Autonomous Region of Ningxia is one of the smallest national administrative entities in China. It is semi-desert, except for the banks of the Yellow River in the north and the Liupan Shan, a mountain range in the south of the region. A focus of human alveolar echinococcosis covering Xiji and Guyan counties, at an altitude of 2,000–2,600 m, was located on either side of the Liupan Shan in the late 20th century. High human prevalences, averaging 2.5%, were found there (Yang *et al.* 2006), but not higher than those found in the more agricultural landscapes of the southern Gansu outbreak, with which this part of Ningxia shares many similarities in both landscape and economy. The residual patches of natural vegetation (e.g. in Huoshizhai Park near Xiji) show that the area belongs to the "sub-alpine" forest belt, the same as the focus of Min and Zhang counties in southern Gansu. It has undergone the same dynamics, with historical deforestation starting here centuries ago and worsening during the Qin Dynasty (1644–1912). Agriculture expanded during this period until the early 1990s, although attempts at reforestation are noted as early as the 1950s in the Xiji County records, but obviously without effect. Since the late 1990s, a firm ban on grazing and cultivation has been enforced on the hills, and reforestation by mass planting in the areas thus abandoned by agriculture and livestock farming was effectively implemented (Figure 4.16).

Eco-epidemiological studies undertaken in 2002–2003 by the research consortium already involved in the other Chinese sites, in collaboration with the Ningxia Medical College, which later became Ningxia Medical University, and then in 2012, in collaboration with an Australian team from the University of Queensland, failed to detect any populations of arvicolidae in the endemic areas, except in the Liupan Shan (Raoul *et al.* 2008; Giraudoux *et al.* personal observations 2012). Thus, southern Ningxia is so far the only place in the world where transmission of *E. multilocularis* occurred in some localities without a single arvicolid rodent species being present in the system. Furthermore, once again, species taxonomy proved to be a difficult issue. For example, a new species of jerboa (of a then unknown genus), *Chimaerodipus auritus*, though common, was not identified and described until many years later in the reference collection assembled on that occasion (Shenbrot *et al.* 2017).

Figure 4.16. *Landscape of cultivated terraces (flax, wheat, rape, potato, etc.) in the vicinity of Xiji, Ningxia Autonomous Region, July 2003. To the left of the road, formerly cultivated areas are left abandoned, then planted with trees; in the distance, the hill heights are being reforested (photo P. Giraudoux)*

Ultrasound screening of 3,205 people, correlated with a land survey conducted using satellite data, showed that pastures on the plains at an altitude of 2,000 m were associated with higher prevalences of the disease than those on higher mountains up to 2,600 m, another feature of this endemic region of Ningxia. Dog ownership, being female, Hui, a farmer and not having a well or running water were major risk factors for human alveolar echinococcosis in this region (Pleydell *et al.* 2008).

Red fox populations, although present and still locally abundant in the 1980s, declined considerably and were virtually absent at the time the Ningxia studies were conducted (Giraudoux *et al.* 2013a). The only possibility for the parasite to maintain itself here would be to circulate in a zokor (*Eospalax* sp.) – dog or hamster (*Cricetulus* sp.) – dog chain of transmission, possibly involving locally other, diverse species such as *Spermophilus dauricus* (the Daurian ground squirrel), and possibly other *Cricetinae*, *Gerbillinae*, *Dipodidae* and *Ochotonidae* (Raoul *et al.* 2008; Yang *et al.* 2012). Zokors are agricultural pests commonly trapped by farmers, as are ground squirrels. Sometimes consumed for human consumption, their viscera and cadavers were regularly fed to dogs, thus increasing the risk of parasite transmission, beyond mere predation by stray dogs. Because of incomplete knowledge of the behavior of domestic dogs with respect to intermediate hosts in the system, as well as the population dynamics of the potential species-rich intermediate hosts, the ecology of *E. multilocularis* transmission in southern Ningxia remains enigmatic compared to that described in all other endemic areas worldwide.

4.5. Alpine meadows from Altai to Pamir

In Xinjiang, for various reasons, studies on the epidemiology of human alveolar echinococcosis could not be extended by the spatial modeling of foci, following the preliminary study conducted by Zhou *et al.* (2000). However, local studies indicated that the prevalence of human alveolar echinococcosis in northern Xinjiang (Altai, Tian Shan) was 10–100 times lower than in the eastern edge of the Tibetan Plateau and its foothills, and similar to the magnitude of the high endemicity foci of the Alpine arc in Europe. Cases were generally concentrated in areas of intermediate altitude (sub-alpine pastures and forests and alpine meadows). In southern Kyrgyzstan, at the foot of the Pamir mountain range, transmission foci of high prevalence, of the order of those observed in Tibetan foci, could recently be identified in areas at 3,000 m altitude in alpine meadows (Raimkulov *et al.* 2015).

The visitor to the Jura or the Alps in Europe will not be disoriented in the sub-alpine forest belt of the Chinese Altai or the Tian Shan, as the landscapes observed there are similar in many respects to those of the European massifs. They are in fact marked by a division between high altitude forests of fir, spruce or larch and meadows, sometimes in wooded pastures similar to the French-Swiss "*prés-bois*", with high altitude alpine pastures occupied in summer by herds of cattle or transhumant sheep (Figure 4.17). They are also physiognomically similar to the forest and pasture types that occupy the eastern foothills of the Tibetan plateau, for example, in the Qilian Range (Gansu) (Han *et al.* 2015). Red foxes are generally common to very common in all of these areas, which the Corsac fox (*Vulpes corsac*), another possible definitive host, does not frequent because of their high winter snow cover (Giraudoux *et al.* 2013a).

a) b)

Figure 4.17. *a) Larch wooded pasture near Lake Kanas, near Baihaba, Altai. b) Columnar spruce forest and grasslands near Narati, in the Tian Shan (photos P. Giraudoux)*

The rodent communities of the highland grassland habitats of the Altai, Tian Shan, southern Kazakhstan and the northern foothills of the Pamir range, although different, are, very surprisingly, structured in the same way as in the Jura Massif: a relatively large (60–90 g), subterranean species, the Zaisan mole vole (*Ellobius tancrei*), which forms mounds by digging its galleries in the manner of the European water vole, and a smaller (about twenty grams) and less subterranean species of the genus *Microtus*. The larval form of *E. multilocularis* of the same genotype as the adult form found in a dog from the same village was identified in *E. tancrei* in Sary Mogul in Kyrgyzstan (Afonso *et al.* 2015). As in Europe, it may also be the common vole (*Microtus arvalis*) or a local sister species, the Altai vole (*Microtus obscurus*) (e.g. in Baihaba and Narati), or the narrow-headed vole (*Microtus gregalis*) (e.g. in Sary Mogul). Each of these species is likely to reach peak densities of several hundred or even several thousand individuals per hectare, in contrast to forest species, which each reaches only a few dozen individuals per hectare (Giraudoux *et al.* 1994, 2008). Landscape modeling based on MODIS and synthetic aperture radar data has enabled the detection of seasonal changes in vegetation cover and the differentiation of, for example, permanent grasslands from cultivated fields, thus improving the identification of intermediate host habitats (Marston *et al.* 2016; Marston and Giraudoux 2019).

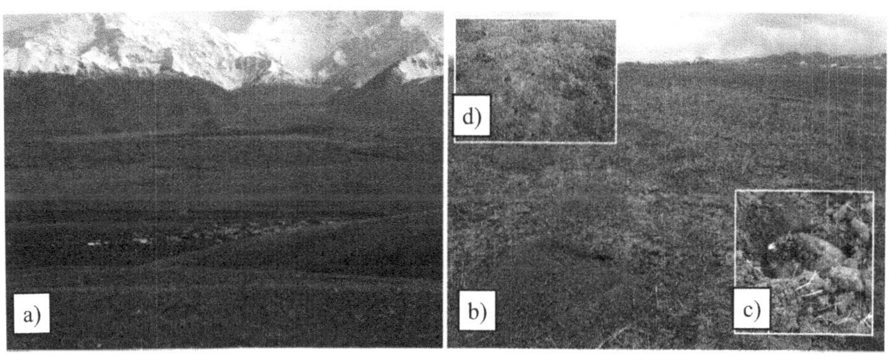

Figure 4.18. *Focus of Sary Mogul, Kyrgyzstan: a) Village and foothills of Abu Ali Ibn Sina Peak (formerly Lenin Peak); b) Ellobius tancrei mound-covered meadow; c) Ellobius tancrei emerging from burrow; d) Microtus gregalis colony (photos P. Giraudoux)*

4.6. Conclusion

Given the complexity of transmission systems (Patz *et al.* 2004; Restrepo *et al.* 2016) and the possible variations in the factors controlling them over time (Craig *et al.* 2019), it is difficult to simultaneously assess in detail, through reductionist approaches, each of the event sequences that lead to changes in *E. multilocularis* transmission intensity over space and time (Figure 4.19). This difficulty is further compounded by the development of the parasite in humans, which is generally slow (several years or decades) but varies from one individual to another, precluding an accurate estimate of the timing of human infection. The reductionist analytical approach is further complicated by the fact that in a socioecosystem where many factors modulate transmission, a factor given as effective at a given time may be ineffective later and compensated by another. In other words, the same effects can be induced by different causes at different times. This is the case, for example, as we have seen above, with intermediate and definitive host densities, one of which may compensate for the other, or with human behavior which varies over time, ecosystems and cultures. Furthermore, some parameters are simply not technically measurable. For example, the low prevalence of *E. multilocularis* in intermediate hosts makes it impossible to establish statistical differences in "natural" infection among the many potential or actual host species. The intensity of transmission also depends on the seasonally variable food preferences of the definitive hosts, which are impossible to quantify in absolute terms.

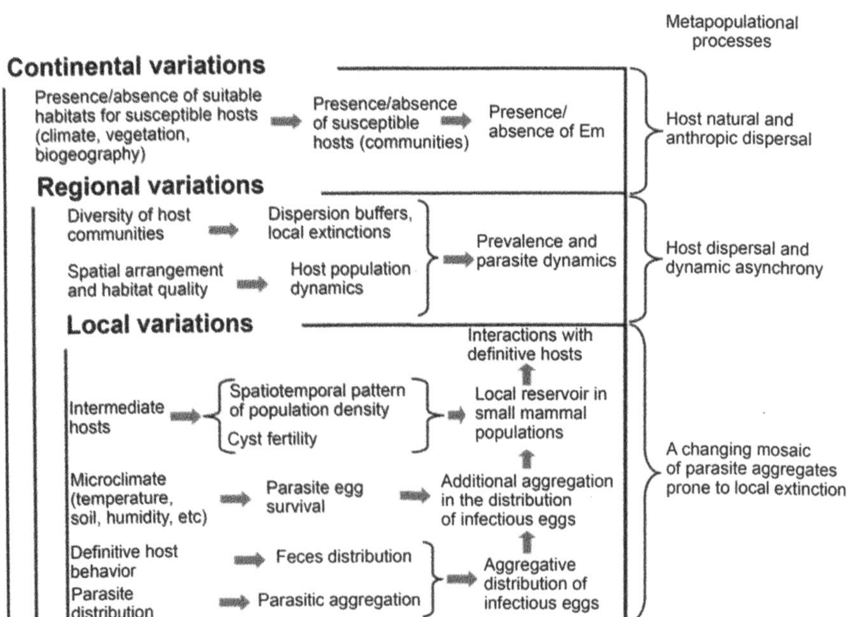

Figure 4.19. *Selected factors affecting the intensity of transmission of Echinococcus multilocularis in its non-human hosts, from Giraudoux* et al. *(2006)*

However, we think that, even if there are major gaps in the details of the processes, an eco-epidemiological approach, combining, in a spatially explicit context, human epidemiology and host and parasite ecology (Figure 4.20), makes it possible to characterize the regional systems that do or do not favor transmission of *E. multilocularis*, as observed on the basis of collected data. This is in line with an approach initiated by Pavlowski (1964), who coined the term landscape epidemiology, and the work of Jean-Antoine Rioux (mentioned in the foreword to this book), who shortly afterward proposed the term eco-epidemiology, as used here.

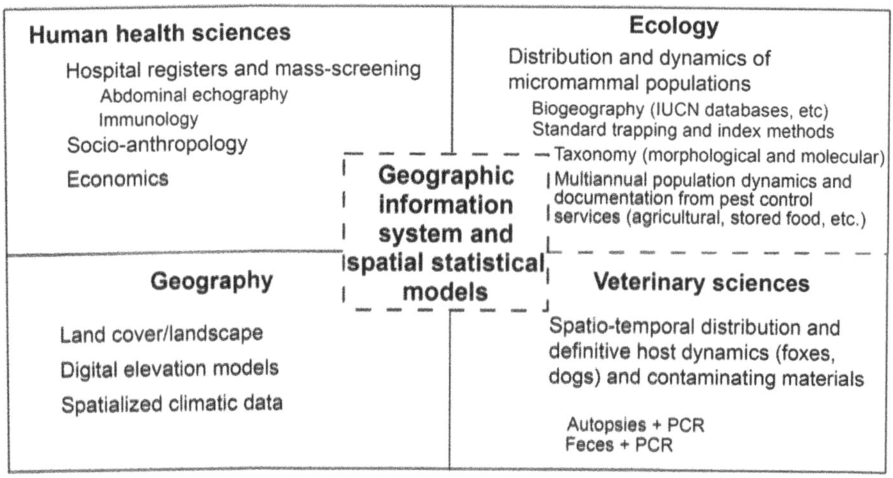

Figure 4.20. *Organization of the disciplines contributing to the eco-epidemiological approach described here. The dotted lines highlight the extreme permeability of the disciplines, both between themselves and through the spatialization methods central to the diagram*

However, it is essential to consider that the list of intermediate host species obtained at a given time is not enough to characterize such a system. Many species show varied population dynamics in different ecosystems, with, for some of them, important seasonal and inter-annual variations (see Chapter 1, Figure 1.6, and previous sections). It is these dynamics, rather than the presence of a particular species, and the resulting prey/predator relationship that are the key factors in transmission of *E. multilocularis*. The apparent absence of one or more host species at a given time, which could be that of a temporary decline in their populations in an inter-annual dynamic, may be completely misleading from this point of view and concern precisely one key species in the transmission. In other words, "fishing" for a rare parasite, over a short period of time, in populations and communities whose spatial and temporal characteristics of abundance fluctuation are unknown, can lead to serious misinterpretation (Giraudoux *et al.* 2007). Here, again, the observation of socioecosystems over a long time and at several spatial scales is essential. This is a lesson to be learned for the study of other emerging zoonoses, whose evolution depends on the dynamics of host populations, which are often unknown or poorly understood over time.

The eco-epidemiological approach results in a proposed classification for China and neighboring countries that recognizes a minimum of four types of "regional transmission system" for *E. multilocularis*, each identified by a flagship species. There are undoubtedly a limited number of other systems that have not yet been studied from this perspective. The flagship species chosen for each transmission system is not necessarily the one that is functionally most important in transmission everywhere and at all times. It was selected because it is widely distributed, overlaps little or not at all with the distribution of other flagship species and best signals the set of animal and human communities collectively involved in the socioecosystem (Giraudoux *et al.* 2013a) (Figure 4.21). This typology continues to be used to support the current national alveolar echinococcosis control program in China.

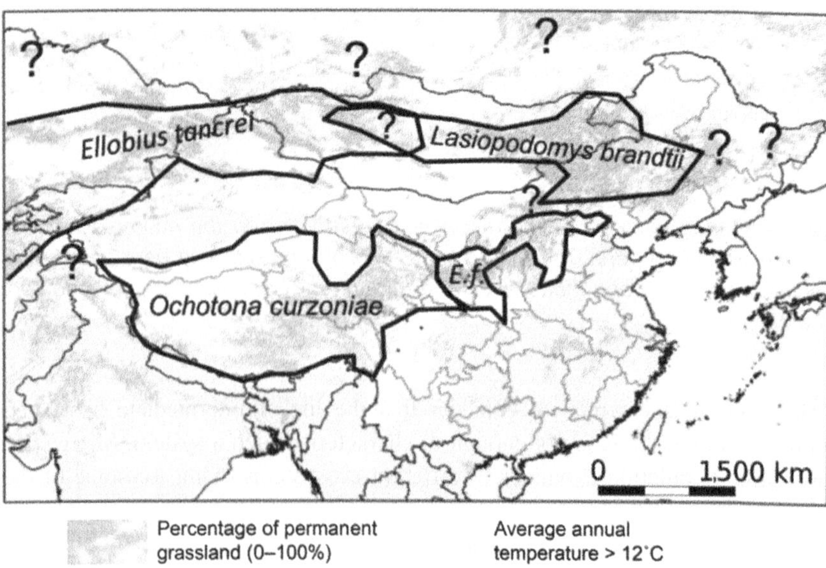

Figure 4.21. *Transmission systems and their flagship species, from Giraudoux et al. (2013a). E.f., Eospalax fontanierii. Question marks are placed on a number of areas with a high proportion of permanent grassland where no baseline eco-epidemiological studies are yet available. Areas with mean annual temperature >12°C are assumed to be too warm to maintain the parasite. For a color version of this figure, see www.iste.co.uk/giraudoux/socioecosystems.zip*

An additional typology of socioecosystems has since been further proposed, based on the consequences of urbanization gradients in transmission on a global scale (Liccioli *et al.* 2015). It is based on the observation that urbanization can alter host communities and the prey/predator relationship, by changing the composition and configuration of the landscape, as well as by directly modifying the definitive host community (e.g. by developing a domestic dog population, by favoring certain wild species or, conversely, by voluntary elimination of species considered as "pests" or involuntary through road mortality).

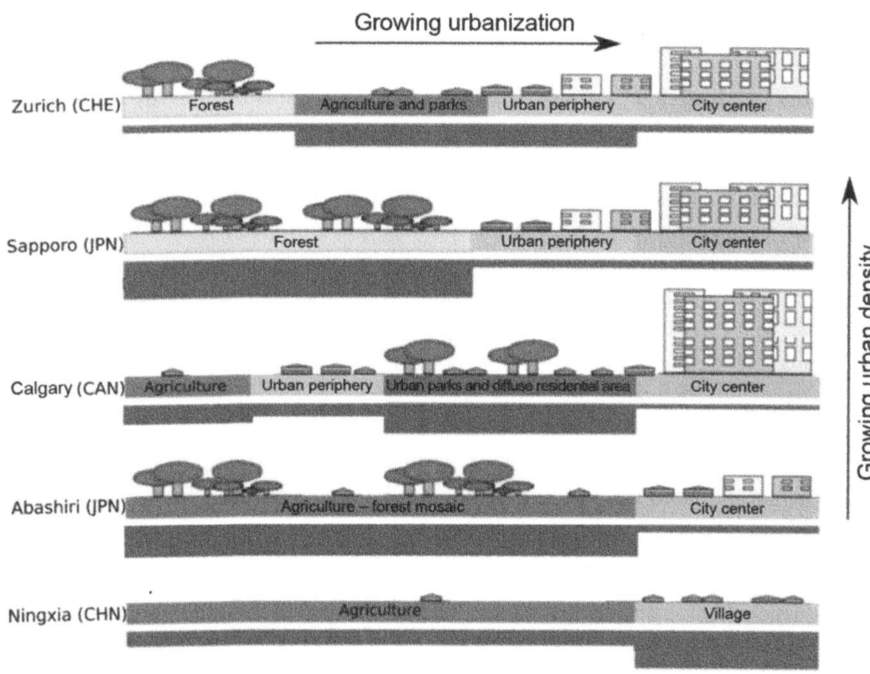

Figure 4.22. *Typology based on the heterogeneity of E. multilocularis transmission in various urban landscapes, from Liccioli* et al. *(2015). The magnitude of the red bars under each landscape represents the relative variation in infection pressure. For a color version of this figure, see www.iste.co.uk/giraudoux/socioecosystems.zip*

Although these "socio-ecological" models represent a certain level of reality regarding the capacity of *E. multilocularis* to come into massive contact with its accidental human host, one should not neglect the "human factors", which may prove to be decisive. The North American example sees a coexistence of fox infection almost equivalent to that measured in Europe in alveolar echinococcosis endemic areas and a near absence of human cases (Massolo *et al.* 2014). Because the risk of infection with a parasite is a combination of hazard (measuring the frequency of the parasite in a given location) and exposure (measuring contact with the parasite), little of this difference can be attributed to the virulence of the local strain of *E. multilocularis* (hazard), or to human behaviors (exposure), or both (Hildreth *et al.* 2000). This observation remains true in the United States, but is beginning to evolve in Canada, where increased contact between humans and infected wildlife in parks near cities, and possibly the importation of a European strain more infectious to humans than the "boreal" strain previously present in animal hosts, appears to be driving clusters of human cases in Alberta and isolated cases in Quebec and Saskatchewan (Massolo *et al.* 2014, 2019; Schurer *et al.* 2020).

In our studies, the sociological aspects of the transmission of the parasite to humans have only been approached by epidemiological techniques based on pre-established questionnaires, thus founded on a limited number of parameters and preconceived ideas, whether in Europe or in Asia. This is not for lack of having proposed research projects, both at the national and European levels, in France and in China. Sociological aspects are not a priority for public and even less for private funders – such as the pharmaceutical industry, which, as with its ecology, is not interested in a rare disease for which the drug market is very limited. Apart from the more detailed sociological study on the perception and representation of this disease by patients and their families, and their consequences in terms of prevention, presented in Chapter 5 of this book, we have only "epidemiological" results. Consistent elements are found, such as living in rural areas and contact with dogs, which in Europe often goes hand in hand with hunting, and, for Chinese households, hygiene and possibly contamination of water from wells or springs, as well as nomadic livestock rearing (with the notable exception of the endemic foci studied in Gansu and Ningxia, where this livestock rearing is reduced) (Craig *et al.* 2000; Kern *et al.* 2003; Budke *et al.* 2005a; Yang *et al.* 2006; Piarroux *et al.* 2013; Kern *et al.* 2017). The human, sociological or other factors (e.g. immune and hormonal) underlying notable and still unexplained differences between endemic areas, within the same large region or across continents, have not yet been finely characterized. This is the case for the sex ratio – with a clear female predominance of cases in Asia, a sex ratio close to 1 is observed in Europe, alternating more "male" and more "female" periods; or for the differences in the average age of patients at diagnosis – close to 60 years in Europe, with a majority of asymptomatic, not very advanced forms, and closer to 40 years in China, with a majority of "advanced" forms and signs of severity (Kern *et al.* 2017; Graeter *et al.* 2020). The role of dogs as definitive hosts, often important, is likely not the only factor behind these differences. The functioning of health systems, the management of alveolar echinococcosis and also of other unrelated chronic diseases, such as cancer or chronic inflammatory conditions, can also interfere in a major way with the diagnostic circumstances and human epidemiology, causing, for example, cases to emerge in Europe in patients treated with potent immunosuppressive drugs, who are almost non-existent in China (outside of the major urban centers hitherto unaffected by this parasitic disease) (Chauchet *et al.* 2014).

Much work remains to be done from a multidisciplinary perspective before both a comprehensive and detailed interpretation can be undertaken. However, the convergent work developed over nearly 30 years has shaped the current global understanding of the development of alveolar echinococcosis from its animal reservoirs (Romig *et al.* 2017) and provided a solid foundation for future work that should certainly give greater prominence to socio-economic and political determinants in the emergence or re-emergence of an admittedly rare but pathogenic zoonotic human disease. Moreover, one advantage of participating in this multidisciplinary research on *E. multilocularis* was a route to greater understanding of emerging diseases and enabling health professionals to become more "alert" to dealing with other diseases resulting from the combined action of environmental upheavals and human behavior (see Chapter 7). Conversely, they have also alerted ecologists, geographers and other non-health professionals and researchers involved in these studies to the sometimes unlikely, often delayed and generally unexpected consequences of environmental change on human health.

KEY POINTS.–

The eco-epidemiological approach, here the joint study of the spatial distribution of human cases of alveolar echinococcosis, fox infection and the distribution and dynamics of small mammal populations, makes it possible to define "regional transmission socioecosystems", characterized by their interactions between landscape and small mammal community dynamics.

The intensity of transmission is most often dependent on the existence of one or more intermediate host species, which can reach high densities over long periods (several hundred individuals per hectare) during multiannual variations in abundance or in open environments (grasslands) easily accessible to their final host predators (foxes, dogs). It can also be locally favored by certain human behaviors (feeding of dogs with infected rodents).

Taking into account landscape and climatic factors, based on knowledge of regional socioecosystems, makes it possible to model and predict areas at risk for alveolar echinococcosis and to orientate operations to prevent this disease.

The eco-epidemiological approach requires permanent collaboration and joint analysis between clinicians, epidemiologists, ecologists, veterinarians, geographers and modelers.

The intensity of human exposure in highly endemic areas for this zoonosis and, even more so, in areas with more limited "parasite pressure" is also modulated by individual and collective behavioral factors specific to each population (gender, age, occupation, customs, etc.) in a given socioecosystem; however, to date, these sociological aspects have not been studied in as much detail and in such an integrated manner as the eco-epidemiological factors.

4.7. Acknowledgment

Thank you to Francis Raoul for proofreading this chapter.

4.8. References

Afonso, E., Knapp, J., Tete, N., Umhang, G., Rieffel, D., van Kesteren, F., Ziadinov, I., Craig, P.S., Torgerson, P.R., Giraudoux, P. (2015). *Echinococcus multilocularis* in Kyrgyzstan: Similarity in the Asian EmsB genotypic profiles from village populations of Eastern mole voles (*Ellobius tancrei*) and dogs in the Alay valley. *Journal of Helminthology*, 89(6), 664–670.

Baudrot, V., Perasso, A., Fritsch, C., Raoul, F. (2016). Competence of hosts and complex foraging behavior are two cornerstones in the dynamics of trophically transmitted parasites. *Journal of Theoretical Biology*, 397, 158–168.

Bresson-Hadni, S., Laplante, J.-J., Lenys, D., Rohmer, P., Gottstein, B., Jacquier, P., Mercet, P., Meyer, J.P., Miguet, J.P., Vuitton, D.A. (1994). Seroepidemiological screening of *Echinococcus multilocularis* infection in a European area endemic for Alveolar echinococcosis. *Am. J. Trop. Med. Hyg.*, 51(5), 837–846.

Budke, C.M., Campos-Ponce, M., Qian, W., Torgerson, P.R. (2005a). A canine purgation study and risk factor analysis for echinococcosis in a high endemic region of the Tibetan plateau. *Veterinary Parasitology*, 127(1), 43–49.

Budke, C.M., Qiu, J.M., Wang, Q., Torgerson, P.R. (2005b). Economic effects of echinococcosis in a disease-endemic region of the Tibetan plateau. *American Journal of Tropical Medicine and Hygiene*, 73(1), 2–10.

Budke, C.M., Jiamin, Q., Craig, P.S., Torgerson, P.R. (2005c). Modeling the transmission of *Echinococcus granulosus* and *Echinococcus multilocularis* in dogs for a high endemic region of the Tibetan plateau. *International Journal for Parasitology*, 35(2), 163–170.

Chauchet, A., Grenouillet, F., Knapp, J., Richou, C., Delabrousse, E., Dentan, C., Millon, L., Di Martino, V., Contreras, R., Deconinck, E., Blagosklonov, O., Vuitton, D.A., Bresson-Hadni, S., FrancEchino Network (2014). Increased incidence and characteristics of alveolar echinococcosis in patients with immunosuppression-associated conditions. *Clinical Infectious Diseases: An Official Publication of the Infectious Diseases Society of America*, 59(8), 1095–1104.

Combes, B., Comte, S., Raton, V., Raoul, F., Boue, F., Umhang, G., Favier, S., Dunoyer, C., Woronoff, N., Giraudoux, P. (2012). Westward spread of *Echinococcus multilocularis* in Foxes, France, 2005–2010. *Emerging Infectious Diseases*, 18(12), 2059–2062.

Comte, S., Umhang, G., Raton, V., Raoul, F., Giraudoux, P., Combes, B., Boue, F. (2017). *Echinococcus multilocularis* management by fox culling: An inappropriate paradigm. *Prev. Vet. Med.*, 147, 178–185.

Courant, F., Brunet-Lecomte, P., Volobouef, V., Chaline, J., Quéré, J.-P., Nadachowski, A., Montuire, S., Bao, G., Viriot, L., Rausch, R., Erbajeva, M., Shi, D., Giraudoux, P. (1999). Identification of *Microtus limnophilus* in a large focus of Alveolar Echinococcosis (south Gansu, China). *C. R. Acad. Sc. III*, 322, 473–480.

Craig, P.S., Liu, D., Ding, Z. (1991). Hydatid disease in China. *Parasitology Today*, 7(2), 46–50.

Craig, P.S., Liu, D., Shi, D., Macpherson, C.N.L., Barnish, G., Reynolds, D., Gottstein, B., Wang, Z. (1992). A large focus of alveolar echinococcosis in central China. *The Lancet*, 340(8823), 826–831.

Craig, P.S., Giraudoux, P., Wang, Z.H., Wang, Q. (2019). Chapter Four – Echinococcosis transmission on the Tibetan Plateau. In *Advances in Parasitology*, 104(2019), 165–246 [Online]. Available at: https://doi.org/10.1016/bs.apar.2019.03.001 [Accessed January 5, 2022].

Craig, P.S., Giraudoux, P., Shi, D., Bartholomot, B., Barnish, G., Delattre, P., Quéré, J.-P., Harraga, S., Bao, G., Wang, Y.H., Lu, F., Ito, A., Vuitton, D.A. (2000). An epidemiological and ecological study of human alveolar echinococcosis transmission in south Gansu, China. *Acta Tropica*, 77, 167–177.

Danson, F.M., Craig, P.S., Man, W., Shi, D., Giraudoux, P. (2004). Landscape dynamics and risk modeling of human alveolar echinococcosis. *Photogrammetric Engineering and Remote Sensing*, 70(3), 359–366.

Davis, A., Pawlowski, Z.S., Dixon, H. (1986). Multicentre clinical trials of benzimidazole carbamates in human echinococcosis. *Bulletin of the World Health Organization*, 64(3), 383–388.

Davis, A., Dixon, H., Pawlowski, Z.S. (1989). Multicentre clinical trials of benzimidazole carbamates in human cystic echinococcosis (phase 2). *Bulletin of the World Health Organization*, 67(5), 503–508.

Delattre, P., Pascal, M., Le Pesteur, M.H., Giraudoux, P., Damange, J.P. (1988). Caractéristiques écologiques et épidémiologiques de l'*Echinococcus multilocularis* au cours d'un cycle complet des populations d'un hôte intermédiaire (*Microtus arvalis*). *Canadian Journal of Zoology*, 66(12), 2740–2750.

Delattre, P., Giraudoux, P., Quéré, J.-P. (1990). Conséquences épidémiologiques de la réceptivité d'un nouvel hôte intermédiaire du Ténia multiloculaire (*Echinococcus multilocularis*) et de la localisation spatio-temporelle des rongeurs infestés. *Comptes Rendus de l'Académie des Sciences, Paris, III*, 310(III), 339–344.

Delattre, P., Giraudoux, P., Pascal, M. (1991). L'échinococcose. *La Recherche*, 230, 294–303.

Deplazes, P., Rinaldi, L., Rojas, C.A.A., Torgerson, P.R., Harandi, M.F., Romig, T., Antolova, D., Schurer, J.M., Lahmar, S., Cringoli, G., Magambo, J., Thompson, R.C.A., Jenkins, E.J. (2017). Global distribution of alveolar and cystic echinococcosis. In *Echinococcus and Echinococcosis, Pt A*, Thompson, R.C.A., DePlazes, P., Lymbery, A.J. (eds) [Online]. Available at: https://doi.org/10.1016/bs.apar.2016.11.001 [Accessed January 5, 2022]

Gillet, M., Miguet, J.P., Mantion, G., Bresson-Hadni, S., Becker, M.C., Rouget, C., Christophe, J.L., Roullier, M., Landecy, G., Guerder, L. (1988). Orthotopic liver transplantation in alveolar echinococcosis of the liver: Analysis of a series of six patients. *Transplantation Proceedings*, 20(1), 573–576.

Gilot, B., Doche, B., Deblock, S., Pétavy, A.F. (1988). Éléments pour la cartographie écologique de l'échinococcose alvéolaire dans le Massif central (France) : essai de délimitation d'un foyer. *Canadian Journal of Zoology*, 66(3), 696–702.

Giraudoux, P. (1991). Utilisation de l'espace par les hôtes du ténia multiloculaire (Echinococcus multilocularis) : conséquences épidémiologiques. PhD Thesis, Université de Bourgogne, Dijon [Online]. Available at: https://tel.archives-ouvertes.fr/tel-00522886.

Giraudoux, P., Delattre, P., Quéré, J.-P., Damange, J.P. (1994). Structure and kinetics of rodent populations in a region under agricultural land abandonment. *Acta Oecologica*, 15(4), 385–400.

Giraudoux, P., Quéré, J.-P., Delattre, P., Bao, G., Wang, X., Shi, D., Vuitton, D., Craig, P.S. (1998). Distribution of small mammals along a deforestation gradient in south Gansu, China. *Acta Theriologica*, 43(4), 349–362.

Giraudoux, P., Raoul, F., Bardonnet, K., Vuillaume, P., Tourneux, F., Cliquet, F., Delattre, P., Vuitton, D.A. (2001). Alveolar echinococcosis: Characteristics of a possible emergence and new perspectives in epidemiosurveillance. *Médecine et maladies infectieuses*, 31, 247S–256S.

Giraudoux, P., Delattre, P., Takahashi, K., Raoul, F., Quéré, J.-P., Craig, P., Vuitton, D. (2002). Transmission ecology of *Echinococcus multilocularis* in wildlife: What can be learned from comparative studies and multiscale approaches? In *Cestode Zoonoses: Echinococcosis and Cysticercosis: An Emergent and Global Problem*, Craig, P., Pawlowski, Z. (eds). IOS Press, Amsterdam.

Giraudoux, P., Craig, P.S., Delattre, P., Bartholomot, B., Bao, G., Barnish, G., Harraga, S., Quéré, J.-P., Raoul, F., Wang, Y.H., Shi, D., Vuitton, D. (2003). Interactions between landscape changes and host communities can regulate *Echinococcus multilocularis* transmission. *Parasitology*, 127, 121–131.

Giraudoux, P., Pleydell, D., Raoul, F., Quéré, J.-P. Wang, Q., Yang, Y.R., Vuitton, D.A., Qiu, J.M., Yang, W., Craig, P.S. (2006). Transmission ecology of *Echinococcus multilocularis*: What are the ranges of parasite stability among various host communities in China? *Parasitology International*, 55, S237–S246.

Giraudoux, P., Pleydell, D., Raoul, F., Quéré, J.-P., Qian, W., Yang, Y., Vuitton, D.A., Qiu, J.M., Yang, W., Craig, P.S. (2007). *Echinococcus multilocularis*: Why are multi-disciplinary and multiscale approaches essential in infectious disease ecology? *Tropical Medicine and Health*, 55, S237–S246.

Giraudoux, P., Zhou, H.X., Quéré, J.-P., Raoul, F., Delattre, P., Volobouev, V., Deforet, T., Ito, A., Mamuti, W., Scheifler, R., Craig, P.S. (2008). Small mammal assemblages and habitat distribution in the northern Junggar Basin, Xinjiang, China: A pilot survey. *Mammalia*, 72(4), 309–319.

Giraudoux, P., Raoul, F., Afonso, E., Ziadinov, I., Yang, Y., Li, L., Li, T., Quéré, J.-P., Feng, X., Wang, Q., Wen, H., Ito, A., Craig, P.S. (2013a). Transmission ecosystems of *Echinococcus multilocularis* in China and Central Asia. *Parasitology*, 140(13), 1655–1666.

Giraudoux, P., Raoul, F., Pleydell, D., Li, T., Han, X., Qiu, J., Xie, Y., Wang, H., Ito, A., Craig, P.S. (2013b). Drivers of *Echinococcus multilocularis* transmission in China: Small mammal diversity, landscape or climate? *PLoS Neglected Tropical Diseases*, 7(3), e2045.

Giraudoux, P., Zhao, Y., Afonso, E., Yan, H., Knapp, J., Rogan, M.T., Shi, D., Jia, W., Craig, P.S. (2019). Long-term retrospective assessment of a transmission hotspot for human alveolar echinococcosis in mid-west China. *PLOS Neglected Tropical Diseases*, 13(8), 1–20.

Graeter, T., Bao, H., Delabrousse, E., Brumpt, E., Shi, R., Li, W., Jiang, Y., Schmidberger, J., Kratzer, W., Liu, W. (2020). Hepatic alveolar echinococcosis: Comparative computed tomography study between two Chinese and two European centres. *Food and Waterborne Parasitology*, 19, e00082.

Han, J., Bao, G., Zhang, D., Gao, P., Wu, T., Craig, P., Giraudoux, P., Chen, X., Xin, Q., He, L., Chen, G., Jing, T. (2015). A newly discovered epidemic area of *Echinococcus multilocularis* in West Gansu Province in China. *PLoS ONE*, 10(7), e0132731.

Hanf, M., Guegan, J.F., Ahmed, I., Nacher, M. (2014). Disentangling the complexity of infectious diseases: Time is ripe to improve the first-line statistical toolbox for epidemiologists. *Infection Genetics and Evolution*, 21, 497–505.

Hildreth, M.B., Sriram, S., Gottstein, B., Wilson, M., Schantz, P.M. (2000). Failure to identify alveolar echinococcosis in trappers from South Dakota in spite of high prevalence of *Echinococcus multilocularis* in wild canids. *The Journal of Parasitology*, 86(1), 75–77.

Houin, R., Deniau, M., Liance, M. (1980). Arvicola terrestris (L.), premier rongeur trouvé naturellement infesté par *Echinococcus multilocularis*, Leuckart, 1863, en France. *Comptes Rendus de l'Académie des Sciences*, 290(series D), 1269–1271.

Kern, P., Bardonnet, K., Renner, E., Auer, H., Pawlowski, Z., Ammann, R.W., Vuitton, D.A., Kern, P. (2003). European echinococcosis registry: Human alveolar echinococcosis, Europe, 1982–2000. *Emerging Infectious Diseases*, 9(3), 343–349.

Kern, P., Menezes da Silva, A., Akhan, O., Müllhaupt, B., Vizcaychipi, K.A., Budke, C.M., Vuitton, D.A. (2017). The echinococcoses: Diagnosis, clinical management and burden of disease. *Advances in Parasitology*, 96, 259–369.

Knapp, J., Bart, J.-M., Giraudoux, P., Glowatzki, M.-L., Breyer, I., Raoul, F., Deplazes, P., Duscher, G., Martinek, K., Dubinsky, P., Guislain, M.-H., Cliquet, F., Romig, T., Malczewski, A., Gottstein, B., Piarroux, R. (2009). Genetic diversity of the cestode *Echinococcus multilocularis* in red foxes at a continental scale in Europe. *PLOS Neglected Tropical Diseases*, 3(6), e452.

Knapp, J., Giraudoux, P., Combes, B., Umhang, G., Boué, F., Said-Ali, Z., Aknouche, S., Garcia, C., Vacheyrou, M., Laboissière, A., Raton, V., Comte, S., Favier, S., Demerson, J.-M., Caillot, C., Millon, L., Raoul, F. (2018). Rural and urban distribution of wild and domestic carnivore stools in the context of Echinococcus multilocularis environmental exposure. *International Journal for Parasitology*, 48(12), 937–946.

LePesteur, M.H., Giraudoux, P., Delattre, P., Damange, J.P., Quéré, J.-P. (1992). Spatio-temporal distribution of four species of cestodes in a landscape of mid-altitude mountains (Jura, France). *Annales de parasitologie humaine et comparée*, 67(5), 155–160.

Liccioli, S., Giraudoux, P., Deplazes, P., Massolo, A. (2015). Wilderness in the "city" revisited: Different urbes shape transmission of *Echinococcus multilocularis* by altering predator and prey communities. *Trends in Parasitology*, 31(7), 297–305.

Marston, C.G. and Giraudoux, P. (2019). On the synergistic use of optical and SAR time-series satellite data for small mammal disease host mapping. *Remote Sensing*, 11(1), 39.

Marston, C.G., Danson, F.M., Armitage, R.P., Giraudoux, P., Pleydell, D.R.J., Wang, Q., Qui, J.M., Craig, P.S. (2014). A random forest approach for predicting the presence of *Echinococcus multilocularis* intermediate host Ochotona spp. presence in relation to landscape characteristics in western China. *Applied Geography*, 55, 176–183.

Marston, C.G., Giraudoux, P., Armitage, R.P., Danson, F.M., Reynolds, S.C., Wang, Q., Qiu, J., Craig, P.S. (2016). Vegetation phenology and habitat discrimination: Impacts for *E. multilocularis* transmission host modeling. *Remote Sensing of Environment*, 176, 320–327.

Massolo, A., Liccioli, S., Budke, C.M., Klein, C. (2014). *Echinococcus multilocularis* in North America: The great unknown. *Parasite*, 21 [Online]. Available at: https://doi.org/10.1051/parasite/2014069 [Accessed January 5, 2022].

Massolo, A., Klein, C., Kowalewska-Grochowska, K., Belga, S., MacDonald, C., Vaughan, S., Girgis, S., Giunchi, D., Bramer, S.A., Santa, M.A., Grant, D.M., Mori, K., Duignan, P., Slater, O., Gottstein, B., Müller, N., Houston, S. (2019). European *Echinococcus multilocularis* identified in patients in Canada. *New England Journal of Medicine*, 381(4), 384–385.

Miguet, J.-P. (1973). L'échinococcose alvéolaire en Franche-Comté (à propos de 20 observations). PhD Thesis, Université de Franche-Comté, Besançon.

Patz, J.A., Daszak, P., Tabor, G.M., Aguirre, A.A., Pearl, M., Epstein, J., Wolfe, N.D., Kilpatrick, A.M., Foufopoulos, J., Molyneux, D., Bradley, D.J. (2004). Unhealthy landscapes: Policy recommendations on land use change and infectious disease emergence. *Environ. Health Perspect.*, 112(10), 1092–1098.

Pavlowski, E.N. (1964). *Natural Foci of Transmission Diseases in Connexion with the Landscape Epidemiology of Zooanthropozoonses*. Nauka, Moscow/Saint Petersburg.

Pesson, B. and Carbiener, R. (1989). Écologie de l'échinococcose alvéolaire en Alsace : le parasitisme du renard roux (*Vulpes vulpes* L.). *Bulletin d'écologie*, 20(4), 295–301.

Piarroux, M., Piarroux, R., Knapp, J., Bardonnet, K., Dumortier, J., Watelet, J., Gerard, A., Beytout, J., Abergel, A., Bresson-Hadni, S., Gaudart, J. (2013). Populations at risk for alveolar echinococcosis, France. *Emerging Infectious Diseases*, 19(5), 721–728.

Pleydell, D.R.J., Raoul, F., Danson, F.M., Graham, A., Craig, P.S., Tourneux, F., Giraudoux, P. (2004). Modeling the spatial distribution of Echinococcus multilocularis infection in foxes. *Acta Tropica*, 91, 253–265.

Pleydell, D.R.J., Yang, Y.R., Danson, F.M., Raoul, F., Craig, P.S., McManus, D.P., Vuitton, D.A., Wang, Q., Giraudoux, P. (2008). Landscape composition and spatial prediction of alveolar echinococcosis in Southern Ningxia, China. *PLOS Neglected Tropical Diseases*, 2(9), e287.

Raimkulov, K.M., Kuttubaev, O.T., Toigombaeva, V.S. (2015). Epidemiological analysis of the distribution of cystic and alveolar echinococcosis in Osh Oblast in the Kyrgyz Republic, 2000–2013. *Journal of Helminthology*, 89(6), 651–654.

Raoul, F., Deplazes, P., Nonaka, N., Piarroux, R., Vuitton, D.A., Giraudoux, P. (2001). Assessment of the epidemiological status of *Echinococcus multilocularis* in foxes in France using ELISA coprotests on fox faeces collected in the field. *International Journal for Parasitology*, 31(14), 1579–1588.

Raoul, F., Michelat, D., Ordinaire, M., Decote, Y., Aubert, M., Delattre, P., Deplazes, P., Giraudoux, P. (2003). *Echinococcus multilocularis*: Secondary poisoning of fox population during a vole outbreak reduces environmental contamination in a high endemicity area. *International Journal for Parasitology*, 33(9), 945–954.

Raoul, F., Quéré, J.-P., Rieffel, D., Bernard, N., Takahashi, K., Scheifler, R., Ito, A., Wang, O., Qiu, J., Yang, W., Craig, P.S., Giraudoux, P. (2006). Distribution of small mammals in a pastoral landscape of the Tibetan plateaus (Western Sichuan, China) and relationship with grazing practices. *Mammalia*, 70(3/4), 214–225.

Raoul, F., Pleydell, D., Quéré, J.-P., Vaniscotte, A., Rieffel, D., Takahashi, K., Bernard, N., Wang, J., Dobigny, T., Galbreath, K.E., Giraudoux, P. (2008). Small-mammal assemblage response to deforestation and afforestation in central China. *Mammalia*, 72(4), 320–332.

Raoul, F., Deplazes, P., Rieffel, D., Lambert, J.C., Giraudoux, P. (2010). Predator dietary response to prey density variation and consequences for cestode transmission. *Oecologia*, 164(1), 129–139.

Rausch, R.L. (1967). On the ecology and distribution of Echinococcus spp. (Cestoda: Taeniidae), and characteristics of their development in the intermediate host. *Annales de parasitologie*, 42, 19–63.

Rausch, R.L. and Schiller, E.L. (1956). Studies on the helminth fauna of Alaska. XXV. The ecology and public health significance of Echinococcus sibiricensis Rausch and Schiller, 1954, on St. Lawrence Island. *Parasitology*, 46, 395–419.

Restrepo, A.M.C., Yang, Y.R., McManus, D.P., Gray, D.J., Giraudoux, P., Barnes, T.S., Williams, G.M., Magalhaes, R.J.S., Hamm, N.A.S., Clements, A.C.A. (2016). The landscape epidemiology of echinococcoses. *Infectious Diseases of Poverty*, 5(13) [Online]. Available at: https://doi.org/10.1186/s40249-016-0109-x [Accessed January 5, 2022].

Romig, T., Deplazes, P., Jenkins, D., Giraudoux, P., Massolo, A., Craig, P.S., Wassermann, M., Takahashi, K., de la Rue, M. (2017). Ecology and life cycle patterns of Echinococcus species. *Advances in Parasitology*, 95, 213–314.

Said-Ali, Z., Grenouillet, F., Knapp, J., Bresson-Hadni, S., Vuitton, D.A., Raoul, F., Richou, C., Millon, L., Giraudoux, P., the FrancEchino Network (2013). Detecting nested clusters of human alveolar echinococcosis. *Parasitology*, 140(13), 1693–1700.

Schurer, J.M., Tsybina, P., Gesy, K.M., Kolapo, T.U., Skinner, S., Hill, J.E., Jenkins, E.J. (2020). Molecular evidence for local acquisition of human alveolar echinococcosis in Saskatchewan, Canada. *The Journal of Infectious Diseases*, 7 August [Online]. Available at: http://academic.oup.com/jid/advance-article/doi/10.1093/infdis/jiaa473/5885121 [Accessed October 7, 2020].

Shenbrot, G., Bannikova, A., Giraudoux, P., Quéré, J.-P., Raoul, F., Lebedev, V. (2017). A new recent genus and species of three-toed jerboas (Rodentia: Dipodinae) from China: A living fossil? *Journal of Zoological Systematics and Evolutionary Research*, 55(4), 356–368.

Thompson, A., Deplazes, P., Lymbery, A. (eds) (2017). *Echinococcus and Echinococcosis*. Academic Press, London.

Vaniscotte, A., Pleydell, D.R.J., Raoul, F., Quéré, J.-P., Qiu, J., Qian, W., Li, T., Bernard, N., Cœurdassier, M., Delattre, P., Takahashi, K., Weidmann, J.-C., Giraudoux, P. (2009). Modeling and spatial discrimination of small mammal assemblages: An example from western Sichuan (China). *Ecological Modeling*, 220(9/10), 1218–1231.

Vaniscotte, A., Raoul, F., Poulle, M.L., Romig, T., Dinkel, A., Takahashi, K., Guislain, M.H., Moss, J., Tiaoying, L., Wang, Q., Qiu, J., Craig, P.S., Giraudoux, P. (2011). Role of dog behaviour and environmental fecal contamination in transmission of *Echinococcus multilocularis* in Tibetan communities. *Parasitology*, 138(10), 1316–1329.

Veit, P., Bilger, B., Schad, V., Schäfer, F.W., Lucius, R. (1995). Influence of environmental factors on the infectivity of *Echinococcus multilocularis* eggs. *Parasitology*, 110, 79–86.

Viel, J.F., Giraudoux, P., Abrial, V., Bresson-Hadni, S. (1999). Water vole (Arvicola terrestris scherman) density as risk factor for human alveolar echinococcosis. *Am. J. Trop. Med. Hyg.*, 61(4), 559–565.

Vuitton, D.A., Bresson-Hadni, S., Liance, M., Meyer, J.P., Giraudoux, P., Lenys, D. (1990). L'échinococcose alvéolaire humaine : hasard épidémiologique ou fatalité immunologique ? *Gastroentérologie clinique et biologique*, 14, 124–130.

Vuitton, D.A., Wang, Q., Zhou, H.-X., Raoul, F., Knapp, J., Bresson-Hadni, S., Wen, H., Giraudoux, P. (2011). A historical view of alveolar echinococcosis, 160 years after the discovery of the first case in humans: Part 1. What have we learnt on the distribution of the disease and on its parasitic agent? *Chinese Medical Journal*, 124(18), 2943–2953.

Vuitton, D.A., Zhang, W., Giraudoux, P. (2017). Echinococcus spp. In *Global Water Pathogen Project*, Rose, J.B., Jiménez-Cisneros, B., Robertson, L. (eds). UNESCO, Michigan State University. doi.org/10.14321/waterpathogens.39.

Wang, Q., Vuitton, D.A., Qiu, J., Giraudoux, P., Xiao, Y., Schantz, P.M., Raoul, F., Li, T., Yang, W., Craig, P.S. (2004). Fenced pasture: A possible risk factor for human alveolar echinococcosis in Tibetan pastoralist communities of Sichuan, China. *Acta Tropica*, 90(3), 285–293.

Wang, Q., Vuitton, D.A., Xiao, Y., Budke, C.M., Campos-Ponce, M., Schantz, P.M., Raoul, F., Yang, W., Craig, P.S., Giraudoux, P. (2006). Pasture types and Echinococcus multilocularis, Tibetan communities. *Emerging Infectious Diseases*, 12(6), 1008–1010.

Wang, Q., Xiao, Y., Vuitton, D.A., Schantz, P.M., Raoul, F., Budke, C.M., Campos-Ponce, M., Craig, P.S., Giraudoux, P. (2007). Impact of overgrazing on the transmission of *Echinococcus multilocularis* in Tibetan pastoral communities of Sichuan Province, China. *Chinese Medical Journal*, 120(3), 237–242.

Wang, Q., Raoul, F., Budke, C.M., Craig, P.S., Xiao, Y., Vuitton, D.A., Campos-Ponce, M., Qiu, D., Pleydell, D., Giraudoux, P. (2010). Grass height and transmission ecology of *Echinococcus multilocularis* in Tibetan communities, China. *Chinese Medical Journal*, 123(1), 61–67.

Yang, Y.R., Sun, T., Li, Z., Zhang, J., Teng, J., Liu, X., Liu, R., Zhao, R., Jones, M.K., Wang, Y., Wen, H. *et al.* (2006). Community surveys and risk factor analysis of human alveolar and cystic echinococcosis in Ningxia Hui Autonomous Region, China. *Bulletin of the World Health Organization*, 84(9), 714–721.

Yang, Y.R., Clements, A.C.A., Gray, D.J., Atkinson, J.-A.M., Williams, G.M., Barnes, T.S., McManus, D.P. (2012). Impact of anthropogenic and natural environmental changes on Echinococcus transmission in Ningxia Hui Autonomous Region, the People's Republic of China. *Parasites & Vectors*, 5(1), 1–9.

Zhou, H.X., Chai, S.X., Craig, P.S., Delattre, P., Raoul, F., Quéré, J.-P., Vuitton, D.A., Wen, H., Giraudoux, P. (2000). Epidemiology of alveolar echinococcosis in Xinjiang Uygur Autonomous Region, China: A preliminary analysis. *Annals of Tropical Medicine and Parasitology*, 94(7), 715–729.

5

"Indigenous" Views of the Disease and Risks Associated with Alveolar Echinococcosis

Dominique JACQUES-JOUVENOT
LASA, Université Bourgogne Franche-Comté, Besançon, France

GUIDING QUESTIONS.–

What are the ways of creating cooperation at work in the case of research on alveolar echinococcosis?

What can socio-anthropology contribute to our knowledge of a disease and the health risks that result from it?

What is the place of indigenous views of disease in preventive behaviors?

5.1. Introduction

This chapter is not intended to add to the long list of works on multidisciplinarity[1], although my purpose is to question the place occupied by sociology within the framework of multidisciplinary research on zoonoses, more specifically on the disease alveolar echinococcosis (see Chapter 4), designated by the acronym "AE". Convinced of the merits of multidisciplinarity, we prefer the

1 The book directed by M. Jollivet, *Sciences de la nature, sciences de la société* (Jollivet 1992), more precisely the chapter entitled "Une histoire récente d'une vieille et grande question: les rapports homme-nature" (pp. 25–39), is a very fine synthesis of multidisciplinary work on the relationship between humans and nature, placed in the political and scientific context of the time.

Socioecosystems,
coordinated by Patrick GIRAUDOUX.

term "scientific collaboration", which is closer to the reality of experience with colleagues from similar or more distant disciplines (medicine and ecology), as in this case.

Indeed, the research project carried out by sociologists, entitled "Perceptions et représentations des risques en relation avec les zoonoses liées à l'alimentation et à l'environnement : l'EA comme modèle" (Perceptions and representations of risks in relation to food- and environment-related zoonoses: AE as a model), formalizes an experience of collaboration that began before this project was submitted and lasted long after[2] with two disciplines, medicine and ecology, represented by Dominique Angèle Vuitton, an internationally recognized physician specializing in AE, and Patrick Giraudoux, an ecologist (Delattre *et al.* 1991). For sociologists, it is a matter of collaborating, in the sense of working with others, in the service of "the production of knowledge for action", as J.-P. Darré (1999) would say, or, following M. Jollivet's example, with the aim of producing "finalized research" (Jollivet 1992). Targeted research is nothing other than research that integrates into its questioning the impact on society of the scientific results produced and, by the same token, integrates those who are called "social partners" so as to demonstrate social usefulness.

> "This is true," says Mr. Jollivet, of all research programs that "have an economic and social purpose, whether it is to find ways of developing or managing fragile French rural areas or areas threatened with abandonment, or to adapt agricultural production systems to the new economic context and to new environmental requirements, to provide knowledge and tools to managers of protected areas, forests or biotopes that are poorly designed, used or managed, or of resources of interest such as water, to develop or improve the development of plant or animal resources that are not or are poorly managed, to set up observatories of ecological, economic and social changes, etc." (Jollivet 1992)

Rather, this article attempts to tell a story: that of the integration of a team of sociologists into a research team that had already been producing a body of knowledge on the subject concerned for a long time (Delattre *et al.* 1991; Bresson-Hadni *et al.* 2000). Sociologists often come later. The relations between the natural sciences and

2 The project was submitted in 2000 and it is with the same colleagues that we are preparing this joint book 20 years later. In the meantime, collaboration with other disciplines took place around the creation of a master's diploma, "Environnement, Santé, Société" (Environment, Health, Society), and also in a university farm project bringing together other social actors, and, finally, some of us found ourselves once again in the program "CARELI" initiated by the ecologists.

sociology are complex, but the institutional disciplinary division is partly the reason for the power relations that have always existed between them (Jollivet 1992), relations that we have tried to overcome in this collective experience, hence the importance of remembering that collaboration begins, in the case that concerns us here, between people who are not, of course, separate from their scientific affiliation, but who contract on the basis of an interpersonal knowledge.

Telling a story means, of course, doing it "in one's own way", that is, by means of a reflexive return that must be understood as a narrative with all the usual precautions that this type of narrative discourse in the field of human sciences entails (Ricoeur 1983). This story begins with a meeting between two people, one a medical doctor and the other a sociologist,[3] and their respective laboratories, SERF (see Chapter 3) and LASA (Laboratory of Sociology and Anthropology), respectively. Dominique Angèle Vuitton's request to sociologists, namely, "how to successfully convey one or more prevention messages related to the risk of human contamination by AE", led us to a reformulation that mobilized the concepts and methods specific to our disciplinary field in order to grasp the "indigenous visions"[4] of disease and health risks related to AE.[5] The collection and analysis of representations – relationship to health, to the animal, to food, and also to disease and health risk – are then correlated to the practices of the actors. The production of scientific knowledge takes shape in a research report (Jacques-Jouvenot *et al.* 2003).[6] Only after this data analysis phase was it possible for us to formulate a

3 The collaboration between LASA-UFC and SERF began 3 years ago, when the research project "Perceptions and representations of risks in relation to food- and environment-related zoonoses: AE as a model" was submitted as part of the "environment and health" program in 2000, financed by the French Ministry of the Environment and Land Planning.

4 Let us consider for the moment that an indigenous vision is a personal conception of the disease, the risk or any other social object, a vision collected by the sociologist, which is a means of access to the understanding of the actors' thought system.

5 *Perceptions et représentations des risques en relation avec les zoonoses liées à l'alimentation et à l'environnement : l'échinococcose alvéolaire comme modèle* (Jacques-Jouvenot *et al.* 2003). Research report for the Ministry of Land Planning and Environment and Inserm, in collaboration with the SERF laboratory (Health and Rural Environment, University of Franche-Comté/WHO Collaborating Center for the Prevention and Treatment of Human Echinococcosis) and the LBE (*Laboratoire de biologie et écophysiologie*).

6 This research is based on 75 interviews conducted with 96 people. Thirty-one interviews were intergenerational; 28 included two generations of the same family (i.e. 14 families contacted more than once); the other three included three generations of the same family (i.e. one family contacted more than once). It should be remembered that, although we sometimes conducted a single interview for couples, we proceeded with separate interviews for people of different generations within the same family.

certain amount of advice on the prevention of health risks related to this disease (Jacques-Jouvenot 2019).

5.2. Building scientific cooperation

It was following a first report to the Commission of the European Communities, entitled “Food-borne Zoonoses”,[7] and the production of a white paper on food safety[8] that our first disciplinary rapprochement took place. The sociological contribution to the reflection on zoonoses is important for understanding the social mechanisms related to these diseases. Indeed, events such as mad cow disease or listeriosis have highlighted the fact that public reactions to these diseases do not necessarily correspond to a rationality of action and cannot therefore be explained solely by taking into account so-called “objective” risks.

The levels of knowledge relating to this disease – scholarly knowledge and lay knowledge – depend on the positions of the actors. At the beginning of this research, sociologists had more or less the same lay knowledge of this disease as the actors concerned by it. The information on AE that we had before taking a scientific interest in it was the same as everyone in Franche-Comté who was “vaguely” aware that doctors had “discovered” a new disease linked to the presence of foxes in the region. This new disease turned out to be AE, of which medical knowledge was, in the year 2000, the beginning of our collaboration, already very significant. Another type of information also constitutes a part of the knowledge of the public. This knowledge comes from the articles produced on the issue by the local and national press.[9] D. Faggianelli, the author of this research, notes that compared to the other two subjects – bromadiolone and bovine spongiform encephalopathy (BSE) – AE is treated a little differently, namely, that the journalistic discourse, most often appearing in the local press, refers directly to the explanations of the scientists interviewed, without any attempt to put the information in law terms, as in the other two cases (Callon 1986). While, as Dominique Angèle Vuitton reminds us in an interview with the *Nouvel Observateur* in April 2001: “Since 1983, 260 cases of human alveolar echinococcosis have been recorded in France, making it a disease

7 Commission of the European Communities – DG SANCO – Scientific Committee on Veterinary Measures relating to Public Health, adopted on April 12, 2000.

8 White Paper on Food Safety, adopted on January 12, 2000 (COM 1999) [Online]. Available at: https://eur-lex.europa.eu/legal-content/EN/TXT/?uri=LEGISSUM%3Al32041 [Accessed January 6, 2022].

9 In a research report complementary to the above-mentioned research (Jacques-Jouvenot *et al.* 2003), D. Faggianelli worked on the construction of journalistic discourse about alveolar echinococcosis, encephalopathy (BSE) and bromadiolone.

that causes about ten deaths each year, i.e., many more than mad cow disease", AE remains a local matter, confined to one French region, and therefore considered an unimportant subject, that is one without journalistic value. An interviewee confirmed this difference in the way AE was treated by the media:

> The media did a lot more in terms of campaigning for everything that happened with rabies, and more fuss was made about that than about echinococcosis. It's true that for that, it's perhaps more localized, but there are people who are dying of echinococcosis whereas at present, no one has died of rabies… well, only one. (Jacques-Jouvenot *et al.* 2003)

However, AE has been known in France for 100 years and studied in Franche-Comté since the 1920s. The process of contamination that leads to the disease takes place in several stages, the details of which the reader will find in Chapter 4. The first step is for sociologists to become acquainted with them. These are consistent both from the medical point of view and from the point of view of the mode of contamination; doctors and ecologists have been working on this for some 30 years (see Chapter 4). Epidemiologists have interviewed sick people using quantitative tools. From the results obtained, the researchers were able to establish a link between food practices and contamination. In his thesis, J.-P. Meyer underlines that

> the consumption of wild berries and dandelions appears to be a major risk (85% of the cases among the sick people)... the consumption of vegetables from an unenclosed vegetable garden, accessible to foxes and dogs, also appears to be a major risk (74% of the cases)... hunting, finally, appears to be a risk as well (30% of the cases), but is often associated with the preceding parameters and with the possession of dogs. (Meyer 1988)

People with such practices are considered as a "population at risk". In the study conducted by J.-P. Meyer on a sample of 71 sick people, it appears that 39% of this population consists of farmers, 25% of craftsmen and tradesmen, 18% of workers and employees and, finally, 15% of housewives. The remaining 3% of the population is made up of teachers, military personnel, religious people and medical professionals. This patient population is composed of both men and women: 53% men and 43% women. These results show that the disease affects a predominantly rural population, with an over-representation of the agricultural population compared to other professions.

The contributions of doctors and ecologists to sociologists are significant. Indeed, whether it concerns the definition of the territories concerned, the differences to be taken into account owing to the different population outbreaks of rodents depending on the altitude of the landscapes, or the definition of the population at risk, which allow us to establish correlations between food behaviors and rates of contamination, the quantitative surveys of our colleagues are precious and, in a way, allow us to sample a corpus of respondents relatively easily, knowing that the health and ecological context is defined beforehand.

5.3. Collaborating: working with others

In fact, the real collaboration between our respective disciplines begins here. The team of sociologists involved in this research works on rural issues, particularly agricultural ones, hence the interest that we quickly showed in the collaboration. It is on the basis of our knowledge of this agricultural public, its professional identity and the ways in which professional knowledge is transmitted, and also its relationship to health, suicide, etc. (Jacques-Jouvenot 1997), that we responded to the request of our colleagues and undertook this research. The population interviewed in the Inserm survey is representative of all the socio-professional categories (SPCs) in our sample with, however, a deliberate over-representation of farmers (35.4% of the total population). The various surveys conducted by sociologists on populations of dairy farmers questioned on their health, in a context of the specific evolution of the profession and professional skills (Dares 1991; Hamon-Cholet 1998; Olm and Simon 2001), show that the economic health of the farms is considered by the actors to be their primary concern, whereas the health of the actors comes far behind.[10] The first observation resulting from these studies is that the health of these professionals is one of the problems underestimated by the professionals themselves (Jacques-Jouvenot and Laplante 2009; Droz *et al.* 2012; Droz *et al.* 2014). If we add to this aspect that social and professional norms are transmitted familially in this profession within which family and work are permanently combined, the likelihood of a relationship to health transmitted within farming families is high. Hence the hypothesis that we are making in the first instance concerning the family transmission of the actors' behavior in the face of health risks in general. The work of P. Peretti-Watel and J.-K. Ward analyzes distrust of vaccines and shows that, as we have already pointed out, "the information given to the public is normatively

10 The CCMSA (*Caisse centrale de la Mutualité sociale agricole*) survey, which concerns the prevention expectations and occupational risks of non-salaried workers in agriculture, was conducted in April 2003. The results confirm that the economic health of the farm is cited as the main concern for 52.3% of the actors questioned, while the health of the actors is a secondary concern (7%) (Jacques-Jouvenot and Laplante 2009).

charged", which explains why certain actors reject information that goes against the collective representations of the group to which they belong (Ward and Peretti-Watel 2020). This is why it is important to take into account the collective systems of thought of agricultural professionals in our research.

What was the sociological knowledge of this relationship to health, the body and risk in the agricultural population before our collaboration? This is where our scientific interests intersect: we want to analyze the relationship between indigenous visions of the reality of illness and risks related to AE and the behaviors adopted in the face of these risks. We are neither health nor risk sociologists, as is customary to place sociologists in one or other of the categories of specialists in different social fields. We therefore claim to be socio-anthropologists who are above all driven by an observational field, based on empirical research and interested in the research being done (Callon 1989; Latour 2010).

The major empirical surveys show a high exposure to physical risks. This is the case of the Dublin Foundation survey of 1996, which showed that farmers were among the self-employed workers most exposed to this type of risk (Fondation européenne pour l'amélioration des conditions de travail 1997). On the other hand, we show in *Les Maux de la terre* that occupational diseases are important (Jacques-Jouvenot and Laplante 2009). Why, then, are the actors so silent on the illnesses of the body? To understand this, we had to make a detour through the actors' representations of the body and of risk, which we were led to do later to understand their relationship to a disease, AE, which has the particularity of being linked to nature.

For example, the results of a comparative research on farmers' health in France/Switzerland/Quebec (Droz *et al.* 2012) show that in all three countries, farmers value their resistance capacities and thus claim "a strong body" as a professional skill. When they mention back problems, accidents or fatigue, the actors refer to the handling of tractors, machines or even fertilizers and phytosanitary products. These problems are the result of the difficult exercise of their profession. These illnesses take a toll on the body but are considered by the farmers to be "normal" consequences of difficult work. This acceptance of illness can be understood as a family norm integrated during family socialization to the profession. A cultural norm, we should say, in the manner of J. Giono: "Working the land is our life, like blood which, until death, whatever happens, must go round a body, everywhere, even if it hurts" (Giono 2013). Their father already had a bad back. Milking hurts the back. The positions to be held on the tractor lead to sciatica. Faced with these various ailments, the farmers cope: "In any case, you have to push the carcass to the end" (Droz *et al.* 2012). The body signals here the trying nature of the job.

Bedridden fatigue is even considered good fatigue! This violence done to the body is therefore evidence of a certain order of things in agriculture. There is no shame in having pain that bears witness to a relationship to suffering that is part of an inherited and shared family culture. This culture of the all-powerful body persists and helps to produce a professional standard that is passed on to young apprentices. The skills valued in the profession are all linked to the body at work: tending, selecting, milking, driving tractors, etc. This passion for animals translates into professional practices that are carried out in the stables and the fields, and that mobilize the body at work. The professional identity of the cattle breeder, which defines a good cattle breeder, is built on the mastery of this activity in its physical dimensions. It is this body that he speaks to us about. Only illness, which prevents him from "being able to do", leads the actor to acknowledge that he is ill: "Only the fever stops us, if we really can't stand up. For the rest, we are used to it. It takes a lot of pain, so that we can't go on. We need more than aches and pains"!

The question that arises from this knowledge acquired in the framework of sociologists' research is the following: what about the relationship of farmers to a disease such as AE, which, beyond the relationship to the body, raises the question of the relationship to the risks of contamination and the resulting behaviors? An initial model developed in the United States in the context of public health campaigns attempted to show that the adoption of a health behavior is explained by "beliefs". According to this model, changes in practices and behaviors are linked to two factors:

– on the one hand, the perception of a threat to health;

– on the other hand, the perception that the adoption of a particular behavior can reduce this threat.

However, as C. Herzlich and P. Adam stated:

> Each of these factors is based on underlying beliefs, in fact on adherence to information provided by the medical profession: belief in the reality of the threat and its consequences, belief in the effectiveness of the preventive measure. (Adam and Herzlich 1994)

Whether we are talking about echinococcosis, other zoonoses or even other diseases – AIDS, for example – we see that this explanatory model does not work. It has a number of limitations:

– it is based on erroneous assumptions: "Individuals are perfectly rational, and for them avoiding health risks is always a primary goal" (Adam and Herzlich 1994);

– it is incomplete insofar as it does not question the basis of beliefs: "It only considers a belief or an isolated piece of information without asking how it fits into the complex sets of representations, and accepts or rejects the new information" (Adam and Herzlich 1994);

– the link between information, beliefs and behaviors is thought of at the individual level, without taking the social context into account.

Before returning to this question, let us pause for a moment on these few discursive aspects relating to the above-mentioned research to make a methodological point which seems to us to be one of the valuable contributions of sociology to the understanding of the object and to scientific cooperation.

5.4. Indigenous visions of social reality

Alongside quantitative methods from the French sociological tradition (Durkheim 1986), sociologists use qualitative methods based on data collected from the social actors concerned by the object of study. These data complement the quantitative studies because of the different materials collected in each piece of research. In the case of this research, these quantitative studies produced by our colleagues have been a precious basis for us both in the sampling of the population and in the knowledge of the disease, its evolution and the actors affected by the disease, considered a population at risk. At the same time, we also note that the over-representation of the agricultural population identified as a population at risk should not lead to false interpretations: it is not so much the profession and/or the place of residence as the fact of having activities linked to nature that makes this population more exposed to the risk of contamination. Picking, gardening or hunting therefore increase the probability of carrying the disease. Picking defined as "at risk" by doctors covers a whole range of plants usually found in fields, forest edges or woods and undergrowth: this ranges from red fruits, such as wild strawberries, raspberries, blueberries and blackberries, to dandelions, including wild lamb's lettuce and mushrooms. In summary, picking and eating raw any plant that may come into contact with tapeworm eggs is considered a risk. The garden is a risk area when it is open; in this case, it becomes easily accessible to wild animals and the cultivated products are then exposed, as are the wild plants. Finally, hunting is a risky practice insofar as any contact with foxes and dogs increases the probability of being contaminated. From these observations stems the methodological choice of creating a sample that takes into account the epidemiological data produced while favoring the analysis of the accounts of agricultural professionals, a population that is the focus of our sociological scientific problems.

The material collected by sociologists, whether it is similar to a life story, a qualitative interview or a biography, is discursive. Indeed, we collect what people say and these words have a meaning, because "discourse is composed of sentences and sentences of signs. The signs must be recognized and the sentence understood. The signs have a meaning, the sentence a meaning" (Compagnon 1979). Moreover, as J.-P. Darré rightly reminds us, it is important to consider actors to be "thinking networks" (Darré 1991), which means that the sociologist must "take people at face value" (Darré 1985). This means, for example, that when an interviewee tells us "that we must push our carcass to the end", he is not saying "we must go as far as the body allows", but he is showing us the farmer's body as a carcass and introducing a temporality into his discourse that will be found as a regularity in some of the discussions held by the farmers, and not in the other interviewees, namely, the prospect of death. A sick body is a useless body. Does this mean that this is true for everyone who practices this profession? No. It is therefore important to collect all the representations of the body in order to reach a threshold of saturation that would allow us to consider the discourses held on the body by our actors to be sufficient to account for the totality of the visions of the body that could be collected from this population if we continued to investigate beyond the number of interviews carried out up to now (Boltanski 1990). This method could be compared to a method used by ecologists when they practice, as P. Giraudoux explains to me, "the inventory of species present in a given space and allowing them to establish a rarefaction curve".

It is only on this condition of a rigorous treatment of the content of the social representations collected that the sociologist can claim access to the indigenous visions of the object studied. In a work entitled *L'Univers de pensée des éleveurs du Ternois*, J.-P. Darré reminds us that even when people use the same words, they do not necessarily attribute the same meaning to them (Darré 1985). To put it more simply, the cow of the agricultural technician is not the same as that of the researcher or the farmer. It is, he says, "through the meaning of words that we can access the meaning of objects for the speaker". But let us stay with AE: talking about "fox disease" has a different meaning from "the disease you get from wild strawberries" or "alveolar echinococcosis". Words organize sentences that make up the speaker's speech. This speech allows the sociologist to access the different systems of thought of the actors and the variations in meaning reveal different points of view or visions of the world. When we speak of an "indigenous vision of disease or health risk", we mean that each person has a personal vision of risk that can be identified in interviews, an indigenous vision rooted in the individual's culture and history. The sociologist is not looking for a true or expert word. The respondent, the doctor and the sociologist have different visions or visions shared with others depending on the object studied, but some are no more valuable than others. Sociological work consists of compiling these different points of view to take account of all the social visions of the same reality and to establish cross-references between the discourse and the social determinants of individuals, which may be SPC, age, education, etc.

From the point of view of the farmers' relationship with AE, one of the results of the research shows that the SPC does not discriminate with regard to the perception of the risk of contamination. However, even if not all farmers share the same vision of nature, they all link it to the risk of contamination, and we can observe behaviors with regard to risk that differ according to whether they see nature as beneficial or dangerous.

Those who consider nature to be generous and healthy are led to believe that there is no risk. What grows "naturally" is good for health and "can do no harm", while products bought in supermarkets are supposed to be less good, even dangerous.

This vision is opposed to another, which is based *a priori* on the idea that nature "is no longer what it used to be", that it only offers consumers dangerous products, "with all the pollution that exists in nature". They pay particular attention to this new food and health risk.

These two types of representation, which construct two different indigenous visions, are linked to the culture of the people, the knowledge inherited from family and culture (Jacques-Jouvenot *et al.* 2010).

On the other hand, farmers are more often than others in denial of the risk. The risk representations that are developed are therefore not the expression of a simple opinion: "There is no risk" or: "It is very dangerous to use this product". Nor are they the result of "irrational" behavior. People mobilize their own knowledge – lay knowledge or thoughts – which are embedded in habits linked to practices that are also inherited. We hear today about glyphosate: "We've always used it and we're not dead" or "My parents never used it and they managed fine without it". We find, similarly, this kind of discourse when faced with the risk of contamination by echinococcosis. What justifies the denial is based on spatiotemporal frameworks specific to the experience of previous generations: "We've always eaten dandelions and we didn't die" or "There have been foxes here for a long time and I've never heard my parents say that they transmitted the disease". These representations of the reality of the disease and the risks incurred are the support for action. They therefore make it possible to understand why stakeholders do or do not change their eating practices and behaviors because of the presence of the disease in the area in which they live.

5.5. Lessons learned allow us to think differently about prevention

Prevention campaigns have been launched by the Doubs *Mutualité sociale Agricole* (MSA) among its members, emphasizing the need to change behavior considered risky. The Doubs MSA appeared quite early in the common research scheme. Indeed, already registered and active in SERF through the commitment of its director of occupational medicine J.-J. Laplante, the MSA is an important player in disease prevention because of its very good knowledge of the agricultural population. From the outset, therefore, the research has brought together health professionals and field workers, researchers in the natural sciences, medical sciences and social sciences. This cooperation between these groups of actors forms a device defined as:

> A set of actions structured around a contingent purpose *[and which]* displays the intention of its bearer to act on society by capturing, orienting, determining, intercepting, controlling and assuring the management, conduct, opinions, and courses of living beings. (Agamben 2007)

In the definition of this cooperative device, we find the same objective of multidisciplinary research already present in the work of M. Jollivet (1992).

However, the expected effects of the prevention campaigns have not really been effective. Indeed, several surveys conducted by doctors show that people who have been informed do not always put into practice what they have read, heard or seen during these campaigns. In other words, some of them do not systematically change their practices, even if they are considered as "at risk" by other prevention actors and in particular doctors. Three types of attitudes to AE-related risks were identified:

– some people change their practices radically once they are informed of this disease. Only the measures recommended in the field of gardening practices are not followed up;

– others are changing some of their activities but are still engaging in risky practices;

– others do not change any of their practices.

The appropriation of information messages gives rise to a skillful alchemy of knowledge among the various individuals, producing meaning, which then offers each one the possibility of structuring his or her life, of acting in an uncertain world which, if it escapes scientific rationality, is nonetheless constructed according to other forms of rationality. All the actors interviewed share a common concern: the distancing of risk. Yet, according to some, modifying practices is not equivalent to preventing risks but, on the contrary, to taking them. This is neither a denial nor a challenge, but rather a questioning of risky practices as defined by scientists. Behavior in the garden speaks for itself. Indeed, a better knowledge of the disease owing to prevention campaigns led people to believe that individuals, whoever they may be, whether farmers or rural dwellers, would fence their garden to prevent the presence of foxes and, consequently, would reduce the risk of contamination. However, this exposure to risk produced very diverse attitudes in relation to the categories that organize the representation of space of the interviewees, namely, the wild and the domestic. Moreover, we must consider that the representations of disease, risk and nature are part of spatiotemporal frameworks that prove to be decisive in the attitude of the actors in the face of the transformation of their gardening practices or their food practices. The long incubation period of the disease makes some people say: "I'm preventing my grandchildren from eating wild strawberries, but at my age, I'm not risking anything..."

The relationship to space can be seen in the symbolic division established between wilderness (the forest) and domestic space (the garden), categories that are very present in the exchanges between interviewer and respondent.[11]

> There's less chance of foxes coming into the garden. We're not going to be slaves to the foxes in our house either.
>
> – And your garden, is it protected?
>
> – No, not even, no. It's behind the house. The risk is the same, if you like, but there are fewer chances, eh? It's not the same risk as in the forest where the foxes go as they please, where there are foxes in freedom. But let's say it's a risk all the same, but we don't make too much of it, we don't treat it like a disease, we don't make a big deal of it… (L.)

The control over what is grown in the garden, as opposed to what is grown in the wild or what is sold in stores, is the reason some people give their garden special

11 All quotes used in the text are taken from the research report *op. cit.* (Jacques-Jouvenot *et al.* 2003).

status. If the garden is not the forest, the strawberries in the woods are not the strawberries in the garden. From the opposition of the wild and the domestic follows that of healthy/unhealthy products. The gardener's knowledge functions here as the guarantor of the healthy product:

> I'm not saying that nothing can happen in my garden, but it's still quite limited compared to what happens in the forest or even in the shops. There, we don't know at all what happens in either case whereas, in my garden, I know.

While some respondents are aware of the risks of consuming produce from their garden and develop a rhetoric that could be described as "reasonable", this does not lead to a change in behavior:

> Yes, that means I hardly ever go to the dandelions anymore, because we are always afraid of picking up this disease. You're going to tell me: "It's also stupid because [your] garden isn't fenced in and foxes can come at night and walk over and defecate on the greens and so on." "It's preconceived ideas to say, 'We don't go to the dandelions and then we don't go to the wild strawberries anymore because of that', because when you look at it from the other side, our garden is not fenced in, the foxes can get to our strawberries, greens and more." (P.)

Claude Fischler's approach to mad cow disease is interesting to consider here because, as he says, "food is not consumption like any other" (Fischler 2001). The risks associated with these two diseases – AE and mad cow disease – are related to the act of eating. And to eat is to incorporate, to put into one's body, here, at the risk of poisoning oneself: "Eating represents a perilous moment, in essence, because it is a moment where the interior of the being is exposed and where something happens between the world and the self" (Fischler 2008). Referring to his attitude toward the consumption of so-called "risky" products, one interviewee makes a very clear separation between the wild and the domestic and stops consuming wild berries, dandelions and mushrooms as soon as he becomes aware of the disease. However, he still consumed dandelions when he was invited to eat at the home of friends who served him some. He explains to his hosts the risks involved, but nevertheless respects the dish. Could we be seeing the construction of a healthy space – the space of the friendship network – free of risk, like the garden?

For this other interviewee, her unenclosed garden and the area around her house – an isolated farmhouse in the middle of a meadow bordering coniferous forests – are not exposed to risk, even though she knows that a fox visits them. Through the account she gives us of an exchange with her son, we can measure the differences in the representation of risk linked to the gap between two generations sharing the same house. For her son, the space in front of the house is at risk, while the garden, which also adjoins the house, is not:

> There was a little piece of garden in front of it. There, I had left to ripen some… I didn't pick it, my lamb's lettuce. Do you know what lamb's lettuce is?
>
> – Yes, yes, of course.
>
> – So it matured, it went to seed, the seeds fell into the garden. I said, "Well, maybe it'll grow back after all." Afterwards, we decided to make a bigger terrace. So with a mini digger, my son, he removed the earth and put it in front. And now, a whole bunch of lamb's lettuce has grown in front of the house. So I said, "Well, I'm gonna trim it." Then my son says, "What's this salad?" I said, "Well, it's salad from the front." He says, "Oh, well, I don't want it." I ask him why: "Oh well, there's a fox that comes around the house all the time. No, I don't want it." "So I tell him, Well, that's the best, the salad right there in front." Finally, he ate it anyway, but he didn't feel good about it.
>
> – But he still ate the salad you planted in your garden?
>
> – Yes.
>
> – And the fox, could it get through in the same way?
>
> – Yes.
>
> – Did you tell him that?
>
> – No. I was really surprised he didn't want to eat salad from the front.

5.6. Proximity to a sick person reinforces preventive behavior

According to C. Herzlich and P. Adam,

> if we really want to understand how views, representations and discourse become meaningful for action, we must always relate them to the daily constraints of people's lives, on the one hand, and to the

characteristics of their relationships, on the other. (Adam and Herzlich 1994)

In the case of AE, research shows that, as in the case of AIDS, "information is less important for the adoption of preventive behaviors than a sense of personal proximity to the disease and the fact of knowing people who have it" (Adam and Herzlich 1994). Knowing someone who has developed the disease – even more so if he or she has died from it and shares the same territory (his or her village, neighboring villages or even department) – leads to changes in disease-related behavior. Spatial proximity to a person with AE plays a decisive role. While prior knowledge of the disease has no impact on the adoption of preventive measures, the change in practices occurs with the knowledge of a person fatally affected by the disease.

The death of the person affected by the disease is clearly correlated with changes in gathering and gardening, as well as in the consumption of products from these activities:

> Because to me, that's maybe the most dangerous thing right now. I could be wrong but that's my point of view. Well, you can die from it already, before your eyes, all that. I saw the person I knew well, well, it's scary, eh? He knew he was doomed, he spoke openly about it, he knew there was no way out. He saw himself decreasing every day. It lasted ten years. Cancer doesn't last that long, it's quicker and you don't even notice it anymore; whereas in this case, he was aware until the end. Right until the end, I mean. No, it's still scarier than anything else. And then, the way to catch it is more dangerous maybe, right? If you go for a walk in the forest, a kid, for example, who is like that on the edge, he can eat fruit… (H.)
>
> Over thirty years ago you heard about it; and it changed your behavior?
>
> – No, not thirty years ago, only nineteen years ago, when we started seeing people sick, well, disappearing, that I became a little more aware. When you're young, you don't believe in it. (B.)
>
> The kid there who died. That's when we stopped eating wild strawberries. Yes, that's when we stopped. (C.)

The development of the disease in a family member always results in a change in eating habits, but in this case, this change is not always imposed over time. M.'s change in eating habits is part of a time frame that follows the evolution of her mother's disease. According to her, the first stage of the diagnosis corresponds to a radical change in the practices of gathering and consuming products at risk:

> At the time, it scared us a lot because we knew that it could lead to cirrhosis of the liver. But afterwards, when she told us it was over, we thought less about it; and then it faded away, when she told us she was cured, well... it was less serious, it seemed less serious… (M.)

S. changed her practices when she learned of her brother-in-law's infection. The fact that the disease did not develop led S. to "pick a few more wild strawberries". If the knowledge of an infected person can have an effect on the modification of practices, the absence of development of the disease leads to a form of relativization of the risk, which leads to the resumption of certain interrupted practices.

This experience of collective research allows us to make a few final remarks on the knowledge produced thanks to or in collaboration. Socio-anthropology shows that in order to understand the behavior of actors in the face of risks, it is no longer just a matter of understanding their representations of the risk, but also the representation they have of the space with which this risk is associated. Without taking into account the representation of the space, the behavior in front of the risks linked to AE or any other zoonosis cannot be understood: when the healthy, natural space is defined on the basis of the control that one has over it (the domestic space), it is difficult for the actors to accept that the use of such a space presents risks, and that the more this space is extended (garden; garden and fields; garden, fields and forest; etc.), the more the practices considered "risky" by medicine are questioned by the actors. In order to be effective, prevention campaigns on the risks related to AE or any other zoonosis should take into account the relationship of the actors to the space.

This sociological knowledge produced from the survey must be fed into the research community, which implies producing a kind of "distributed cognition" and only collaboration or cooperation can make this possible. Researchers do not know the same things, and different types of knowledge must be combined, respecting what is really specific to each, in order to better understand what is going on and act with relevance. Knowledge therefore remains partly monodisciplinary, constructed or manufactured within a discipline.

For example, sociology and its knowledge are already evolving within the same discipline, creating new capacities for expertise, sometimes new paradigms that reorient the work of researchers and produce new skills, acquired after several years of research on the health of the populations concerned. These are precise and rich expertises, acquired over time in other fields, which are adjusted to those of the doctors and ecologists in this collaboration.

KEY POINTS.–

Scientific cooperation between several disciplines consists of sharing the same object of work, while each of the scientists remains anchored in his or her specific theories and methods. The dialog is then based on a distributed cognition between all the researchers.

Gathering indigenous views of the disease allows for a better understanding of the barriers and drivers for prevention of a disease such as AE.

The representations of the risks linked to AE zoonosis are linked to the representations of space and the categories that organize it – wild/domestic – of the spatial proximity of a person who is ill or has died from this disease. All these representations are the basis for action, that is, for changing or not changing food practices and behavior.

5.7. References

Adam, P. and Herzlich, C. (1994). *Sociologie de la maladie et de la médecine*. Nathan, Paris.

Agamben, G. (2007). *Qu'est-ce qu'un dispositif ?* Payot & Rivages, Paris.

Boltanski, L. (1990). *L'Amour et la justice comme compétences. Trois essais de sociologie de l'action*. Métailié, Paris.

Bresson-Hadni, S., Vuitton, D.A., Barthomot, B., Heyd, B., Godard, D., Meyer, J.-P., Hruvosky, S., Becker, M.C., Mantion, G., Lenys, D., Miguet, J.-P. (2000). A twenty-year history of alveolar echinococcosis: Analysis of a series of 117 patients from Eastern France. *Eur. J. Gastrendeterol. Hepatol.*, 12, 327–338.

Callon, M. (1986). La domestication des coquilles Saint-Jacques et des marins pêcheurs dans la baie de Saint-Brieuc. *L'Année sociologique*, 1(36), 169–208.

Callon, M. (ed.) (1989). *La Science et ses Réseaux : genèse et circulation des faits scientifiques*. La Découverte, Paris.

COM (1999). White Paper on food safety. [Online]. Available at: https://eur-lex.europa.eu/legal-content/EN/TXT/?uri=LEGISSUM%3Al32041 [Accessed January 6, 2022].

Compagnon, A. (1979). *La Seconde Main ou le travail de la citation*. Le Seuil, Paris.

Dares (1991). Enquête conditions de travail. Report, Ministère du Travail, Paris.

Darré, J.-P. (1985). *La parole et la technique l'univers de pensée des éleveurs du Ternois*. L'Harmattan, Paris.

Darré, J.-P. (1991). Les hommes sont des réseaux pensants. *Sociétés contemporaines*, 5(1), 55–66.

Darré, J.-P. (1999). *La Production de connaissance pour l'action : arguments contre le racisme de l'intelligence.* Éditions de la Maison des sciences de l'homme/Institut national de la recherche agronomique, Paris.

Delattre, P., Giraudoux, P., Pascal, M. (1991). L'Échinococcose alvéolaire. *La Recherche*, 22(230), 294–303.

Droz, Y., Mieville-Ott, V., Jacques-Jouvenot, D. (2012). Les conséquences sociopsychologiques des politiques agricoles. Étude interdisciplinaire et comparative Suisse-France-Québec. Report, IHEID, LASA-UFC, Agridea, MSA, La Coop Fédérée.

Droz, Y., Mieville-Ott, V., Jacques-Jouvenot, D., Lafleur, G. (2014). *Malaise en agriculture : une approche interdisciplinaire des politiques agricoles France-Québec-Suisse.* Karthala, Paris.

Durkheim, É. (1986). *Le Suicide.* Presses universitaires de France, Paris.

Fischler, C. (2001). *L'Homnivore : le goût, la cuisine et le corps.* Odile Jacob, Paris.

Fischler, C. (2008). Les ressorts des paniques alimentaires. Rencontre avec Claude Fischler. *Sciences Humaines* [Online]. Available at: https://www.scienceshumaines.com/les-ressorts-des-paniques-alimentaires-rencontre-avec-claude-fischler_fr_27585.html.

Fondation européenne pour l'amélioration des conditions de travail (1997). Les conditions de travail dans l'Union européenne. Report, Fondation européenne pour l'amélioration des conditions de travail, Dublin.

Giono, J. (2013). *Lettre aux paysans sur la pauvreté et la paix.* Héros-Limite, Geneva.

Hamon-Cholet, S. (1998). Les conditions de travail des non-salariés sont proches de celles des salariés. *Économie et statistique*, 319(1), 113–125.

Jacques-Jouvenot, D. (1997). *Choix du successeur et transmission patrimoniale.* L'Harmattan, Paris.

Jacques-Jouvenot, D. (2019). Prévention et perception du risque en milieu rural. *Soins*, 64(832).

Jacques-Jouvenot, D. and Laplante, J.-J. (2009). *Les Maux de la terre : regards croisés sur la santé au travail en agriculture.* Éditions de l'aube. Paris.

Jacques-Jouvenot, D., Gillet, M., Guigon, S. (2003). Représentations des risques en relation avec les zoonoses liées à l'alimentation et à l'environnement : l'échinococcose alvéolaire comme modèle. Report, Ministère de l'Aménagement du territoire et de l'Environnement, Inserm, Université de Franche-Comté.

Jacques-Jouvenot, D., Guigon, S., Gillet, M. (2010). La transmission d'une culture familiale de la prudence face au risque alimentaire. In *Famille et santé*, Cresson, G., Mebtoul, M. (eds). Presses de l'EHESP, Rennes.

Jollivet, M. (1992). *Sciences de la nature, sciences de la société.* CNRS, Paris.

Latour, B. (2010). *La Science en action.* La Découverte, Paris.

Meyer, J.-P. (1988). L'échinococcose alvéolaire humaine : hasard épidémiologique ou fatalité immunologique ? PhD Thesis, Université de Franche-Comté, Besançon.

Olm, C. and Simon, M.-O. (2001). Les conditions de vie des exploitants agricoles à faibles revenus. Report, CREDOC, Ministère de l'Agriculture et de la Pêche, Paris.

Ricoeur, P. (1983). *Temps et récit. Tome 1 : L'Intrigue et le récit historique.* Le Seuil, Paris.

Ward, J.K. and Peretti-Watel, P. (2020). Comprendre la méfiance vis-à-vis des vaccins : des biais de perception aux controverses. *Revue française de sociologie*, 243–273.

6

Conservation of the Black-and-White Snub-nosed Monkey

Eve AFONSO[1], Cécile CALLOU[2], Céline CLAUZEL[3], Patrick GIRAUDOUX[1,4] and Li LI[4]

[1] *Chrono-environnement UFC-CNRS, Université Bourgogne Franche-Comté, Besançon, France*

[2] *MNHN, Paris, France*

[3] *LADYSS UPD-CNRS, Université Paris Diderot, France*

[4] *Yunnan University of Finance and Economics, Kunming, China*

GUIDING QUESTIONS.–

How does fragmentation of the habitat of the black-and-white snub-nosed monkey lead to the genetic isolation of sub-populations, and how can it be measured?

How can historical and museological records help to document this isolation?

How can the combination of genetic studies and spatial modeling help to identify key areas of habitat reconnection?

What are the benefits and risks of ecotourism?

6.1. Introduction

Chapters 1 and 4 show, through two examples, how the eco-epidemiological approach has led to improved control of organisms with undesirable effects, whether they are agricultural pests or pathogenic organisms. It shows that sustainable control of these organisms requires action on the socioecosystems that favor them, rather than direct control of the target population, which is often ineffective or even counterproductive (Jiguet 2020). The European hamster (*Cricetus cricetus*) provides a striking example, in Alsace, of a species whose habitat modification has turned it from an intensively destroyed agricultural pest in the 1960s to a critically endangered species today. The expansion of maize cultivation at the expense of other cereals and farming techniques, leading to the disappearance of weed cover in winter, leave the population without food and highly vulnerable to predation when it emerges from hibernation in April (O'Brien 2015). From the 1960s to the present day, the water vole (*Arvicola amphibius*), as discussed in Chapter 1, has undergone a reverse trajectory in the Jura Mountains owing to the expansion of permanent grassland and the increase in its productivity. The principles of the approach presented in this book, potentially, apply as much to species whose population size is to be reduced as to species whose population size is to be maintained or even increased. We present here the research path used to resolve questions related to the second category. It concerns the conservation of a species of highland primate endemic to southeastern Tibet and Yunnan in China, the black-and-white snub-nosed monkey (*Rhinopithecus bieti*) (Figure 6.1).

Figure 6.1. *Young male black-and-white snub-nosed monkey (photo Li Li)*

6.2. Historical context and issues

As victims of deforestation and subsequent habitat fragmentation and poaching (meat, trafficking for private zoos, etc.), 60% of the world's primates are currently threatened with extinction and 75% have declining populations (Estrada *et al.* 2017). Endemic to South and Southeast Asia, all species of "snub-nosed monkeys" (*Rhinopithecus sp.*) are unique in being folivorous, with prestomachal fermentation functionally analogous to that of ruminants (Yu *et al.* 2016) (Figure 6.2). All, over the long term, have undergone a reduction in habitat and a decline in numbers.

Figure 6.2. *Lichens on which black-and-white snub-nosed monkeys mainly feed (photo P. Giraudoux)*

Several *Rhinopithecus* species occur in the Three Parallel Rivers region of China. They diverged in the early Pleistocene (Li *et al.* 2004), isolated by the barriers that these rivers represent (Xing and Ree 2017) (Figure 6.3).

– The golden snub-nosed monkey (*Rhinopithecus roxellana*), or golden monkey, inhabits the eastern Yangtze River. It was discovered for science by Father Armand David and described in 1870 by Alphonse Milne-Edwards of the National Museum of Natural History (Paris).

– The black-and-white snub-nosed monkey (*R. bieti*) or Yunnan snub-nosed monkey survives between the upper reaches of the Yangtze River (called Jinsha Jiang, in China) and the Upper Mekong River (called Lancang Jiang, in China). It is still confusingly named (see following) black snub-nosed monkey in Wilson and Reeder (2005). It was discovered by Jean-André Soulié, of the Paris Foreign Missions, and described in 1897 by Alphonse Milne-Edwards.

– No *Rhinopithecus* species are known between Lancang and Salween (Nu Jiang, China), but a very small and isolated population of Burmese snub-nosed monkey (*Rhinopithecus strykeri*) was discovered west of Salween, first in 2010 in Burma (Myanmar) (Geissmann *et al.* 2011) and then in 2011 in China (Long *et al.* 2012). It was commonly named later Myanmar snub-nosed monkey or black snub-nosed monkey based on the color of its fur.

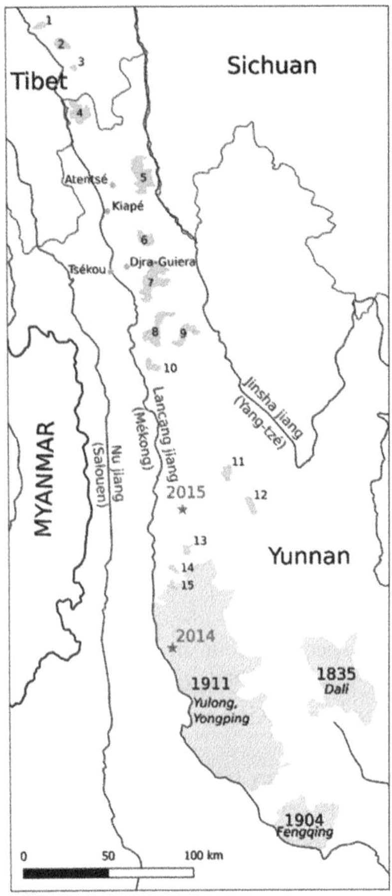

Figure 6.3. *Historical and current distribution of black-and-white snub-nosed monkeys. For a color version of this figure, see www.iste.co.uk/giraudoux/socioecosystems.zip*

COMMENT ON FIGURE 6.3.– *Beige are historical locations (Li* et al. *2002); green are known modern groups, with the number assigned to them (Wong* et al. *2013); red stars are recently rediscovered groups, with the year of discovery (Wang* et al. *2019). The orange circles are the sites near which holotypes and paratypes of species were captured, probably in 1895, and described in more detail at the* Muséum national d'histoire naturelle, *Paris (Milne-Edwards 1897; Rode 1938).*

The biology and ecology of the black-and-white snub-nosed monkey has, since its discovery by the French in the 19th century, remained virtually unknown to the scientific world for nearly 70 years. The first research outcomes were published in 1962 (Peng *et al.* 1962), but research did not intensify until the very late 1970s[1]. While in the past, local communities generally did not hunt monkeys, as they were considered closely related to humans, their collection seems to have started in the 1970s. The bones, brains and other body parts of black-and-white snub-nosed monkeys were then used to prepare medicinal remedies, which amounted to several dozen animals per year in the late 1970s and early 1980s (Yang 1988). The species itself was protected by Chinese law from 1975 onward,[2] turning hunting into poaching. Some poachers, accustomed to tracking black-and-white snub-nosed monkeys in the dense, steep forests of the Baima Xueshan, the "Snowy Mountains of the White Horse", had acquired enough knowledge to outstrip that of scientists (Yang 1988). Beginning in the summer of 1992, Long Yongcheng (龙勇诚), a Chinese researcher at the Kunming Institute of Zoology affiliated with the Chinese Academy of Sciences, and Craig Kirkpatrick of the University of California, USA,

1 In 1962, Peng *et al.* (1962) of the Kunming Institute of Zoology and the Chinese Academy of Sciences reported collecting eight black-and-white snub-nosed monkey skins from Deqin County, thus confirming its existence. Since then, the Kunming Institute of Zoology has organized several scientific surveys: (1) from October to November 1979, a first expedition was conducted to the Adong Mountains in the hinterland of Deqin County, and three specimens were collected (Li *et al.* 1981); (2) the eastern Baima Xueshan in Deqin County was explored from April to July 1981 (Mo and Yang 1982); (3) the geographical distribution and numbers of the black-and-white snub-nosed monkey were then studied from 1985 to 1986 (Bai *et al.* 1988); (4) the behavior of a group of monkeys on the eastern slope of the Baima Xueshan was studied from 1986 to 1988 (Wu *et al.* 1988; Zhao *et al.* 1988). This body of work paved the way for the work of Long Yongcheng and Craig Kirkpatrick and their successors. In May 1993, the photographer Xi Zhinong was able to photograph the species at less than 200 m, after 3 months of approach, thus giving the world the first document of its kind.

2 China created the Baima Xueshan Nature Reserve in 1983. It was the first nature reserve to house a population of black-and-white snub-nosed monkeys. In 2009, the "Yunnan snub-nosed monkey" National Park, located within the reserve, was officially opened to the public in Xiangguqing, near Tacheng. The national park is operated and managed by Weixi Yunnan Snub-nosed Monkey National Park Management Co., Ltd. The objective assigned to this company is to develop tourism, protect the ecological environment and conduct environmental education. The park hired local farmers to be the guardians who track and protect the monkeys. Their monthly salary was 1,500 yuan (about 195 euros) in 2020.

prospected for 3 years in the mountains of Yunnan and Tibet with the help of local farmers to search for and identify populations of this species (Long *et al.* 1994, 1996). They confirmed some 15 groups, totaling between 2,000 and 3,000 individuals, living in high-altitude evergreen forests between 1,800 and 4,600 m above sea level, making the species the only one among non-human primates living in such extreme altitude conditions (Figure 6.4).

Figure 6.4. *Typical biotope of black-and-white snub-nosed monkeys at 3,700 m altitude, in the Laojun Shan, the "Jade Mountains" (group no. 11, see Figure 6.3) (photo P. Giraudoux). For a color version of this figure, see www.iste.co.uk/giraudoux/socioecosystems.zip*

Beyond these 15 groups, there are likely still other unrecorded groups, either relics or recolonizations, as evidenced by recent findings in the south of its historical range (Wang *et al.* 2019). The species is threatened by habitat alteration, including expansion of agricultural and ranching areas, logging and mining, and collection of firewood. Current guarding limits the poaching of which it was a victim until recent years, but some individuals are still found unintentionally trapped in snares for musk deer (*Moschus chrysogaster*).

The fragmentation of the black-and-white snub-nosed monkey population into 15 physically disconnected sub-populations or groups therefore raises the question of the viability of each of these and the consequences of the limitation of genetic exchanges that it could entail. This is already underway: the genetic structuring of current populations is largely determined by the fragmentation of their habitat (Liu *et al.* 2009). It is therefore clear that reconnecting currently disjointed habitats is a priority for the conservation of the species.

From an operational point of view, a measure of the genetic cost of habitat loss and population isolation could be envisaged by comparing the diversity of alleles in the current population with those that could be expected to be preserved among the specimens collected in the Museum National d'Histoire Naturelle in Paris in the 19th century. It was essential at the time to be able to locate the places of capture of the museum's specimens as precisely as possible in order to compare them with the closest current groups. Names were inscribed on some of the bases of the naturalized animals and the brief comments of the collector, Jean-André Soulié, reported in Rode's catalog (1938), but these names do not correspond to the current place names of this region of Upper Mekong. Finally, in addition to the establishment of national nature reserves, such as the Baima Xueshan, and the care of other populations, part of the conservation effort by Chinese NGOs, private tourism agencies and the government is based on the belief that the visibility of the usually very discreet and elusive monkeys will attract tourists and familiarize them with the need to protect the species, while at the same time allowing the local population to benefit from the income generated. A group of around 40 monkeys from sub-population no. 9 (Xiangguqing, near Tacheng), whose total number is estimated at several hundred individuals, was therefore isolated without a barrier in 2008, simply by providing them with regular lichen for food near the village.

From the global to the local, it was therefore necessary to: (1) establish which areas of the population distribution chain, which extends over 300 km of mountains that can exceed 4,000 m in altitude, have reconnectable habitats; (2) establish which current populations can be compared with the ancient specimens in the museum; and (3) explore the possible consequences of the relative isolation of a fraction of the Xiangguqing population for ecotourism purposes. Ecological, geographical, historical, museological and, transversally, spatial modeling expertise were convened and combined to answer these questions.

6.3. Habitat connectivity and population genetics

The mapping materials available to us were (1) the vegetation map (2.5 × 2.5 km resolution) derived from an assemblage of 13 SPOT5 satellite images (November 2004, 60 × 60 km) covering the study area, with ground-truthing by the Conservation Information Center of The Nature Conservancy of China and

(2) extraction of road density, distance to the nearest road, human population density and distance to the nearest towns and villages from 1:250,000 topographic maps from the China Mapping Center. A least-cost path analysis consists of assigning each habitat a resistance factor named "viscosity" representing the difficulty of crossing that habitat for a given species and modeling the least-cost paths between two groups on this basis. The estimation of the viscosity of each habitat is a critical point as there is a lack of field data to support it. Values were calculated for each habitat individually and iteratively, progressively optimizing the correlation between genetic distances and least-cost distances between groups. These calculations indicated that the genetic distances[3] observed were better explained by these least-cost paths, combined with human disturbance, than by the Euclidean distance between groups, thus indicating the importance of habitat and the barriers that certain environments represent in individual movements. They allowed a first identification of potential corridors and proposed seven sensitive areas for restoration (Li *et al.* 2014).

A spatial modeling of the ecological network of black-and-white snub-nosed monkeys was then carried out in order to evaluate the degree of connectivity of its habitats and the impact of different development scenarios (construction of a motorway, reforestation, conversion of agricultural plots, etc.). The method used, graph theory, is an interesting compromise between cartographic approaches requiring little data, but limited to a structural analysis of connectivity, and individual-centered simulations, which are more realistic, but more complex and more data-intensive. From a landscape map (generally a land use map, Figure 6.5(a)), the ecological network of a species is modeled in the form of a landscape graph in which the nodes represent the optimal habitats of the species considered and the links represent the potential connections between these habitats (Figure 6.5(b)). An ecological network thus translates the living and movement space of a species on a specific spatiotemporal scale, corresponding to the ecological process under study: foraging (daily movements), migration (seasonal movements), dispersal (annual movements) and modification of the distribution area (intergenerational movements). The graph created, a mathematical topological object, is then used to calculate connectivity indices that allow nodes (patches of favorable habitats) and links (movement paths) to be prioritized according to their importance in terms of

3 Genetic distance quantifies the difference in allele frequencies between two sub-populations. A large distance, corresponding to differences in the distribution of allele frequencies, is generally interpreted as the result of a barrier to the dispersal of individuals that are not able to join and reproduce in sub-populations other than those where they were born. Conversely, when the genetic distance between two sub-populations is small, it is considered that enough individuals can move between the sub-populations and reproduce without constraint, resulting in genetic mixing and homogenization of genetic diversity in a large population encompassing all sub-populations.

reachable habitat surface, potential interaction with other nodes or even connector role within the network (Foltête *et al.* 2014).

A landscape map at 50 m resolution was constructed from supervised classification of 13 SPOT5 images from November 2004 (Zhang *et al.* 2015). Five categories were identified according to the ecological preferences of the black-and-white snub-nosed monkey: old-growth coniferous forests (optimal habitat); mixed forests (suboptimal habitat); deciduous forests or shrub layer (movement-friendly features); Yunnan pine forests (movement-unfriendly features); and urban or agricultural areas (barrier features). The nodes of the graph were defined as the heart of old-growth coniferous forests, eliminating edges that are avoided by the species owing to human disturbance. Links were calculated in least-cost distance, taking into account the permeability of each landscape category (Li *et al.* 2014) weighted by the slope between two pixels. This weighting, suggested by the primatologists, was added to avoid the creation of paths that would cross the valley to reach a habitat located on the other slope, and thus favor its alternative route by following contour lines. Only the links in the graph whose distance was less than a presumed dispersion distance were kept. This assumption shows one of the limitations of modeling when biological data are missing. The resulting graph (Figure 6.5(b)) reveals a highly fragmented ecological network owing to high altitudes in the north (>4,000 m) that limit tree growth, and human activities in the south. The central part, with five sub-populations in the Baima Xueshan Nature Reserve and two sub-populations in the Laojun Shan Nature Reserve, appears to be much better connected owing to a large area of optimal habitat. On the other hand, the graph highlights a potential disconnection of these two nature reserves because of less optimal habitat and a landscape that is more restrictive to movement. Finally, the geographical isolation of the three sub-populations of black-and-white snub-nosed monkey remaining in the south was confirmed by the very high fragility of the network in this area.

After this initial diagnosis of habitat connectivity, the impact of several real or imaginary scenarios was assessed. A first set of scenarios (Clauzel *et al.* 2015) concerned the construction of the G214 highway linking Xining (Qinghai Province) to Jinghong (southern Yunnan). The section crossing the northern mountainous part of the black-and-white snub-nosed monkey's range consists of viaducts and tunnels. These structures, by shifting traffic from the conventional road to underground or overhead sections, could potentially contribute to improved connectivity. The question was thus to assess whether the barrier effect of the conventional road could be partially or totally compensated by this traffic diversion associated with reforestation operations. The results revealed that only targeted habitat restoration could potentially increase connectivity. Strategically positioned within the network, these microhabitats would reconnect habitats located on either side of the highway by constituting intermediate steps (stepping-stone function) (Saura *et al.* 2013). The scenarios including traffic diversion alone or restoration of environments favorable to movement (deciduous forests, grasslands) would reduce the cost of movement between existing habitats, without creating new connections.

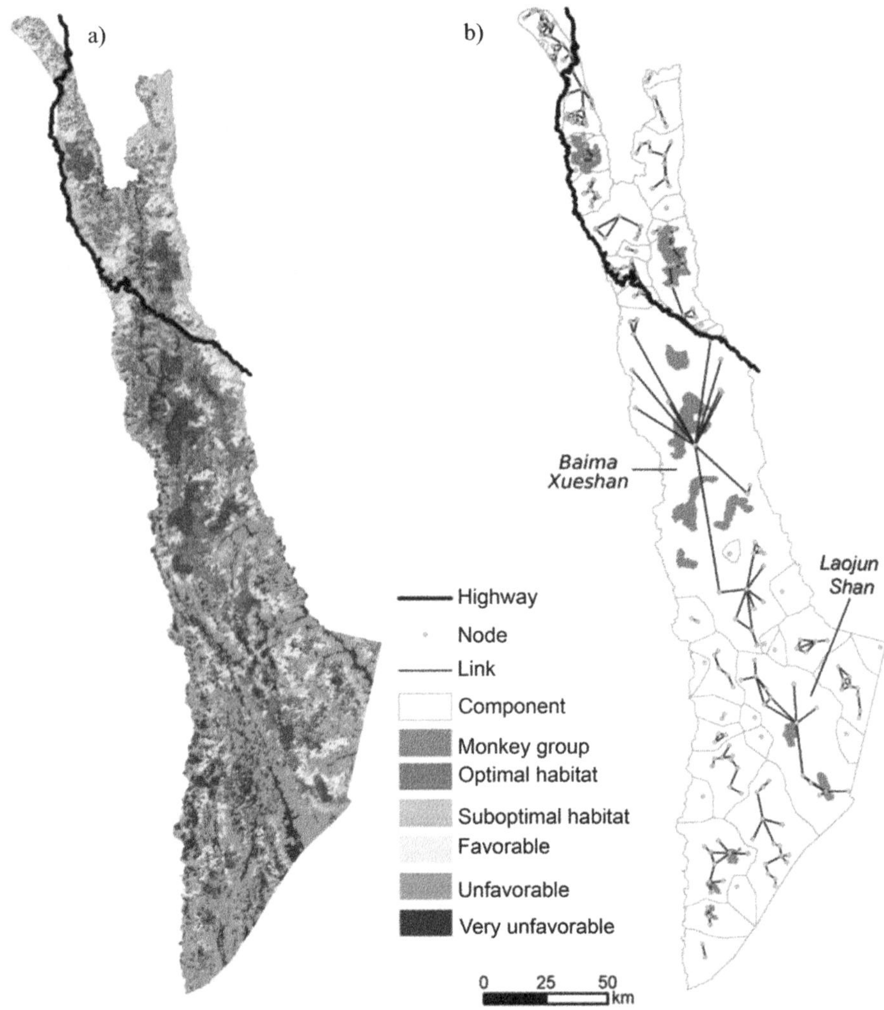

Figure 6.5. *Using a land-use map adapted to the ecology of the black-and-white snub-nosed monkey (a), a graph was constructed to model its ecological network at the dispersal scale (b). Its analysis revealed a high degree of fragmentation and potential disconnection of the Laojun Shan groups with those of the Baima Xueshan Reserve. For a color version of this figure, see www.iste.co.uk/giraudoux/socioecosystems.zip*

A second set of scenarios (Li *et al.* 2017) was built around the conversion of agricultural plots to improve the functionality of corridors for snub-nosed monkeys (Figure 6.6(a)). The impact on connectivity was measured at different scales depending on the type of vegetation favored (favorable or unfavorable environments for movement, optimal or suboptimal habitat). At an overall scale (whole network), only the restoration of suboptimal or optimal habitat increased connectivity by 2.4% and 24%, respectively. In the first case, the scenario led to an improvement in the permeability of the landscape allowing existing habitats to be reconnected. In the second case, the scenario led to the creation of new habitats, as well as the extension of some existing habitats. At a local scale, the 1,482 agricultural parcels were prioritized according to the gain in connectivity they would provide if they were converted to optimal habitat. While the overall increase in connectivity was small (+3.5% for 10 parcels converted), the "connector" quality of the parcels is clearly evident at the local scale. The majority of them are located in the largest component corresponding to the Baima Xueshan Reserve (Figure 6.6(b)). Plots 1 and 2, located at its periphery, would allow it to be reconnected with the subnetwork to the south. Parcels 3–6, as well as parcel 10, would improve existing connections in the central and northern parts of the reserve. Finally, parcels 7–9 are of strategic importance as they allow for a potential reconnection of the two nature reserves.

Spatial modeling of the black-and-white snub-nosed monkey ecological network therefore sheds some light on the degree of fragmentation of its habitats and the actions to be considered in order to reduce it. The various simulations revealed that only the creation of new optimal habitats would make it possible to significantly increase connectivity.

Despite its interest, this scenario appears highly unlikely, as the habitat of black-and-white snub-nosed monkeys consists of old-growth coniferous forests. These results underline the importance of anticipating landscape transformations and focusing conservation actions on these old-growth forests before they disappear. A similar approach was implemented to assess the impact of global warming on habitat connectivity of the golden snub-nosed monkey (*R. roxellana*) in Shennongjia National Park and Badong Nature Reserve, Hubei (Zhang *et al.* 2019). Beyond these results, landscape graphs, like any spatial model, have a didactic function. The schematic representation of the ecological network is sufficiently synthetic to be understandable and appropriable by non-specialists. The cartographic results thus constitute an interesting support for discussion and consultation between stakeholders.

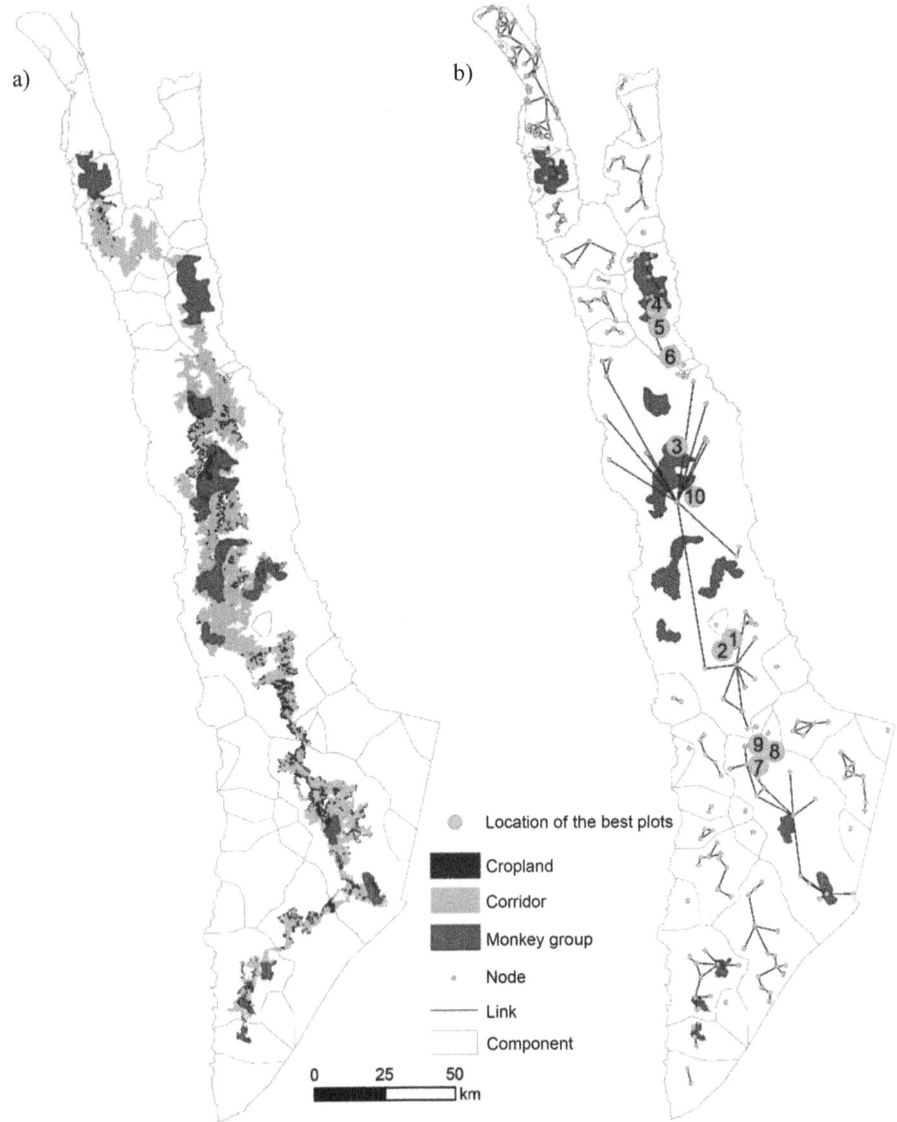

Figure 6.6. *Location of the 1,482 agricultural parcels located within the corridors (a) and the 10 parcels of greater interest for conversion to optimal habitat (b). The number indicates the prioritization of parcels based on the gain in connectivity provided by habitat creation. For a color version of this figure, see www.iste.co.uk/giraudoux/socioecosystems.zip*

6.4. In search of lost place names

Restoring connectivity among black-and-white snub-nosed monkey sub-populations is difficult without understanding what forces contributed to the fragmentation of the population, and whether they persist today. One answer can be found in the study of mitochondrial DNA (transmitted exclusively through the mother) and the modeling of the appearance of genetic mutations, which accumulate in the genome over time. Such studies have estimated that black-and-white snub-nosed monkey sub-populations began to diverge between 950 and 21,550 years ago, a time period that appears to point to both biogeographic and anthropogenic changes in the landscape as the cause of this divergence (Liu *et al.* 2009). The timing of the initial separation of the two most distant groups may have coincided with the Last Glacial Maximum: snow-covered valleys and high mountain forests may have served as a refuge for sub-populations disconnected by glaciations (Jablonski 1993; Zhang *et al.* 2002). Archeological studies show that human colonization of the Tibetan plateau began 10,000 years ago (Liu *et al.* 2009). Anthropogenic fires then played an important role in the dynamics of forest landscapes in southern Yunnan from 4,300 years ago (Xiao *et al.* 2017). Post-glacial climate warming then forced coniferous forests to persist only at higher elevations (Jablonski 1993). The rapid increase in human population density from the 17th century onward and its acceleration in the 19th century resulted in an expansion of agricultural land and an increase in economic activities related to wood exploitation, to which was added the development of road and then motorway infrastructures and the residential and commercial expansion of cities, leading to significant deforestation. Mainly exploited as energy resources, the forests were also destroyed during several periods of conflict (first opium war between the Chinese and the British, 1840–1842; Taiping revolt, 1850–1864), following the setting up of camps and the deliberate burning of wooded areas to repel the enemies. Persisting in fragmented and increasingly unconnected forest habitats, the black-and-white snub-nosed monkey has also been hunted due to popular beliefs about the medicinal properties of some of its body parts (mainly bone and skin) and trafficked to private zoos. Numbers during the 20th to 21st centuries were estimated at between 1,000 and 1,500 individuals (Long *et al.* 1996; Li *et al.* 2002; Xiao *et al.* 2003).

The initial population size of black-and-white snub-nosed monkeys before it began to be fragmented is not known, but it is clear that the species has declined dramatically over the past four centuries, particularly over the last two centuries, both in terms of spatial distribution and numbers (Li *et al.* 2002). While the population is currently stable or even increasing (Long *et al.* 2020), and numbers close to 3,000 individuals, past population variations and habitat fragmentation have made it possible for one or more genetic bottlenecks to occur in past centuries. This phenomenon is observed when a population decreases in size and fragmentation occurs. They lead to the rarefaction or even disappearance of rare alleles, and thus a reduction in genetic diversity. This reduction is generally accompanied by a decrease in the capacity of populations to adapt to new environmental constraints. It then takes millennia before, through the appearance of new mutations, a new genetic diversity similar to the initial one is restored. To corroborate this scenario with evidence, it is necessary to compare ancient genetic diversity with that observed today. Techniques for purification and amplification of ancient DNA now make it possible to make use of museum specimens. Although museums house millions of animal specimens, most of which have been collected since the 19th century, black-and-white snub-nosed monkey specimens from this period remain extremely rare. Seven individuals described by Alphonse Milne-Edwards in 1897, then chair of vertebrates and director of the Muséum national d'histoire naturelle, were nevertheless used to describe and name *R. bieti*. In systematics, a type is the reference element attached to a scientific name from which a species has been described (International Commission on Zoological Nomenclature 1999). These primates thus constitute the reference series of the species (holotype and paratypes) composed of:

– a "very adult" male taken at Kiapé, one day journey from Atentsé, on the left bank of the Mekong; "Tchru-tchra" (MNHN-ZM-MO-1897-72; holotype);

– an "adult" male, taken in the vicinity of Atentsé, on the left bank of the Mekong (MNHN-ZM-MO-1897-71);

– a "semi-adult" male, of the same provenance as the old male (MNHN-ZM-MO-1897-73);

– a "very adult" female, killed at Djra-Gniéra, not far from Tsékou, left bank of the Mekong (MNHN-ZM-MO-1897-74);

– a young infant of undetermined sex, of the same origin as the female, his or her mother (MNHN-ZM-MO-1897-75);

– two young males, of the same provenance as the female (MNHN-ZM-MO-1897-76 and 77).

The skin of all specimens was stuffed, often staged, and the complete skulls with jaws were carefully preserved (Figure 6.7). Their excellent general state of preservation thus allowed us to hope for a good quality of DNA preservation for comparison with present-day groups from the same localities, provided we could locate them precisely. It was therefore essential to know the locations, as precisely as possible, of their capture in order to deduce their geographical proximities with the present groups.

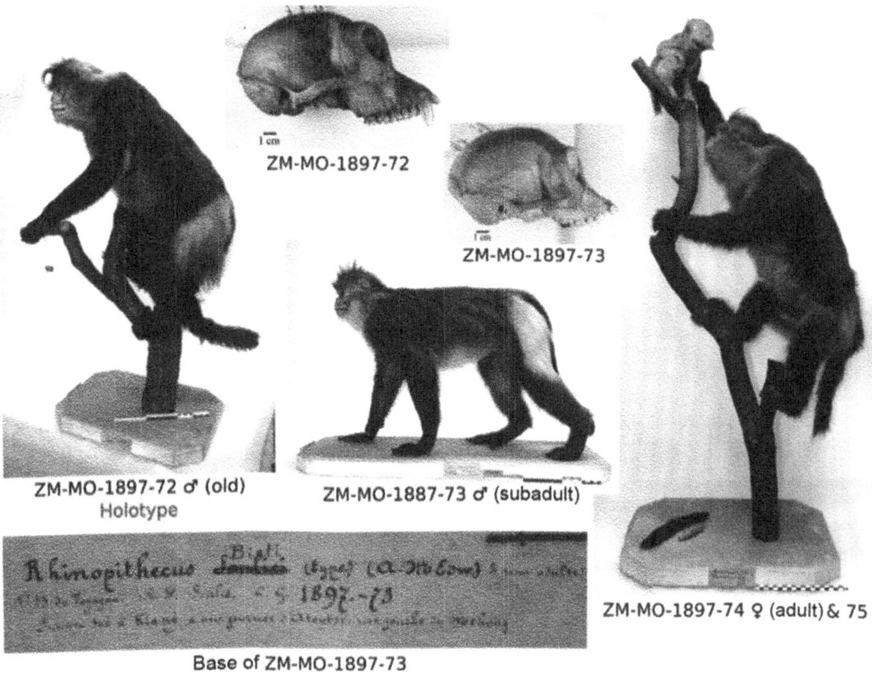

Figure 6.7. *Specimens from the collection of the Muséum national d'histoire naturelle and inscription of the holotype base (photos C. Callou)*

These seven specimens are among the thousands of plants and animals that the Catholic missionaries of the Paris Foreign Missions (*Missions étrangères de Paris*, MEP), encouraged by modest funding (Garnier 2017), sent to the National Museum of Natural History in Paris. Jean-André Soulié, for his part, on mission in the region from 1885 to 1905, sent over 7,000 species of plants! He stayed in "Tsékou" from December 8, 1894 to July 20, 1896, with Father Dubernard who had built a church there and gathered a small community converted to Catholicism, from a Tibeto-Burman ethnic group, the Lisu people (Espinasse 1998).

It is from there that the specimens collected in the surroundings were sent to Paris. This church was destroyed in 1905, during the "lama revolt", and Jean-André Soulié, then ministering in "Yaregong" (Yarigong, 亚日贡), was killed during these events. It was rebuilt "not far away" in Cizhong (茨中), where it still stands today and brings together Catholics in peaceful coexistence with local Tibetan Buddhists, all of whom are mainly Lisu (Figure 6.8).

Figure 6.8. *Church of Cizhong and vineyard planted by missionaries in the 19th century (grape variety: baco, honey rose). In the background, the snowy mountains of the Baima Xueshan (photo P. Giraudoux)*

On the base of one of the type specimens (Figure 6.7), it can be seen that the original intention was to name the species *Rhinopitecus souliei*, but the collector's name, for some unknown reason, was crossed out and replaced by that of Biet, his direct superior, hence *Rhinopithecus bieti*. Many of the new species were dedicated at this time to Felix Biet who, as bishop *in partibus* of Diana[4], and later "vicar apostolic of Thibet", based at Tatsienlou, now Kangding, in south-eastern Sichuan, administered all the missionaries in the region. The inscriptions on the bases of the specimens and Rode's catalog, which includes information directly related to the specimens (marking of the skulls, labels, publication of origin), are the only sources available to identify the most exact place of their capture. They report them in a phonetic writing that, presumably, has more or less altered the names heard by the

4 The Diana is a kingdom that existed between 2400 BCE and 109 CE, partially encompassing the present provinces of Yunnan, Sichuan and the Tibet Autonomous Region. *In partibus (infidelium)* means "in the lands of the infidels", clearly illustrating the evangelizing purpose of his appointment on July 23, 1878. The Mandarin name for black-and-white snub-nosed monkey, diān jīn sī hóu (滇金丝猴), literally means "golden-hair monkey of Diana". It alludes to this extinct kingdom and to the fact that the species is related, as a *Rhinopithecus*, to that of the golden monkey.

missionaries in Lisu (a Tibeto-Burmese language) and in local Mandarin. These are not present in the Chinese toponymy of the current maps. It took 3 years of documentary research and three on-site missions, from 2016 to 2018, to get to the bottom of the mystery.

An initial correspondence between ancient names and modern Chinese toponyms was established from an appendix to the list of Eastern Palaearctic Sphingidae, discovered on the Internet (Kitching and Pittaway 2000). This appendix was a great help, but it was not sufficient. Some writings of Jean-André Soulié (Soulié 1904), available on the Internet, made the existence of the MEP archives known. They were then kindly opened to us. A map of the "Thibet mission", drawn up in 1889, included the toponyms of the bases and of Rode's catalog (Figure 6.9).

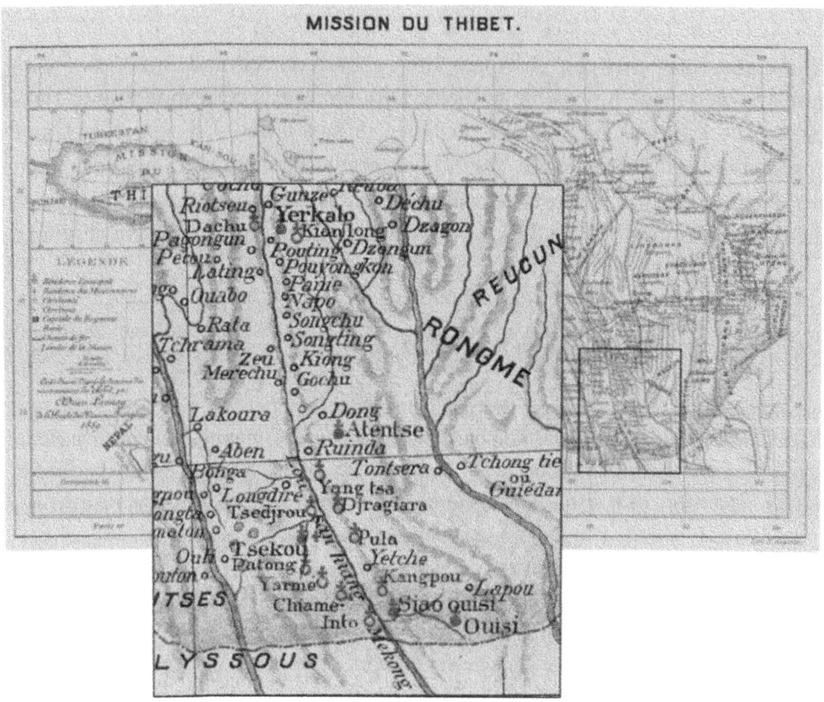

Figure 6.9. *Old map from the archives of the Foreign Missions of Paris, drawn up by Adrien Launay in 1889 (MEP archives). The toponymy used by the missionaries is recorded (© Institut de recherche France-Asie/MEP). For a color version of this figure, see www.iste.co.uk/giraudoux/socioecosystems.zip*

It shows that these are not just Lisu, but also include altered Mandarin place names of the time. For example, "Siao Ouisi" is positioned at the relative location of the present Xiao Weixi (小维西). This approach shows the type of alteration practiced and identifies a safe point of georeferenced location here. Supplemented by other less precise maps from Taiclet (1998), a georeferencing of each identifiable toponym was thus possible and was carried out. It then appeared that Tsékou was located a few kilometers southwest of the present-day Cizhong church and corresponded to the hamlet of Cigu (茨姑). An on-site survey in April 2017 located the church at the current site of a farm and a walnut plantation, whose pro-owner Lisu, a Catholic, had saved some crosses and instruments of worship from the ill-treatment during the Cultural Revolution.[5] This location corresponds perfectly to that which could be established from old engravings in the archives of the MEP (Figure 6.10).

Back and forth between old maps thus georeferenced and interviews with the inhabitants then made it possible to locate Djra-Gniéra as corresponding to Jia Kong (贾空) on the left bank of the river, between groups 6 and 7. Kiapé corresponds to the present-day village of Jia Bi (佳碧), at an intermediate level between groups 5 and 6. Finally, Atentsé is close to group 5 and corresponds to the historical districts of the city of Deqin (德欽), still known in Lisu by this name at present (Figure 6.3).

In parallel with the collection of hundreds of feces from current populations and their analysis by the Laboratory for Conservation and Utilization of Bioresources

5 The success of these surveys owes everything to the hospitality of the people of Cizhong, Cigu and Jiabi, whom we thank here, combined with often very improbable circumstances. For example, we were in contact, for our work in the Baima Xueshan Nature Reserve, with Ms. Shi Fangqing (施方勤), of the Yunnan Department of Wildlife and Forestry. On the first mission to Cizhong where she accompanied us, she revealed that her father and uncle, orphans, had been raised by the missionary fathers in Cizhong (her father spoke and wrote French) and that part of her family was settled in Xiao Weixi. Her father's grave was located there, in a small cemetery, not far from the graves of French and Swiss missionaries. She thus established for us a unique connection with the priests of the churches in the area. As a result, in November 2016, we had the privilege of meeting Mr. Xiao Jieyi (肖杰一) in Cizhong. A churchman in his nineties, he had faithfully known the French priests of the first half of the 20th century, spoke French and knew the Catholic hymns in Latin, which he sang perfectly. It was he who *finally* introduced us to Taiclet's (1998) biography of Father Georges André. It includes maps that were useful for georeferencing. We were totally surprised to discover that the birthplace of Georges André was Ronchamp, a small town in the Franche-Comté region at the foot of the Vosges mountains, a few kilometers from the town of origin of one of the authors of this chapter (P.G.), and that the publishing house was located in Vesoul. How did this book get to Xiao Jieyi in Cizhong? This remains a mystery.

at Yunnan University, DNA analysis of museum specimens is underway. Their relatively precise location, obtained through a historical-geographical approach, justifies the relevance of the genetic comparisons that will be made with the cousins and descendants of the specimens captured in the 19th century.

a) b)

Figure 6.10. *a) Early engraving (MEP archives, © Institut de recherche France-Asie/MEP). b) Location of the former church of Tsékou (yellow arrow) in April 2017 (photo P. Giraudoux). In both images, one can see the ridge of hills and the mountains in the background. For a color version of this figure, see www.iste.co.uk/giraudoux/socioecosystems.zip*

6.5. Animal tourism: what are the consequences for the black-and-white snub-nosed monkey?

The work described in the previous sections is based on an approach that takes into account all of black-and-white snub-nosed monkey sub-populations. Based on the genetic distances between sub-populations, it allows us to determine the fragmentation of the population, which is directly related to the connectivity of the landscape. Isolation of small sub-populations in a fragmented landscape exposes them to an extinction vortex, a process by which a disconnected small group loses its genetic diversity and its ability to cope with stresses on individual survival, leading to a reduction in group size in a spiral leading to group extinction. The loss of genetic diversity and the rate of inbreeding are elements that indicate the vulnerability of a sub-population to such risks, and therefore the need to implement conservation measures. In order to study the feasibility of molecular analyses to acquire these data at the scale of each sub-population, our interest was first focused on a sub-population that is easy to access and allows for rapid collection of feces (from which genetic analyses can be performed). This sub-population is one of five identified in the Baima Xueshan National Nature Reserve, near the villages of Xiangguqing and Gehuaqing (group 9). Reserve rangers have set up feeding sites near Xiangguqing where tourists can easily observe snub-nosed monkeys in their natural environment. Although these sites respond to a growing need for tourists to interact with nature, they are far from the idealized image of wilderness and are usually part of complex socioecosystems that are heavily dominated by human activities. As we sought to develop methods for genetic analysis, we quickly realized that our pilot site represented a population-wide singularity: the creation of a small, fed group created the potential to affect both the genetic diversity of this sub-population and the health status of individuals overexposed to the risk of pathogen transmission.

6.5.1. *Ecotourism feeding sites: ready-to-see animals*

Ecotourism is defined by The International Ecotourism Society (2015) as "responsible travel to natural areas that preserves the environment, supports the well-being of local people, and helps to increase tourists' awareness of the political, environmental, and social climates of host countries". This form of tourism has grown rapidly around the world in recent decades (Balmford *et al.* 2015). Ecotourism responds to a demand for interaction with nature, while developing biodiversity conservation programs and increasing financial and educational benefits for local communities (Orams 2002). Organized wildlife watching is one of the most common forms of ecotourism and exists in a variety of formats, ranging from itinerant safaris to permanent viewing sites. It involves a wide variety of taxonomic groups around the world, including birds, cetaceans, terrestrial herbivores and carnivores (Orams 2002; Okello *et al.* 2008; Penteriani *et al.* 2017).

Observing wildlife requires the ability to travel over large vegetated areas, which may be difficult to access and/or walk through, and the necessary naturalist knowledge of the ecological and behavioral requirements of the species to locate individuals. Passionate naturalists readily accept that observing wildlife in its habitat is an unpredictable and uncertain activity, and enjoy the experience, even when they fail to see the animal. Wildlife viewing takes place over a long period of time and is more valuable the less frequently it occurs. Conversely, tourists have a strong time imperative in their schedules, as they plan their arrival, their presence at the site and their departure. They want to be able to enjoy these observations without spending the time and the search for knowledge required. Organizing a systematic and regular supply of food in fixed locations creates the conditions that make a group of individuals observable at predictable times and places. As there is ready-to-wear in fashion, Knight (2010) draws the comparison with the "ready-to-see" animal.

6.5.2. *Ecotourism in Xiangguqing: what are the consequences for the black-and-white snub-nosed monkey?*

The Baima Xueshan Nature Reserve was established in the 1980s and became a national park by the end of the decade. Local farmers were recruited as rangers for the reserve, and the joint efforts of the reserve and a private company led to the creation of education programs around the conservation of black-and-white snub-nosed monkeys. The development of monkey tourism has provided funding for scientific studies and conservation programs. A meeting in March 2019 with Yu Jianhua, head ranger of the Baima Xueshan Reserve, allowed us to trace the genesis of the creation of the fed group. Starting in 2008, the guards of the Baima Xueshan reserve began to attract black-and-white snub-nosed monkeys to feeding sites and gradually accustom them to human presence. Over a period of 7–8 months, the rangers gradually managed to habituate about 80 individuals, testing different kinds of food. At first, the snub-nosed monkeys refused the food offered by the guards, which consisted mainly of foodstuffs that the guards could easily obtain (peanuts and apples, for example). The rangers then observed the behavior of the monkeys at length to understand which food resources were most attractive to them. The most successful food sources were lichens, which are the species' preferred resource, but which required a lot of work to collect in the surrounding highland forests. During this period, monkeys were free to move in and out of the area where feeding sites were distributed, and it was not possible to differentiate between a fed subgroup and the rest of the sub-population. The effort required to feed the individuals at the feeding sites soon became considerable and the wardens decided to limit the number of individuals fed. Forty individuals were chased away from the feeding sites by the wardens. As a result, at the end of 2008, 40 individuals made up the fed group. The food supply had a positive effect on the number of births and the fed group quickly grew to over 50 individuals. This threshold represented a limit for the wardens above which the efforts to maintain the sub-group were no longer sustainable. When the size of the fed group exceeded 50 individuals, the wardens chased individuals away from the feeding sites and prevented new individuals from accessing the feeding sites as much as possible.

Since 2009, the reserve rangers have maintained a sizable sub-population of about 50 individuals near Xiangguqing village (Figure 6.11). However, in the absence of a physical barrier, individuals can move freely in the area and it cannot be excluded that they still maintain contact with the rest of the sub-population.

Feeding sites are supplied daily (Figure 6.11(a) and Figure 6.11(b)) and guards move from one feeding site every two to three days to another one located a few hundred meters away. This simulates the natural movements of a group in search of food and semi-normalizes behavioral impacts. During feeding sessions, tourists can easily observe monkeys (Figure 6.11(c) and Figure 6.11(d)). The remainder of the sub-population is distributed across an estimated range of 82.9 km^2 (Wong *et al.* 2013). Long hikes up steep mountains are required to observe wild individuals and/or evidence of their presence.

a) b) c) d)

Figure 6.11. *Tourist visits to black-and-white snub-nosed monkey feeding sites in Xiangguqing, Baima Xueshan Nature Reserve (Yunnan, China), November 2016*

COMMENT ON FIGURE 6.11 *"Ready-to-see" animals.– a) Lichens are collected in the rainforests and kept in bags by the reserve officers. b) A black-and-white snub-nosed monkey eating lichens. c) Tourists observe monkeys at a feeding site; a simple rope separates them from them, but it is not permitted to feed them or approach them. d) Monkeys may occasionally approach tourists, but reserve guards usually discourage this behavior (photos E. Afonso and P. Giraudoux).*

6.5.3. *Genetic consequences of human intervention*

From December 2016 to May 2017, rangers from the Baima Xueshan Reserve actively assisted us in collecting black-and-white snub-nosed monkey feces from feeding sites and the rest of the group 9 range (Figure 6.12). A total of 91 feces were collected, each of which was geo-located and assigned to either a fed individual (as located at the feeding sites) or a wild individual. Back in the laboratory, DNA was extracted from these samples, in order to determine a genetic fingerprint for each sample based on the genotypic characterization of 10 microsatellite markers. The genotype thus obtained made it possible to assign each sample to an individual and to identify cases where several samples of feces had been collected for the same individual. Bayesian statistical methods, with no preconceived notions about the distribution of allele frequencies, were used to determine that the sub-population was structured into two genetic clusters with different allele frequencies. For each individual, we had the probability that it belonged to one or the other of the two clusters: the majority of individuals identified as fed according to field observations belonged to one cluster and individuals identified as wild belonged to the other cluster. Depending on the assignment threshold chosen, a variable number of individuals did not belong to either cluster, which could suggest the possibility of genetic profiles resulting from genetic mixing between the two clusters. The genetic differentiation between the fed and wild subgroups was surprising: although moderate, it was greater than that observed between sub-populations several tens of kilometers apart. Human intervention to create the fed subgroup thus appeared to act more strongly than geographical distance and/or landscape fragmentation (Afonso *et al.* 2021a).

There were several indications of a loss of genetic diversity in the fed subgroup: allelic richness was lower in the fed subgroup than in the wild subgroup, and the fed subgroup had excess homozygosity. Samples collected from the fed subgroup had a significantly higher coefficient of relatedness than those collected from the wild: on average, relatedness measures in the fed individuals were close to the values observed in half-siblings, whereas in the wild individuals these values were more characteristic of unrelated individuals. The small size of the fed subgroup (about 50 individuals) may explain the risk of loss of genetic diversity due to both genetic drift and inbreeding effects. However, the history of the creation of the fed group suggests strong hypotheses that human intervention may have had an impact on the fed subgroup as a founder effect, which occurs when a new population is established by a very small number of individuals from a larger population. Even if the individuals that initially created the fed subgroup were randomly selected, they represent only a fraction of the wild source genetic diversity.

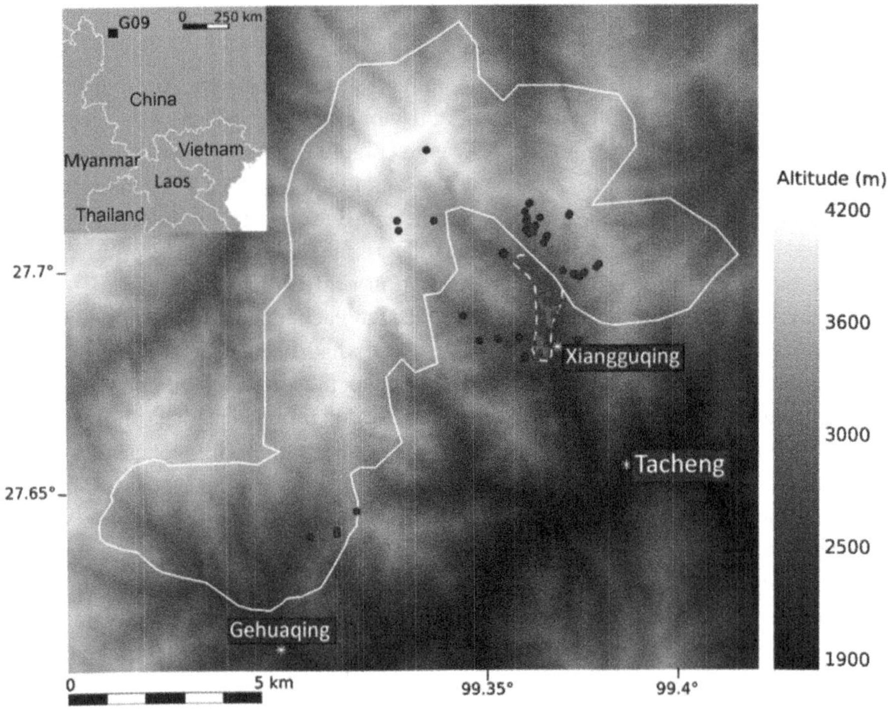

Figure 6.12. *Location of black-and-white snub-nosed monkey feces collected in the Baima Xueshan Nature Reserve. Blue circles represent feces collected and attributed to "wild" individuals, red circles represent feces collected and attributed to fed individuals. The solid white line represents the distributional boundaries of the wild group, from Wong* et al. *(2013). The dashed white line represents the area in which feeding sites are distributed. For a color version of this figure, see www.iste.co.uk/giraudoux/socioecosystems.zip*

Regardless of the evolutionary processes that led to the differentiation between the fed and wild subgroups, the loss of genetic diversity in the fed subgroup is an element that could justify conservation measures on its own. We explored different genetic restoration scenarios as a possible way to ensure the coexistence of the two subgroups by minimizing their genetic differentiation. Based on the current distribution of allelic frequencies, we ran several simulations to estimate the evolution of different measures of genetic diversity by imagining the translocation of m females from the wild to the fed subgroup (Figure 6.13). These simulations were based on the limited knowledge of the pairing system, social relationships and demographic parameters of the species available in the scientific literature (Kirkpatrick *et al.* 1998; Liu *et al.* 2007; Cui *et al.* 2008; Ren *et al.* 2011; Xia *et al.*

2019). Other demographic parameters were estimated from known data for other species. The results of the different simulations suggest that genetic differentiation between fed and wild subgroups may increase in the future if no conservation measures are undertaken (Figure 6.13). Even a one-time translocation of 10 adult females would not restore genetic diversity in the fed subgroup in a sustainable manner. These simulations show that it is unlikely to rely on a one-time solution and that other scenarios involving regular efforts by feeding site managers should be explored.

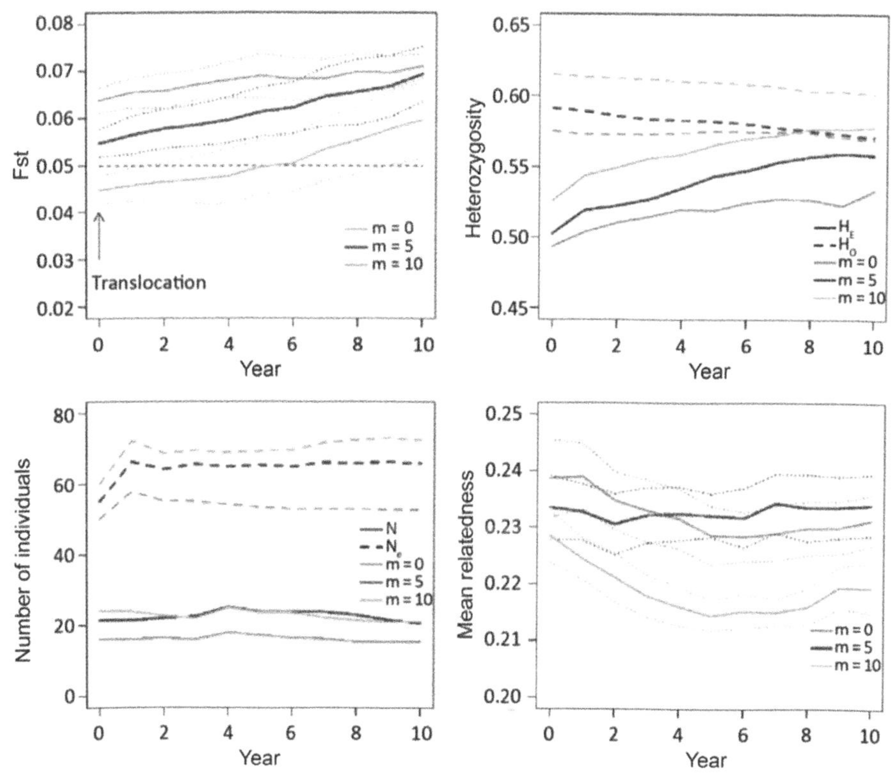

Figure 6.13. *Simulated temporal evolution of genetic diversity in the black-and-white snub-nosed monkey sub-population in Xiangguqing/Gehuaqing, from Afonso* et al. *(2021a). At time t = 0, a number m of females from the wild subgroup virtually introduced once from the wild subgroup to the fed subgroup. For a color version of this figure, see www.iste.co.uk/giraudoux/socioecosystems.zip*

COMMENT ON FIGURE 6.13.– *a) Genetic differentiation between the fed and wild subgroups (FST) and a 95% confidence interval. b) Expected (HE) and observed*

(HO) heterozygosity in the fed subgroup. c) Subgroup size (N) and number of individuals actually participating in reproduction (Ne) in the fed subgroup. d) Mean relatedness coefficient in the fed subgroup.

6.5.4. *Feeding sites: hotspots of parasitic transmission?*

During our various visits to the feeding sites, we found that the proximity of these sites to the village of Xiangguqing was likely to create favorable conditions for the transmission of pathogens between monkeys, humans (villagers and tourists) and domestic wildlife. The local domestic fauna (mainly pigs, cows and chickens) is highly concentrated in the village and is also present more than 1 km around the village. Pigs and cows are present at the feeding sites, where they feed on the ground (Figure 6.14). Pigs may even consume monkey dung (Fu *et al.* 2018), and return to the village for the night. Feeding sites are therefore a potential relay for pathogen transmission between the village and the surrounding mountains.

Figure 6.14. *Cattle (a) and pigs (b) grazing near feeding areas (photos P. Giraudoux)*

Using the same samples that we used to study the genetic diversity of the Xiangguqing/Gehuaqing sub-population, we looked for the presence of protozoan parasites that are widespread in primates, humans and domestic fauna: amoebae of the genus *Entamoeba*. These parasites colonize the digestive system of hosts and include an environmental free phase (contained in soil and water) before being transmitted to a new host. Molecular methods based on the amplification of a region of DNA common to all *Entamoeba* species (here, the 18S rDNA gene) and the massive sequencing of the DNA thus amplified in each positive sample generate an unbiased screening of the parasitic species contained in the feces. Parasite prevalence was high in the fed subgroup (89% of individuals had at least one *Entamoeba* DNA sequence in their feces) and significantly higher than in the wild subgroup (33%). Based on this finding, we conducted systematic sampling of domestic feces (pigs, cows, chickens and dogs) in Xiangguqing village, as well as at the feeding sites in May 2018. We also sampled two latrines used by villagers.

Across all hosts, we detected 13 different DNA sequences that could all be assigned to an *Entamoeba* species or lineage. The most common DNA sequences found in feces belonged to the species *Entamoeba polecki* and were present in both fed and wild snub-nosed monkey, pigs and humans. Several DNA sequences assigned to *Entamoeba*, usually found in cattle, were also present in snub-nosed monkey feces. Pigs, cows and monkeys in the fed group were not only the hosts with the highest parasite prevalences, but they also had the highest numbers of parasite co-occurrences (Figure 6.15).

In snub-nosed monkeys the probability of finding *E. polecki*-positive feces was directly related to distance from domestic fauna and to the individual level of heterozygosity, both of which are representative of the fed subgroup. The closer an individual was located to domestic fauna, the greater the chance of finding the parasite in their feces (Figure 6.16).

The circulation of *E. polecki*, an amoeba common to humans, domestic wildlife and black-and-white snub-nosed monkey, and the high parasite prevalence observed in the fed subgroup raise questions about the possible implication for the conservation of fed monkeys. The pathogenicity of the amoebae found in this study is poorly documented and has never been explored in snub-nosed monkeys. It is not impossible that the body condition of the most fragile individuals is altered by digestive symptoms observed in other host species, which remains to be confirmed by in-depth examinations of the health status of the individuals making up the fed subgroup. In addition, parasitic exposure to amoebae could be representative of exposure to other pathogens common to both domestic and wild fauna, and circulating at feeding sites, which are suitable for the concentration of different hosts.

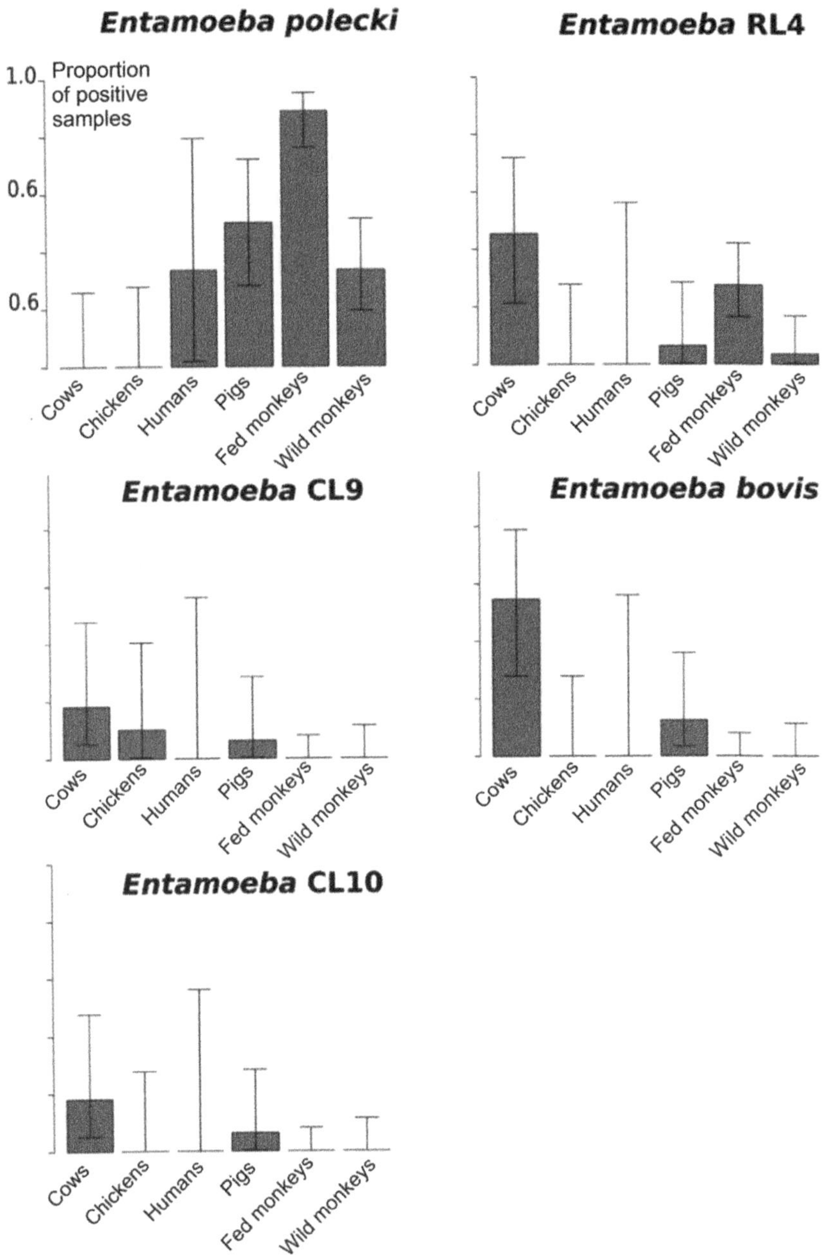

Figure 6.15. *Proportion of fecal samples positive for Entamoeba spp. by species and lineage of Entamoeba. S., monkeys from Afonso* et al. *(2021b). The scale is the same for all graphs*

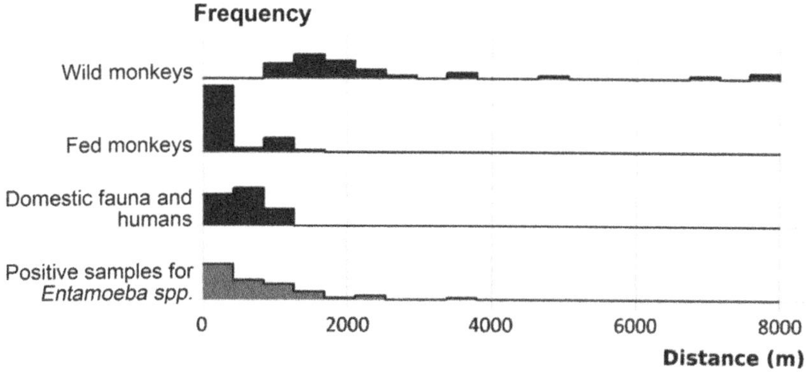

Figure 6.16. *Location of fecal samples collected for the different host categories and samples in which Entamoeba DNA was detected (positive for Entamoeba spp.) according to the distance from the centroid of the area where humans and domestic fauna are located, from Afonso* et al. *(2021b)*

While the various studies presented above show some of the negative impacts of feeding sites on the conservation of black-and-white snub-nosed monkeys, they do not take into account the socio-economic benefits that wildlife tourism represents for local communities. These benefits are essential if local communities are to coexist without conflict with an endangered species for which every individual counts. Wildlife tourism in Xiangguqing provides sufficient financial resources for rangers to provide surveillance to limit snub-nosed monkey poaching. The creation of a fed subgroup is perhaps a price to pay for improving the living conditions of local communities, and for funding research and awareness-raising programs for the conservation of this species, which is now a major national cause in China. A center for injured individuals has also been created and collects several individuals each year that are then released. It is therefore possible that the benefits for the species exceed the costs incurred by 50 or so fed individuals. However, different arrangements could be considered that take into account the health of the fed monkeys, as well as that of humans and wildlife, without placing too great a burden on the rangers. This includes the following recommendations: (1) expand the distribution of feeding sites to as wide an area as possible; (2) encourage contact and/or rotate feeding and wild individuals to promote genetic mixing; and (3) avoid the presence of domestic wildlife within the range of feeding sites. Moreover, at a time when the global biomass of humans and their domestic animals, which accounts for 96% of the mammalian biomass (Bar-On *et al.* 2018), puts them at risk from many wildlife-derived pathogens, as shown by the growing number of epidemics, some of which have triggered pandemics (Morand 2020), it is certainly essential to be very vigilant about artificial situations that create privileged contacts with wildlife.

> KEY POINTS. –
>
> Landscape graphs provide a simulation tool that can be used to prioritize options for reconnecting fragmented habitats based on population genetic structure and habitat mapping.
>
> Museological data combined with historical documentation from archives allow comparison not only of the present and past distributions of a species, but also of its genetic polymorphism.
>
> Isolation of wild populations for ecotourism purposes, while it has the advantage of raising public awareness of the existence and preservation of a species, can lead to genetic impoverishment of the isolated group and increase the circulation of potentially pathogenic organisms between wildlife, domestic animals and humans.

6.6. References

Afonso, E., Fu, R., Dupaix, A., Goydadin, A.-C., Yu, Z., Li, D., Giraudoux, P., Li, L. (2021a). Creating small food-habituated groups might alter genetic diversity in the endangered Yunnan snub-nosed monkey. *Global Ecology and Conservation*, 26, e01422.

Afonso, E., Fu, R., Dupaix, A., Goydadin, A.-C., Yu, Z., Callou, C., Villette, P., Giraudoux, P., Li, L. (2021b). Feeding sites promoting wildlife-related tourism might highly expose the endangered Yunnan snub-nosed monkey (Rhinopithecus bieti) to parasite transmission. *Scientific Reports*, 11(1), 15817. [Online]. Available at: https://doi.org/10.1038/s41598-021-95166-5 [Accessed January 7, 2022].

Bai, S.C., Zhou, S.Q., Lin, S., Tuo, D., Tu, Z., Zhong, T., Wang, X.H. (1988). Population distribution and feeding habits of *Rhinopithecus Bieti*. *Zoological Studies*, 9(suppl.), 67–75.

Balmford, A., Green, J.M.H., Anderson, M., Beresford, J., Huang, C., Naidoo, R., Walpole, M., Manica, A. (2015). Walk on the wild side: Estimating the global magnitude of visits to protected areas. *PLOS Biology*, 13(2), e1002074.

Bar-On, Y.M., Phillips, R., Milo, R. (2018). The biomass distribution on Earth. *Proceedings of the National Academy of Sciences*, 115(25), 6506–6511.

Clauzel, C., Deng, X.Q., Wu, G.S., Giraudoux, P., Li, L. (2015). Assessing the impact of road developments on connectivity across multiple scales: Application to Yunnan snub-nosed monkey conservation. *Biological Conservation*, 192, 207–217.

Cui, L.-W., Huo, S., Zhong, T., Xiang, Z.-F., Xiao, W., Quan, R.-C. (2008). Social organization of black-and-white snub-nosed monkeys (*Rhinopithecus bieti*) at Deqin, China. *American Journal of Primatology*, 70(2), 169–174.

Espinasse, J. (1998). *Tibet, mission impossible : Lettres du père Dubernard, 1864–1905.* Fayard-Le Sarment, Paris.

Estrada, A., Garber, P.A., Rylands, A.B., Roos, C., Fernandez-Duque, E., Di Fiore, A., Nekaris, K.A.-I., Nijman, V., Heymann, E.W., Lambert, J.E. *et al.* (2017). Impending extinction crisis of the world's primates: Why primates matter. *Science Advances*, 3(1), e1600946.

Foltête, J.C., Girardet, X., Clauzel, C. (2014). A methodological framework for the use of landscape graphs in land-use planning. *Landscape and Urban Planning*, 124, 140–150.

Fu, R., Li, L., Afonso, E., Yu, Z.H., Giraudoux, P. (2018). Spatial and temporal distribution of Yunnan snub-nosed monkey, *Rhinopithecus bieti*, indices. *Mammalia*, 83(2), epub.

Garnier, E. (2017). Une nouvelle espèce de rhinopithèque. *Revue des Missions étrangères de Paris*, 530, 43–46.

Geissmann, T., Lwin, N., Aung, S.S., Aung, T.N., Aung, Z.M., Hla, T.H., Grindley, M., Momberg, F. (2011). A new species of snub-nosed monkey, genus Rhinopithecus Milne–Edwards, 1872 (Primates, Colobinae), from northern Kachin state, northeastern Myanmar. *American Journal of Primatology*, 73, 96–107.

International Commission on Zoological Nomenclature (1999). *International Code of Zoological Nomenclature*, 4th edition. The International Trust for Zoological Nomenclature, London.

Jablonski, N.G. (1993). Quaternary environments and the evolution of primates in East Asia, with notes on two new specimens of fossil cercopithecidae from China. *Folia Primatologica*, 60(1/2), 118–132.

Jiguet, F. (2020). The fox and the crow. A need to update pest control strategies. *Biological Conservation*, 248, 108693.

Kirkpatrick, R.C., Long, Y.C., Zhong, T., Xiao, L. (1998). Social organization and range use in the Yunnan snub-nosed monkey *Rhinopithecus bieti. International Journal of Primatology*, 19(1), 13–51.

Kitching, A.R. and Pittaway, I.J. (2000). Gazetteer of the Sphingidae of the eastern Palaearctic – Species list [Online]. Available at: http://tpittaway.tripod.com/china/chigaz.htm [Accessed September 13, 2020].

Knight, J. (2010). The ready-to-view wild monkey: The convenience principle in Japanese wildlife tourism. *Annals of Tourism Research*, 37(3), 744–762.

Li, Z.X., Ma, S.L., Hua, C.H., Wang, Y.X. (1981). The distribution and habit of Yunnan snub-nosed monkey. *Zoological Research*, 2(1), 9–16.

Li, B., Pan, R., Oxnard, C.E. (2002). Extinction of snub-nosed monkeys in China during the past 400 years. *Internat. Journal of Primatology*, 23(6), 1227–1244.

Li, M., Wei, F., Huang, C., Pan, R., de Ruiter, J. (2004). Phylogeny of snub-nosed monkeys inferred from mitochondrial DNA, cytochrome B, and 12S rRNA sequences. *International Journal of Primatology*, 25(4), 861–873.

Li, L., Xue, Y., Wu, G., Li, D., Giraudoux, P. (2014). Potential habitat corridors and restoration areas for the black-and-white snub-nosed monkey *Rhinopithecus bieti* in Yunnan, China. *Oryx*, 49(4), 719–726.

Li, W., Clauzel, C., Dai, Y., Wu, G., Giraudoux, P., Li, L. (2017). Improving landscape connectivity for the Yunnan snub-nosed monkey through cropland reforestation using graph theory. *Journal for Nature Conservation*, 38, 46–55.

Liu, Z., Ding, W., Grueter, C.C. (2007). Preliminary data on the social organization of black-and-white snub-nosed monkeys *Rhinopithecus bieti* at Tacheng, China. *Acta Theriologica Sinica*, 27(2), 120–122.

Liu, Z.J., Ren, B.P., Wu, R.D., Zhao, L., Hao, Y.L., Wang, B.S., Wei, F.W., Long, Y.C., Li, M. (2009). The effect of landscape features on population genetic structure in Yunnan snub-nosed monkeys (*Rhinopithecus bieti*) implies an anthropogenic genetic discontinuity. *Molecular Ecology*, 18(18), 3831–3846.

Long, Y.C., Kirkpatrick, C.R., Zhongtai, Xiaolin (1994). Report on the distribution, population, and ecology of the Yunnan snub-nosed monkey (*Rhinopithecus bieti*). *Primates*, 35(2), 241–250.

Long, Y.C., Kirkpatrick, C., Zhong, T., Xiao, L. (1996). Status and conservation strategy of the Yunnan snub-nosed monkey. *Chinese Biodiversity*, 4(3), 145–152.

Long, Y.C., Momberg, F., Ma, J., Wang, Y., Luo, Y.M., Li, H.S., Yang, G.L., Li, M. (2012). Rhinopithecus strykeri found in China! *American Journal of Primatology*, 74(10), 871–873.

Long, Y., Bleisch, W., Richardson, M. (2020). *Rhinopithecus bieti*. The IUCN Red List of Threatened Species 2020: e.T19597A8986243 [Online]. Available at: http://dx.doi.org/10.2305/IUCN.UK.2008.RLTS.T19597A8986243.en [Accessed November 21, 2019].

Milne-Edwards, M.A. (1897). Note sur une nouvelle espèce du genre Rhinopithèque provenant de la Haute Vallée du Mékong. *Bull. Mus. Hist. Nat. Paris*, 5, 156–160.

Mo, W.W. and Yang, D.H. (1982). A preliminary observation of *Rhinopithecus bieti* population, movement routes and feeding habits in the eastern slope of Baima Snow Mountain. *Journal of Veterinary Science*, 2(2), 125–131.

Morand, S. (2020). *L'Homme, la faune sauvage et la peste.* Fayard, Paris.

O'Brien, J. (2015). Saving the common hamster (*Cricetus cricetus*) from extinction in Alsace (France): Potential flagship conservation or an exercise in futility? *Hystrix, the Italian Journal of Mammalogy*, 26(2), 89–94.

Okello, M.M., Manka, S.G., D'Amour, D.E. (2008). The relative importance of large mammal species for tourism in Amboseli National Park, Kenya. *Tourism Management*, 29(4), 751–760.

Orams, M.B. (2002). Feeding wildlife as a tourism attraction: A review of issues and impacts. *Tourism Management*, 23(3), 281–293.

Peng, H.S., Gao, Y.T., Lu, C.K., Feng, J.X., Cheng, Q.X. (1962). Taxonomic study of animals from Southwest Sichuan and Northwest Yunnan. *Acta Zoologica Sinica*, 14(suppl.), 105–132.

Penteriani, V., López-Bao, J.V., Bettega, C., Dalerum, F., del Mar Delgado, M., Jerina, K., Kojola, I., Krofel, M., Ordiz, A. (2017). Consequences of brown bear viewing tourism: A review. *Biological Conservation*, 206, 169–180.

Ren, B., Li, D., He, X., Qiu, J., Li, M. (2011). Female resistance to invading males increases infanticide in langurs. *PLoS ONE*, 6(4), e18971.

Rode, P. (1938). *Catalogue des types de mammifères du Museum national d'histoire naturelle. Ordre des Primates, sous-ordre des Simiens.* Museum national d'histoire naturelle, Paris.

Saura, S., Bodin, Ö., Fortin, M.-J. (2013). Stepping stones are crucial for species' long-distance dispersal and range expansion through habitat networks. *Journal of Applied Ecology*, 51(1), 171–182.

Soulié, J.-A. (1904). Géographie de la principauté de Bathang. *La Géographie*, 9, 87–104.

Taiclet, G. (1998). *Le père Georges André, missionnaire comtois du bout du monde. Thibet 1920–1952.* Les éditions de Haute-Saône, Vesoul.

The International Ecotourism Society (2015). What is ecotourism? [Online]. Available at: https://ecotourism.org/what-is-ecotourism/ [Accessed January 27, 2020].

Wang, H., Xu, H., Li, Y., Xu, Z., Ding, W., Xiao, W., Ye, H. (2019). New distribution records for the endangered black-and-white snub-nosed monkeys (*Rhinopithecus bieti*) in Yunnan, China. *Journal of Vertebrate Biology*, 68(2), 79–85.

Wilson, D.E. and Reeder, D.M. (2005). *Mammals Species of the World: A Taxonomic and Geographic Reference*, 3rd edition. Johns Hopkins University Press, Baltimore.

Wong, M.H.G., Li, R.Q., Xu, M., Long, Y.C. (2013). An integrative approach to assessing the potential impacts of climate change on the Yunnan snub-nosed monkey. *Biological Conservation*, 158, 401–409.

Wu, B.Q., Zhong, T., Wu, J. (1988). Preliminary observations of the ecological behavior of a *Rhinopithecus bieti*. *Zoological Studies*, 9(4), 373–382.

Xia, W., Ren, B., Zhou, H., Feng, H., He, X., Krzton, A., Hu, J., Aouititen, M., Luan, X., Li, D. (2019). Reproductive parameters of wild *Rhinopithecus bieti*. *Folia Primatologica*, 91, 202–218.

Xiao, W., Ding, W., Cui, L.W., Zhou, R.L., Zhao, Q.K. (2003). Habitat degradation of *Rhinopithecus bieti* in Yunnan, China. *International Journal of Primatology*, 24(2), 389–398.

Xiao, X., Haberle, S.G., Shen, J., Xue, B., Burrows, M., Wang, S. (2017). Postglacial fire history and interactions with vegetation and climate in southwestern Yunnan Province of China. *Clim. Past*, 13(6), 613–627.

Xing, Y. and Ree, R.H. (2017). Uplift-driven diversification in the Hengduan Mountains, a temperate biodiversity hotspot. *Proceedings of the National Academy of Sciences*, 114(17), E3444–E3451.

Yang, D. (1988). Black-and-white snub-nosed monkeys in China. *Oryx*, 22(1), 41–43.

Yu, L., Wang, G.-D., Ruan, J., Chen, Y.-B., Yang, C.-P., Cao, X., Wu, H., Liu, Y.-H., Du, Z.-L., Wang, X.-P. *et al.* (2016). Genomic analysis of snub-nosed monkeys (Rhinopithecus) identifies genes and processes related to high-altitude adaptation. *Nature Genetics*, 48(8), 947–952.

Zhang, B., Chen, X., Li, B.-L., Yao, Y.-H. (2002). Biodiversity and conservation in the Tibetan Plateau. *Journal of Geographical Sciences*, 12(2), 135–143.

Zhang, Y., Li, L., Wu, G., Zhou, Y., Qin, S., Wang, X. (2015). Analysis of landscape connectivity of the Yunnan snub-nosed monkeys (*Rhinopithecus bieti*) based on habitat patches. *Acta Ecologica Sinica*, 36(1), 51–58.

Zhang, Y., Clauzel, C., Li, J., Xue, Y., Zhang, Y., Wu, G., Giraudoux, P., Li, L., Li, D. (2019). Identifying refugia and corridors under climate change conditions for the Sichuan snub-nosed monkey (*Rhinopithecus roxellana*) in Hubei Province, China. *Ecology and Evolution*, 9(4), 1680–1690.

Zhao, Q., He, S., Wu, B., Nash, L.T. (1988). Excrement distribution and habitat use in *Rhinopithecus bieti* in Winter. *American Journal of Primatology*, 16, 275–284.

7

Cholera in Africa, from Fatalism to the Hope of Elimination: The Story of the Revolt Against a *Status Quo*

Didier BOMPANGUE[1,2]

[1] *Service d'Écologie et de Contrôle des Maladies Infectieuses, Université de Kinshasa, Democratic Republic of the Congo*

[2] *Chrono-environnement UFC-CNRS, Université Bourgogne Franche-Comté, Besançon, France*

GUIDING QUESTIONS.–

How can the drivers of the spread of an epidemic infectious disease be understood from the field?

In what way can discoveries depend on the sociological conditions of a personal pathway and the serendipities that result from it?

How can concepts of ecology be combined with those of epidemiology, for a global understanding of the conditions of emergence and adjustment of approaches to control cholera epidemics?

What is the nature of the obstacles encountered in implementing such approaches?

How is indiscipline a practical necessity for removing obstacles to understanding and controlling epidemics?

7.1. How can we engage in a collective approach to action research in health?

Sometimes it is very difficult to start writing a story. This is especially true when the story is a combination of a professional and personal adventure, the story of the fight against a disease and the story of an investment in a paradigm shift, both scientific and operational. Understanding the ecology of a pathogen's transmission and putting in place the conditions to control it requires immersion in a dual system of thought. First, it is a matter of understanding the socio-epistemic conditions that need to be met, in this case from a "classical" medical training, to test the challenges faced by a "doctor at the end of the trail", to think about the socioecosystem as such and then discover, by mobilizing other disciplines, the processes at work. The coordinator of this book encouraged me to take an unusual stylistic detour – and I thank the field directors and the publisher for accepting it – which is partly subjective, and which will facilitate the sharing of experience with young researchers and the assessment of the scientific, human and institutional challenges to be met. It is essential that everyone be able to measure how a medical questioning can evolve toward a socioecosystemic questioning, thus resolving the problems encountered, including in the sometimes extremely difficult working conditions encountered, for example, in Africa, when the "traditional" approaches have been failing for years. The reader will forgive me for the relative intimacy of this account and its digressions, which are only seeming. They provide material for a socio-ecological analysis at two levels: the epistemic one of the conditions of a study through a personal itinerary and the one of the functioning of a socioecosystem of transmission of a pathogen, in this case, cholera.

Brazzaville, Republic of Congo, early 1980s. I was still in elementary school. One day I was brought to a clinic for pain in the inner part of one of my ears. Just as I arrived, the doctor was leaving the consulting room: "I'm not welcoming any more, it's time to go", he told us without batting an eyelid. It was at that moment that I decided to go into medicine one day to take care of people who were not welcome in hospitals. When I became a doctor in the late 1990s, I was torn by the institutionalization of discrimination in the reception and care of patients. These practices, common in our hospitals, were to form the basis of my orientation toward so-called "humanitarian" medicine. A single sentence convinced me that this was what I had to do: "We care for those who are abandoned, we go to those who are forgotten, we go where no one wants to go". I had read it in a document of the NGO *Médecins sans frontières* (MSF). Since then, I have dreamed of nothing else: to be at the service of destitute populations, of these children, these women and men whose social appearance and precariousness exacerbate their exclusion from ever more unequal health systems.

Then came the war of 1997. From June 5 that year, we were scattered everywhere like the shrapnel thrown by Russian helicopter gunships piloted by Ukrainian soldiers (that is how they were known). Fleeing the civil war in Brazzaville, I found myself across the river in the Democratic Republic of the Congo, first in Kinshasa and then as a bush doctor in Inongo, in a rural health area nearly 600 km northwest of Kinshasa (Figure 7.1). In these rural areas, the doctor is the one who performs surgery. The nurse is the one who only prescribes. It was therefore necessary to move quickly into surgery. However, it was the nurses who taught me the most important surgical techniques that would make me famous in this bush hospital. We started from nowhere, in a hospital deserted by the population, with no state subsidy. No NGO was interested in it. It is a pity we were not in Kivu, where NGOs and all the international agencies rush to support the smallest dispensary, but, with effort from the heart, it is possible to change many things. Three years later, during one of his short visits to the town of Inongo, the late Minister of Health, Leonard Mashako Mamba, made a statement as a reward for the efforts of all the courageous and generous staff I had: "In all my visits to rural areas, I had never seen a hospital as clean as this one". However, at the end of 2000, I had to leave Inongo, harassed from all sides for not having accepted a deal to embezzle funds from the national immunization days against polio. It was a great experience to work in this health zone of Inongo. It was there that I first saw and treated cases of profuse diarrhea accompanied by vomiting, causing severe dehydration in a very short time. I remember examining the first cases without knowing that it was cholera. They had arrived through the lake port of Inongo, coming from the Mushie area, itself connected to the town of Bandundu, where there was a major cholera epidemic of which we had not even heard.

Cholera. I could not even remember if we had really spent any time on it in our infectious diseases or gastroenterology classes in medical school. What I did know for sure was that I had never seen a real case until the outbreaks that occurred in those very remote villages in the Basengele sector of the Inongo health zone between March and April 2000. The information had arrived in Inongo through the traders. Every *pirogue* crossing Lake Mai-Ndombe and bringing back agricultural products brought back the same stories, with several cases of unusually severe diarrhea having already claimed the lives of adults in the very remote villages on the other side. Fortunately, we had rehydration solutions left over from the research teams that had come a few months previously, long before that, to deal with a whooping cough epidemic.

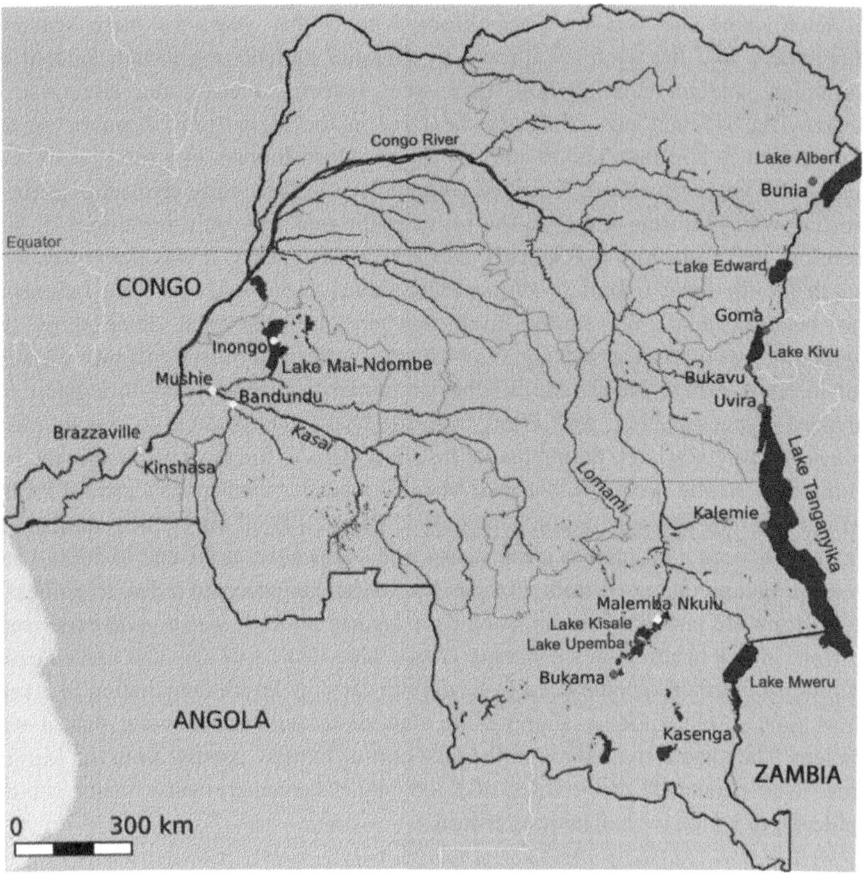

Figure 7.1. *Map of the Democratic Republic of the Congo, with the main places mentioned. In yellow, the first medical posts occupied by the author. In red, the main towns of the seven health zones targeted by the strategic plan for cholera elimination in the Democratic Republic of the Congo (see section 7.4), according to Muyembe* et al. *(2013). For a color version of this figure, see www.iste.co.uk/giraudoux/socioecosystems.zip*

There was no known history of cholera in this area located in the central basin of the Congo. I had never seen a case of cholera in the Inongo health zone; the health personnel were not familiar with this disease. In terms of infectious diseases, it was an area of measles, bloody diarrhea and monkeypox and, because of low vaccination coverage, cases of polio and whooping cough, also called pertussis, were also reported. It was precisely to investigate and organize the response to a whooping cough epidemic that a large delegation accompanied by the then Minister of Health,

Professor Leonard Mashako Mamba, visited Inongo between February and March 2000. UNICEF (the United Nations Children's Fund) and the NGO MSF Belgium had provided most of the logistics for the response. The NGO Épicentre, led by Dr. Sirenda Vong, was responsible for the epidemiological study to determine the evolution of the epidemic and to guide the interventions.

Before the team returned to Kinshasa, Dr. Sirenda Vong called me for a discussion at the lake. He wanted to put me in charge of coordinating the follow-up of pertussis cases in the health zone, in order to continue monitoring the evolution of the epidemic after the interventions that had been carried out there, including a large vaccination campaign. For about 30 minutes, Dr. Vong told me what field epidemiology was and how I would benefit from being involved in this field rather than spending my time in the operating room. He then asked me to manage a system for monitoring cases of whooping cough in the community, in the city of Inongo, and to report these cases every Tuesday to Kinshasa via a UHF radio. The level of functioning of the health structures in the villages of the zone was very low, less than 15% of the rate of use of health services in the health areas still operational. Dr. Vong explained to me that we could not rely on the health system reporting to monitor the epidemic. We needed to set up a monitoring system within the community. This monitoring system relied on local block and avenue officers to keep track of the number of children identified in their community with this particular rooster-like cough. It was easy to recognize by the people, especially since it had a local name, *mbo lokesu*, which, translated into English, means "the big cough". The big cough, unlike the other more traditional coughs. So it was through community surveillance that I began to study the diseases, through this monitoring and reporting to the Épicentre project, based in Kinshasa within the Directorate for Disease Control (*Direction nationale de lutte contre la maladie*, DLM). Every Tuesday, I began to exchange with Épicentre and DLM, as well as the department of the Ministry of Health in charge of monitoring epidemic trends, diseases with epidemic potential and the organization of the fight against some of them, like cholera. At the end of the pertussis epidemic, around April, I received a letter of invitation from Épicentre to participate in a training session organized in Kinshasa on epidemiological disease surveillance and response to epidemics. At the end of the training, Dr. Vong told me about the possibility of extending the epidemiological surveillance project monitoring site, which had been in place in Kinshasa for 2 years, to Bandundu. This is how I found myself in Kinshasa at the end of December 2000, within the DLM, in a disease surveillance project financed by the European Union and implemented by the NGO Épicentre, the operational research branch of the NGO MSF.

Each doctor involved in the Épicentre project at DLM was responsible for one or more diseases. One had to be found for “the villager” (as one of the team doctors called me when I arrived at the office), who did not even have a public health degree. The country had only one school of public health at the time, at the Faculty of Medicine at the University of Kinshasa. The process for admission was long, procedural, not always transparent and not always equal and, above all, the school was very expensive and offered few scholarships. Dr. Vong had therefore broken a code by including me in this project. I was not welcome, especially according to the head of the epidemiology division of DLM, so I had to be assigned to a disease that nobody was interested in. After some thought, the decision was quickly made: “the villager” would work on cholera.

This is how I found myself tracking down the last cases of cholera in the city of Kinshasa, after the very large epidemics of 1996–1999, the last cases of which were reported in Kinshasa and in the western provinces of the Democratic Republic of the Congo (including Bandundu) until 2002. I had a vehicle at my disposal to drive around the sentinel sites in Kinshasa, investigating the smallest cases of diarrhea that would make me suspect the persistence of cholera cases in the city. I found this exciting. It reminded me of another one of the jobs I had also dreamed of doing: detective.

From December 2000 to December 2001, under the leadership of Dr. Vong, I arrived at the office first and often came back last. I did not get tired of investigating cases of cholera or other diarrheal diseases in the city. At the end of the month, I was not paid. Unfortunately, I had arrived at the wrong time. The assassination of Mzee Laurent-Desiré Kabila on January 16, 2001, had brought several international collaboration projects to a halt. The project for which I was recruited, the extension of the Kinshasa project to Bandundu, was put on hold. We had the choice between waiting at home, doing something else or staying and continuing to learn by working, without being paid until the international cooperation resumed. On the advice of Dr. Vital Mondonge, one of the Congolese doctors involved in the Épicentre Kinshasa project, I opted for the third option, to continue learning, working under the supervision of the sentinel sites in Kinshasa, investigating outbreaks in the city and managing databases. My female colleagues, Hélène Chenge, Liliane Diatezulwa and, most importantly, Annie Mutombo, helped me to integrate into the team. They were in charge of database management. They introduced me to the computer.

In the meantime, Dr. Vong had completed his mission with Épicentre in Congo. He was replaced by Dr. Xavier de Radiguès, a Belgian living in France and born in Congo’s Katanga province. Xavier continued in the same vein as Dr. Vong, with the added emphasis of training the Congolese involved in the Épicentre project. He was demanding on the quality of reports, which did not give him the opportunity to make many friends. It was with Xavier that I began to be paid by the Épicentre project.

In 2002, when the project in the Democratic Republic of the Congo came to an end, I had the choice between becoming head of the epidemiological surveillance office at DLM and joining an international NGO. I decided to join an international NGO: MSF.

What I needed was an environment where I could continue the work I had started with Épicentre. I had developed a taste for tracking epidemics. Identifying and managing them had become a second passion, after having mourned the death of the operating theater. MSF, and especially its Congo emergency pool project, specifically dedicated to early detection and response to epidemics, was the perfect place to continue the adventure I had started in February 2000, following the meeting with Dr. Vong during the great whooping cough epidemic in Inongo.

7.2. The first years of the fight against cholera and the disillusionment of humanitarianism

At the end of 2002, I found myself in the Congo emergency pool of MSF Belgium after a brief but very instructive stint with MSF France in a project to respond to epidemics and malnutrition problems in the north of the then Grand Katanga, now Haut-Lomami province. In 2002 and 2003, the country was still divided into several geographical blocks, occupied on one side by the central government in Kinshasa and on the other side by the rebel groups that had re-formed after the assassination of Laurent-Désiré Kabila. We were working in the government part, between Malemba-Nkulu and Kamina Lenge. The aim was to provide overall support, mainly nutritional, to these health zones in the form of logistical and technical support to the general hospital in Malemba-Nkulu, in the Kamina Lenge zone. Indeed, there was a high proportion of malnutrition in this whole region located not far from the front line between the government forces and the rebels of the RCD Goma (Congolese Rally for Democracy). For several weeks, the population had not been able to harvest what they had grown, forced each time to flee to the other side of the combat zones. When they returned to their villages, priority was given to rebuilding their houses, which were often burnt down. They lived mainly on the help of UN agencies and NGOs. It was during this period that we first heard the name *Mayi-Mayi*, armed indigenous people, fighting first against the rebel forces and then later against the government forces, whom they sometimes accused of abuses against the local population. Haut-Lomami was at the time a very isolated region and remains so today. It is from here that the Congo River, which rises much further south, began to form its first lakes. There were several of them, two of them very important, Lake Kisale and Lake Upemba. How could there be such high rates of malnutrition on the shores of such fishy lakes? Insecurity was the main cause. The many islets that emerged in the middle of the lakes, once seasonal fishing grounds, had become camps for local and foreign armed groups rather than the traditional fishing grounds they had always been. The years of war had aggravated the poverty of the population and there was no indicator to predict the end of this situation following this infernal cycle: war-displacement of populations-malnutrition-epidemic-poverty-recurrence of old endemics. Indeed, this area was, and still is, one of the most affected by cases of multidrug-resistant tuberculosis. Today, 18 years later, the situation has not changed much; part of this region is in what is called, in the jargon of humanitarian crisis managers, "the triangle of death". Part of the Butumba area, in the Haut-Lomami, is still today the stronghold of local warlords. The latter shares the record of social devastation with herds of elephants, which also spreads terror in the villages of the Butumba area, all around the Upemba Park Nature Reserve. The herds are also one of the causes of the recurrent malnutrition in this region, owing to the ravages on the meager crops of a population already exhausted by years of bearing the burden of a war that has only stopped in the media. It was enough to set up the slightest surveillance system in this region to realize the abundance of known and unknown pathologies circulating there, but the most emblematic of all these diseases is cholera, with its recurrent epidemics around the lakeside villages (Figure 7.2).

At the time, a major cholera epidemic was raging throughout the region, mainly between the localities of Malemba-Nkulu and Kinkondja. MSF France, which was working in the region, had requested the expertise of Épicentre to conduct an epidemiological study to guide field interventions. No one was available to come from Paris. The coordinator in Kinshasa was on medical leave in the capital, so a solution had to be found with the team in Kinshasa. Against all odds, it was the last person to arrive, the "villager" condemned to work on cholera, who had been designated to carry out this mission. This was an affront to the elders, and another reason for my rejection by the others. This mission would reinforce my investment in cholera.

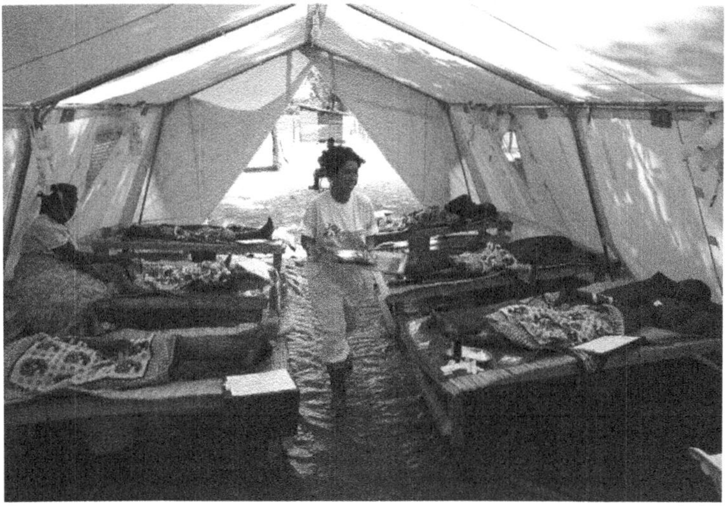

Figure 7.2. *Temporary cholera treatment center in rural areas (photo D. Bompangue)*

I arrived in Malemba-Nkulu in July 2001 as part of a very large MSF team of several nationalities, people from all walks of life, and we all had one goal: to stop the cholera epidemic. We had a small treatment facility, not quite up to standard given the sudden influx of cases. It took the outcry of one of MSF's Parisian headquarters managers, visiting the field, to launch the major construction of what would soon become one of the best cholera treatment centers in the region. The center was packed. Day and night, men, women and children arrived in a state of advanced dehydration, vomiting in spurts and emitting whitish liquid stools. In the context of my mission, apart from the classic description of the epidemic, it seemed essential to solve two questions in order to feel useful: how to stop the epidemic quickly? How to limit the number of deaths? The most urgent thing was to save lives. So our priority was to understand why people were dying and how we could improve the situation. Reviewing patient registers, consultation forms and conversations with caregivers were the means used to research the profiles of people who were dying in the care center. This initial rapid investigation revealed two important pieces of information. The first was that a large number of people died at night. The second was that many of the people who were dying very quickly or arriving almost dead were all from a few distinct locations in the records. As in a police investigation, this was a lead that could help us reduce the number of deaths in the treatment center considerably. To understand the causes of the nightly deaths, I decided to make surprise visits to the treatment center at night.

According to the organization of the service, the expatriate and local supervisor teams did not work at night. There were on-call teams and a local on-call manager. You had to be on night duty, with a clear question in your mind, to understand how difficult it is to find a vein to stick in the black skin of a dehydrated, agitated subject. From there, we could understand why people left at the last evening visit (often between 6 and 7 pm), at a moderate level of dehydration, presented in the morning a severe level, and for those who were at a severe level, it was not surprising to find empty beds in the morning. In addition to the difficulty in finding infusion lines, we were repeatedly surprised to find that the flow rates of infusion fluids were reduced during nighttime supervision. Upon investigation, we realized that night staff, in order to avoid having to get up or wake up during the night to change infusions, were reducing the flow rate in the hope of making them last as long as possible. However, this practice unbalanced the rehydration balance of the patients who, in the end, lost more liquid through vomiting, and especially through diarrhea, than they gained through rehydration. These two explanations, even if they were certainly not the only ones, were already sufficient to allow us to understand the causes of nocturnal deaths of patients. All that remained was to review the situation with the MSF coordination teams and put in place corrective measures. This was done immediately. The lighting in the treatment center was improved, the night shifts were refocused and strengthened, and night supervision was made more systematic. We even had to spend entire shifts with night teams. This experience would follow me throughout my years of work on cholera and would prove very useful, almost 10 years later, across the Atlantic, in one of the metastases of Africa in the Americas, Haiti. We will come back to that.

The enigma of very rapid deaths on arrival and late arrivals of patients in a state of incipient death remained to be resolved. All these people had one thing in common: they came from a group, from a few localities or neighborhoods well identified by the local staff who helped us, through the answers to our questions, to understand the local context. We wanted to know what was specific about the places where the patients who arrived late with very advanced dehydration were being cared for. We wanted to go there to understand what was going on. The answer was clear:

> You can't go there.
>
> – Why?
>
> – It's far away, it's not here.
>
> – This is the city of Malemba-Nkulu. Where is it, then?

– It's on the lake.

– Really? On the lake? On the lake how?

– Yes, on the lake, there are camps there where people are fishing.

But there you are! Maybe that was just the problem. I hurried to find out why we could not go there. The answer from the MSF colleagues was: "It's dangerous there. We are not allowed to go outside the city. We run the treatment center here in the city. We are not allowed to go out of the city, unless we get permission from Paris if we really have to, and then only then". So a visit to the lake was out of the question. It was dangerous – the storms in the lake, the *Mayi-Mayi*, the insurance problems, etc. – and then the cholera is managed by the government. Then cholera is managed in a cholera treatment center. The management protocol, which was included in the MSF/WHO "Cholera Treatment Guide", did not include activities outside the treatment center. We could not do anything that the protocol did not say. In addition, we were not allowed to go into the restricted areas. Things could have ended there, but no, I had not come this far to be locked into instructions I did not understand. I was not interested in the protocols, the instructions from headquarters and the restrictions. I had a mission, with clear objectives, and I wanted to see it through.

Every evening in the mission, we debriefed the day. The base manager (where the mission team resided – mostly expatriates and locals from Kinshasa) would give the floor to all the teams – medics, logisticians, hygiene and sanitation officers, and administrators. I waited for my turn to present my finding and ask the field mission to visit the lake to explore the camps where the dehydrated cases were coming from. When I was given the floor, I took my time to explain how I had done it and what I had found. I concluded by drawing a parallel with treating an abscess on a leg. As long as one did not treat the abscess deeply, being sure to cure possible secondary abscess pockets, one could be sure that the abscess would last for a long time. So I explained to them that if we wanted to stop this epidemic quickly and not keep the treatment teams under stress by receiving cases in an advanced state of dehydration, we had to understand and manage what was happening in these camps in the middle of the lake. It seemed obvious, since the people had been there for months; everyone saw them coming, and I was surprised that they did not know they were fishermen or traders from the fishing areas, but I could not understand why it had never been obvious that we had to integrate the management of these areas of origin of severe cases in the control strategies. It was not written into the protocol and it was Paris that decided. Fortunately, there were a few decision-makers from Paris in the mission, normally based at the organization's headquarters, but present in Malemba-Nkulu to bring their expertise to the response to this major epidemic. It was these executives that I could rely on. The greatest support would come from Francisco

Diaz. Francisco was a logistician, well versed in water and environmental sanitation issues. He was an outstanding worker, exemplary, with a calmness that I still remember today, attentive, and he exuded real experience. It was Francisco who surprised me by taking the decision, against all odds, against the advice of the majority of the group and against Paris, to authorize an investigation of the camps on the lake. To make sure that everything would go well and knowing that he was making a difficult decision, Francisco had decided to be part of the mission himself, which left the next morning. How could we have missed this all this time? Everything was happening in the camps (Figure 7.3). This was the bottom of the abscess; without "cleaning up" the camps, there was no hope of ending the epidemic.

Figure 7.3. *Fishing camp on an island in Lake Upemba, upper Congo River basin. Fishermen and their families typically spend several weeks each year in camps like this, where the lake is the only source of water. Because there is no firewood in these areas, campaigns to promote boiling water are futile (photo D. Bompangue)*

Fishing activities took place between the beginning of May and the end of October, between the end of the rainy season and the beginning of the initially intense phase of the following rainy season. This was the great fishing season. Fishing was the main source of income for the people living along the lake. This activity attracted many traders coming from the big cities of Katanga (Likasi, Kolwezi and Lubumbashi) and Kasaï (Mbujimayi, Ilebo, city connected to Kinshasa by the river). During this period, almost the entire population of the villages and localities located on the lakes' shores moved to the camps floating on the lake. There is no drinking water; they drink the water from the lake. The same water is used just a short distance from the latrines (Figure 7.4).

Figure 7.4. *Water collection point in the Bukama area (photo D. Bompangue)*

In spite of these very precarious living conditions, with the regular occurrence of diarrhea cases, it was out of the question to miss the fishing season. How could we convince these populations that the camps are a place of transmission of cholera cases?

> "What cholera are you talking about? These camps have always been here every dry season. For years, our grandparents have always fished here, these camps belong to clans, families, it's all our life, it's a big part of our history", that's what the fishermen we met in these camps say.

Even the death of a few family members, brought late by dugout to the treatment center, did not change anything. In any case, they had an explanation for these deaths:

> It is not cholera. These people are dying because the water genies, who own the wealth of the waters and therefore also the fish, cannot give us a good season without sacrifice. So those who die, it's sad, but that's how it is, the water genies need sacrifices to grant us a generous fishing season.

It is this cocktail of beliefs that needed to be addressed through the implementation of a specific project targeting the lake camps. This project, implemented by MSF France, the NGO Action Against Hunger (United States section) and the Malemba-Nkulu health zone, combined early detection of diarrhea through the setting up of surveillance posts in the camps. These posts were also used for the early management of potential cases of diarrhea. A mobile emergency response and sensitization team in a motorized dugout canoe traveled to the various camps scattered around the lake. This mobile team was also in charge of rowing away the affected people very quickly and of starting first aid for complicated cases of severe dehydration identified in the camps. Finally, the last activity consisted of the distribution of utensils for the storage and treatment of large quantities of water, soap and water disinfection products. Fishing nets were also distributed – the idea was to increase the daily fish catching capacity, in order to reduce the time spent by the fishermen in the camps. This whole mission was a great revolution, it went very fast: getting MSF out of the treatment center, getting them to accept a mission in an area not authorized by Paris and then starting a project that was not specifically focused on curative care for patients.

Other revolutions were to take place, thanks to Francisco's presence on the ground. This revolution would take place following the identification of one of the explanations for the persistence of the cases coming from certain districts of the city of Malemba-Nkulu. There was an old pipe for the drinking water distribution network in the city. The areas from which the cases originated in Malemba-Nkulu were those where the distribution was very irregular, the leaky pipe was the cause of low water pressure, loss during distribution and certainly of contamination, especially at the time of the resumption of rains. The chronic lack of fuel made the operation of the generator that allowed the distribution of water in the city irregular. The periods of interruption in water distribution coincided with an increase in the number of cases of diarrhea admitted to the cholera treatment center. We were faced with two problems to solve in the city if we were to stop the succession of cholera cases. As with the lake, I prepared to report back to the team during the evening debriefings. And, as with the decision to explore the lake camps, Francisco agreed with me and we would have to start implementing another very unusual project for MSF. Indeed, repairing the pipes of a water network in a city, donating barrels of fuel to a water distribution plant, for anyone who knows MSF, even today, almost 20 years later, it remains surreal. Yet we did it. Yes, we did it because of the strength of conviction, and we did it because there was a person at the decision-making level, who was at the level where decisions are made, the level where problems occur. The water supply had been improved in Malemba-Nkulu thanks to a specific project targeting the lake camps, which had been put in place to improve the water supply and with a view to being renewed (through the NGO Action Against Hunger, United States section, which was working in the area on a long-term basis), and the conditions for night work at the cholera treatment center had been improved. The result was the end of the cholera epidemic in Malemba-Nkulu, while cases continued to be reported in the south, in the Bukama area. As evidence that this was exactly what needed to be done, this area of Malemba-Nkulu remained free of cholera cases for 5 years after these interventions – in fact, for as long as it was possible to maintain prevention activities in the lake camps.

It was a memorable experience. Looking back, I think it was the founding act of my investment in cholera. I understood that it was possible to find solutions to this disease, and above all I learned to trust myself. I had understood that it was not necessary to feel blocked by the protocols, not to let yourself be impressed by those who were there long before you, because people can evolve in an environment or in a system for years and not see anything, not understand anything, because they are too locked into the protocols, in the rigidity of the instructions received from headquarters. This type of intervention in the context of cholera epidemics will be repeated on other occasions, often in the same periods, with the same lessons each time: the role of the lakes, the role of the people in the camps and the impact of improving the quality of drinking water, even with the simple distribution of water purifiers.

With all these field observations, how can we be convinced of the unpredictable nature of cholera? This is, indeed, what the ancients said. How can we accept the idea that it is impossible to maintain sustainable actions against cholera, knowing that the disease is a consequence of poverty in Africa? I still wonder if it is not from this *preconception* that the choice was made to have protocols centered on the setting up of care centers for the sick, without asking the slightest question about what was happening outside these centers.

From my experience in Malemba-Nkulu with Francisco, I was convinced that one should always try to propose solutions when one is sure of the rationality and the correctness of one's ideas. So in 2003, after observing the same phenomenon several times, I decided to suggest to MSF Belgium a project to anticipate the period of cholera epidemics in the region of lakes Upemba and Kisale, in the Upper Lomami. This project was strongly inspired by the project implemented in

Malemba-Nkulu in 2001. To be accepted and implemented in the field as an MSF project, it had to pass two stages. After passing the assessment stage by the MSF Democratic Republic of the Congo office, the project had to be sent to Brussels to be reviewed by headquarters. For me, this was just a formality. It all seemed so obvious and I did not imagine that rational, intelligent people could invalidate such an obvious fact. However, Brussels never followed up on our project. What is more, there was not even a discussion about it. Because of a long silence, I finally understood that the project had been rejected. Silence was the answer. I was very shocked. I wanted to understand why the project had not been validated. We were spending a lot of money every year in our recurrent interventions in the heart of these regions. I had learned phrases at MSF such as, "Money should not be an obstacle to our actions". I was convinced that we were there to relieve those who were suffering, and I was convinced that we could more sustainably improve the well-being of these populations in the lake areas by intervening as we had done with Francisco, rather than by repeating short interventions focused on the management of cases confined to treatment centers, as we had been doing with the MSF Belgium emergency pool. Why was not my project accepted by Brussels? Was it not well enough written? Was it not relevant enough? I tried to understand the reason. At one point, I even wondered if being black was the reason for this lack of consideration. Then I quickly pulled myself together, refusing to find an easy excuse like racism. I began to think, "You weren't convincing enough, the idea is certainly good, but you probably lack the credibility to make such a proposal and have it accepted by an institution like MSF." At that moment, I remembered that, despite all the trust I had gained from my Épicentre coordinators, despite a privileged position and the trust of my head of mission at MSF, I had neither a degree in public health nor a degree in epidemiology. I felt that this lack of an appropriate degree, in relation to what I wanted to do, that is, provide guidance on how to better manage epidemics, certainly limited the credibility of my proposals. I then began to perceive the urgency to invest myself in an academic project aimed at improving my knowledge on the problems of epidemics, in general, and cholera, in particular. I was convinced that this was the best approach to become scientifically credible when making proposals for control projects. This is why I decided, at the beginning of 2004, to prepare myself for enrollment in an English-language university in order to take a degree in public health or epidemiology, ideally with an environmental or sociological component. I wanted to study epidemics but in the context of the global environment in which they occur. The Malemba-Nkulu experience had definitely left its mark.

At this point, my decision to study was made, but English universities were financially out of my reach. Sirenda and Xavier had done their masters in epidemiology in London, at the London School of Hygiene and Tropical Medicine.

If I had been able to pay for the course myself, I would have done it. Unfortunately, I could not afford it and there were no scholarships available to me. So I decided to apply to the School of Public Health of the Free University of Brussels. Deep down, I was not convinced that this school would answer my questions. Not because the skills were lacking; no, it is a very respectable institution. I learned a lot and I do not regret my time there, but the public health studies were designed to manage the existing health system, not to provide answers to my questions about epidemics, especially cholera.

Then I remembered an encounter I had during my years of work at Épicentre, supporting the disease control department of the Ministry of Health of the Democratic Republic of the Congo. In February 2002, during an accompaniment mission to Bandundu city, I had the opportunity to talk for several hours with Professor Renaud Piarroux. We talked about our passions and our objectives. It was there that Renaud told me for the first time about Besançon and the master's degree in Environment, Health and Society (*Environnement, Santé, Société*, ESS) that was offered there. Renaud told me that the ESS master's degree in Besançon could help me to find answers to the questions I was asking myself. However, it was a second-year master's degree and, in order to enter, you had to have completed a first-year master's degree. So Brussels was just a stepping stone to Besançon. This scenario was the alternative to my first choice.

7.3. From unpredictable to predictable cholera

I spent the academic year 2004–2005 at the School of Public Health of the Free University of Brussels, but some of my ideas were already in Besançon. In February 2005, I took advantage of a few days off in Brussels to go to Besançon and attend the ESS master's conference. These were two days of presentation of the research themes of the Master's students, the aim of which was to receive the opinions of the supervisors and the teaching staff on the methodological aspects. It was an opportunity to become aware of the level of requirement expected, and also of the variety of research topics and methods, and to understand the need to anticipate data collection for the work that awaited me the following year. Collecting data on cholera epidemics was the basis of the work. I already had collected some information, but I needed more structured information covering a longer period (Figure 7.5). I had not anticipated this and had to rely on the two organizations that could provide the data, MSF and DLM. I had written a note to the head of mission in the Democratic Republic of the Congo requesting the use of this MSF data and permission was granted. I was also able to access the DLM databases. Indeed, I had initiated a new relationship with DLM at the time I decided to embark on this study project. For the work that was to be done later, I needed to stay in touch with the DLM data managers. Two people played a major role in achieving this goal, Hélène Chenge at the beginning and Annie Mutombo later. Thanks to Annie and Hélène, I was able to continue to update the databases with information on the context of the epidemics (population movements, trade activities, etc.) found in the investigation reports they sent me, as well as information on the organization of the epidemics, the types of interventions that were put in place and the profile of the actors involved in the field.

a) b)

Figure 7.5. *Sources of epidemiological data in paper and electronic formats (photos D. Bompangue)*

COMMENT ON FIGURE 7.5.– *a) Search for old data in paper format, compiled and stored in the offices of the provincial health inspectorates in Bukavu, South Kivu. Despite the conflicts in the region, this data had been kept in the cellars of the provincial health inspectorates and in the health zones. b) Extraction of data in electronic format from the hard drives of disused computers.*

The most difficult data to find were the rainfall data. They were generated in Kinshasa by the Binza meteorological station. It was not easy to get them, but we needed them. We had become aware of their importance during work done in 2001 with Sirenda in monitoring persistent cholera cases in Kinshasa. Sirenda sent me to collect rainfall data from the Binza weather station. I would also go to the *Régie des voies fluviales* (RVF) in Limete to retrieve data from the Congo River water level monitoring. RVF monitors the rising water of the river and is able to anticipate the identification of areas likely to be affected by flooding in Kinshasa. I was piling up information, data of many types and from many sources. I knew that all of this would be useful in understanding the cholera epidemics, but I had no idea how complex the work would be to make it all work out the way I wanted it to. How should the data be presented? Had I anticipated everything? Did I have all the data? Without answers to these questions, I just hoped I had anticipated enough. This was not a typical way of working. In articles on cholera, it was often a question of describing the epidemic, studying reported cases, analyzing case-controls or even looking for risk factors. In this case, we envisaged studying the functioning of a system and all its complexities, in its spatial and temporal dynamics.

Professor Piarroux and I started with what was available to us. In the year of the Master's degree, we essentially worked on the clinical data of the patients, that is, the series of numbers of cases and deaths of people admitted to different cholera treatment centers in the areas and provinces that reported cholera cases in the Democratic Republic of the Congo during the last 4 years.

This is where I was going to face one of the first big challenges, which is organizing and cleaning up a large database. Getting a database ready to use is a time consuming task. We were not used to dealing with long series of several years. Several chunks of data were missing and we had to go back and forth with Hélène first, then with Annie, to improve the completeness of the databases. Much of the data were aberrant; you had to know the local context, at least have followed the progress of some of these epidemics and have worked in database management to get out of the labyrinth. I called it data archeology. We would clean, we would check, we would recheck, until the graphs that appeared on our dynamic cross tabs would no longer allow us to detect unexplained outliers or absences, guided by what we knew of the reality of a field where we had been tracking everything closely for the last 4 years. As we saw and deciphered the first curves from the cleaned data files, we began to see the first information emerge and, above all, the first hypotheses that we would have to prepare to verify.

Some areas were affected almost all the time: they were all lacustrine, mainly around Lake Kivu and Lake Tanganyika. Some areas had short periods of interruption of no more than 2 months: these were also lacustrine, located around the Upper Lomami lakes, some around Lakes Albert, Edward, Kivu and Tanganyika. Some areas had interruptions of more than 2 months or even more than a year. These were mostly located far from the lakes, such as the urban areas of Katanga and Kasai. Some areas had hardly been affected at all. These were located far from the lakes and without much connection to the lake areas. In areas where epidemics persisted throughout the year, the periods of increase and decrease in cases seemed to be the same each year.

When we took the time to follow the evolution of an "epidemic wave" week after week, we realized that epidemics often started from lake areas and then seemed to spread to new, more remote areas that were previously free of cases.

Two major results appeared to me: the almost certain role of the lake areas in the persistence of cholera cases and the existence of a certain periodicity of the cholera cycle in these lake areas. All these observations, impressions and intuitions were not sufficient to give rise to a full scientific knowledge. A further step was to cross-reference and exploit all the data using concepts, methods and tools derived from eco-epidemiology. Traditional approaches to the analysis of epidemiological patient data were no longer sufficient. It was to learn these concepts, methods and tools that I joined the ESS Master's degree in Besançon. However, for a doctor with no knowledge of ecology, with only about 60 hours of biostatistics under his belt, it was difficult to be sufficiently autonomous in all the analyses necessary to explore the hypotheses that had emerged from the exploration of time series of cholera cases. We had to turn to the ecological sciences, which we lacked – this is where the support of Professor Patrick Giraudoux would prove crucial. Here, I include in ecology everything that is geography, sociology, anthropology and biostatistics. I first met Patrick in the context of the ESS Master's degree. A few years earlier, in 2000, Patrick and a few other colleagues, including Professor Jean-François Viel, had created the ESS Master's program with the aim of tackling societal issues from a global perspective. We were not yet talking about socioecosystems! In a world that wants to be hyperspecialized, this degree was quite atypical. However, it is this "atypicality" that we need today to face the increasingly atypical and complex situations that never cease to challenge us.

I was learning as the work progressed. At each stage, I had to integrate more analytical tools, concepts of ecology, geography and biostatistics. Ideally, one would have known the necessary ecological concepts and tools early enough to avoid being confronted with all this so late and abruptly. Something needs to change in the health sciences curriculum if we want to better prepare health actors for the challenges of today and, above all, of tomorrow. This idea is the basis of the master's project in infectious disease ecology that we will create later, in 2014, at the Faculty of Medicine of the University of Kinshasa. We were convinced that health science curricula needed to be adjusted to include notions related to the understanding of how ecosystems work. The teaching of epidemiology cannot continue to focus solely on humans. The environment in which the individual evolves should become the main focus of investigation in the new way of thinking about health issues. For the moment, far from being autonomous in all these analytical aspects, I could count on Patrick's unfailing investment to explore all our hypotheses. Two other people also helped me a lot during my first years of the master's degree and thesis: Dr. Martine Piarroux, a physician and geographer, and Professor Pascal Handschumacher, a health geographer. Both introduced me to geographic information systems.

Thus, during the 3 years of the thesis, from September 2006 to October 2009, with the help of colleagues, under the direction of Professor Renaud Piarroux and the support of Professor Patrick Giraudoux in France, and that of Professor Jean-Jacques Muyembe in the Democratic Republic of the Congo, we gradually arrived at results that have considerably changed the understanding of cholera epidemics and raised hopes to the point where we can see the possibility of eliminating cholera as a public health problem in the countries where this disease is still endemic. This involved reconstructing cholera case data from the DLM databases, which contained information from nearly 500 health zones in the country's 11 provinces. The information from these databases was cross-checked with the databases obtained from MSF during the field missions. Having been able to personally follow several cholera epidemic sequences during this period and knowing the contexts of occurrence and propagation facilitated the cleaning of the databases by identifying, for example, outliers, differentiating the gaps corresponding to "zero reported cases" from those corresponding to "missing data".

These results have been published in international peer-reviewed scientific journals, thus bringing scientific credibility which, until then, had not been granted to us.

7.3.1. *A metapopulation mode of operation linked to the Great Lakes*

An initial study of weekly cholera cases in Katanga and Kasai Oriental, compiling a database of all cases recorded from 2000 to 2005, showed that the lakes were the source of cholera outbreaks long before the epidemics reached the major cities, where they were late in alerting the authorities and public opinion (Bompangue *et al.* 2008). The entire region of the western shore of the Great Lakes showed a continuity in the reporting of cholera cases. However, this continuity consisted of a set of discontinuous and asynchronous time series at the level of the different health zones composing this region. The Kalemie zone was the only one to present a continuous time series, but this apparent continuity was no longer apparent with great clarity at the level of the health areas composing the Kalemie zone, which themselves presented discontinuous and asynchronous reporting (Figure 7.6).

We could therefore conclude that the notion of endemicity was a function of the geographical scale to which it was related. The next step of investigation was to explore the details of the temporal component of these epidemics.

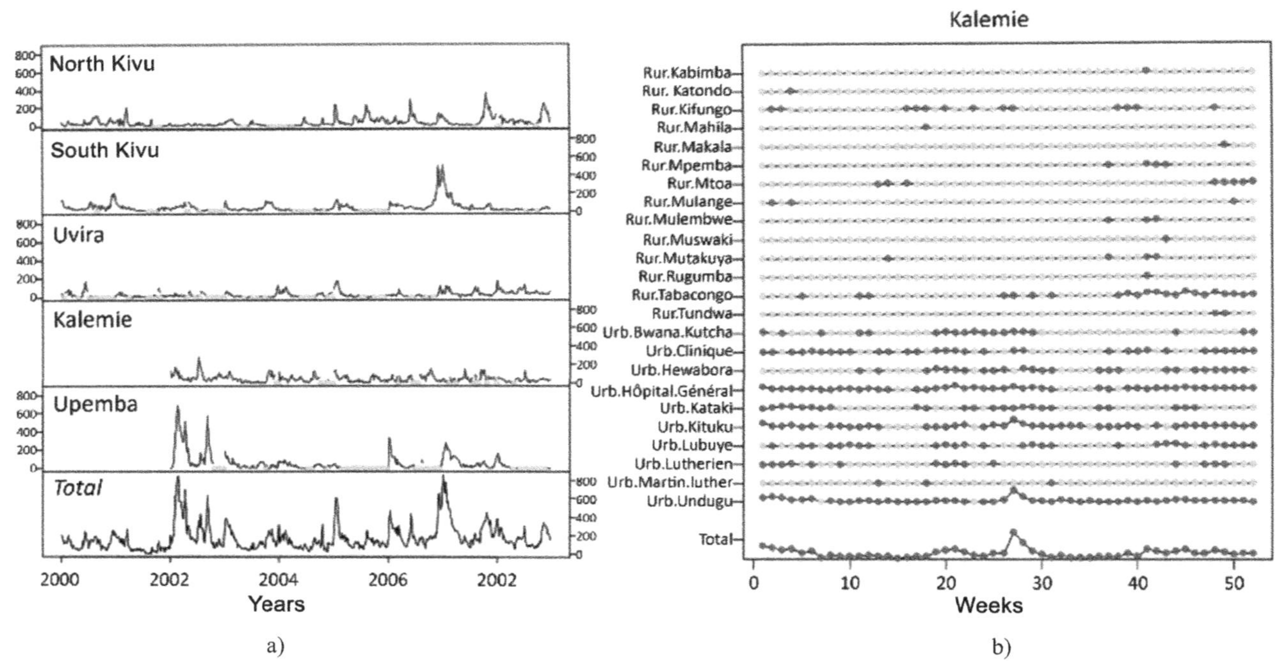

Figure 7.6. *Weekly reporting of cholera cases in the health zones of the Great Lakes region (a) and in the health zones of the city of Kalemie (b). The periods in green are periods without reported cholera cases. We can see that the overall stability of the disease is because of asynchronous local emergence-extinction successions at both spatial scales, characteristic of a metapopulation mode of operation. For a color version of this figure, see www.iste.co.uk/giraudoux/socioecosystems.zip*

7.3.2. *Seasonality of cholera epidemics in lake areas*

The search for periodicity in a time series consists of decomposing it into a periodic curve (seasonality), independent of a general trend assumed to be non-periodic, and residuals (non-seasonal variations not linked to the trend). Most of this analysis was based on the Cleveland *et al.* (1990) algorithm using local regressions (loess). The potential temporal autocorrelation, often inherent to time series, was also taken into account in the statistical analyses. Figure 7.7 shows an example for South Kivu.

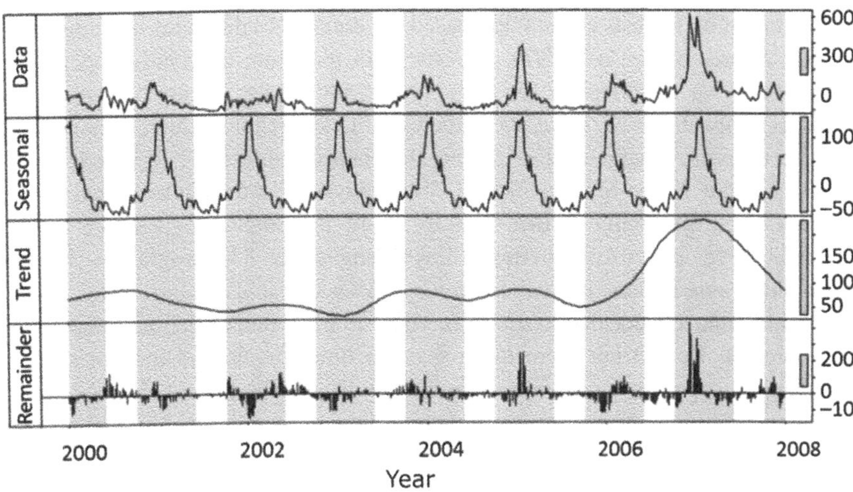

Figure 7.7. *Time series of weekly cholera cases from 2000 to 2007 in South Kivu. For a color version of this figure, see www.iste.co.uk/giraudoux/socioecosystems.zip*

COMMENT ON FIGURE 7.7.– *Yellow: dry season; blue: rainy season. The time series (data, number of weekly cases) are broken down into components – seasonal, trend and remainder. The sum of these three components is equal to the weekly number of cases (data). To facilitate comparison of effect sizes, the gray bars on the right of each graph correspond to the same absolute number of cases. From Bompangue* et al. *(2009).*

The resolution of this first decomposition was then refined, focusing on health zones and looking for a possible link between cholera outbreaks and excess rainfall. The analysis was performed using the residuals from the decomposition of the time series of cholera cases in the five lake areas. The rainfall time series were decomposed in the same way to extract the residuals. A correlation between the residuals of the two time series, thus made independent of seasonal (dry season-rainy season alternation) and interannual variations, could then be sought. It was then shown that a significant part of cholera outbreaks, beyond seasonal rainfall variations, were correlated to abnormally rainy episodes, notably by following the abnormally warm episodes of the El Niño Southern Oscillation (Bompangue *et al.* 2011). The detailed mechanisms by which variations in cholera incidence were linked to seasons and rainfall were still unclear, however, although it could be assumed that leaching from bacteria-impacted soils and latrines could facilitate the subsequent contamination of drinking water.

A common hypothesis at the time, based on studies of epidemics in Bangladesh, was that the association of *Vibrio cholerae* with plankton, especially copepods in algal blooms, was responsible for the sporadic and erratic occurrence of cholera epidemics (Colwell 1996). What about the Great Lakes region, which does not have the same estuarine characteristics as the Bangladesh outbreaks? A search on the Internet for the keywords "plankton", "Tanganyika" and "cholera" led to the discovery of a program, CLIMFISH, to study the impact of climate on the sustainable development of fisheries in Lake Tanganyika. The program was led by hydrobiology researchers from the Tervuren Museum in Belgium. After several unanswered emails, I decided to go to Tervuren to meet one of the main instigators of this project. He did not seem to have much time for me and did not seem to understand why a doctor was coming to him to discuss the project. I asked him to take a moment to look at my graphs of cholera cases in the Great Lakes region. I had seen their time series of algal blooms and fish catches, and was struck by the similarity to our time series of cholera cases. I was pretty sure that our time series would come as a shock to him. That is exactly what happened when I opened my computer. He was stunned for a few seconds; he could not believe it: the person who said he did not have time for me then offered me a coffee to talk. This is how my collaboration with the Royal Museum of Tervuren was born, which would lead to the submission of a joint project, called CHOLTIC, to study a possible link between cholera around Lake Tanganyika and algal blooms, based on a collaboration between the Museum of Tervuren, the Chrono-environment Laboratory of Besançon, the Faculty of Medicine of the University of Kinshasa and the Center for Research in Hydrobiology of Uvira in South Kivu, situated on the shores of Lake Tanganyika. The main difficulty consisted of matching series collected with temporal resolutions that were not exactly the same and that were derived from several sources:

– weekly time series of cholera cases from the Ministry of Health of the Democratic Republic of the Congo;

– chlorophyll concentration data collected specifically for this study over the course of 3 years at three stations located on the left bank of Lake Tanganyika, near Uvira and Kalemie;

– available data on algal blooms from the CLIMFISH project, from 2002 to 2006, monitored through the measurement of chlorophyll concentration to an indicator of phytoplankton abundance detectable using MODIS satellite images;

– precipitation data obtained from the International Research Institute for Climate and Society (IRI/LDEO Climate Data Library), providing daily precipitation from the U.S. National Oceanic and Atmospheric Administration's Climate Prediction Center.

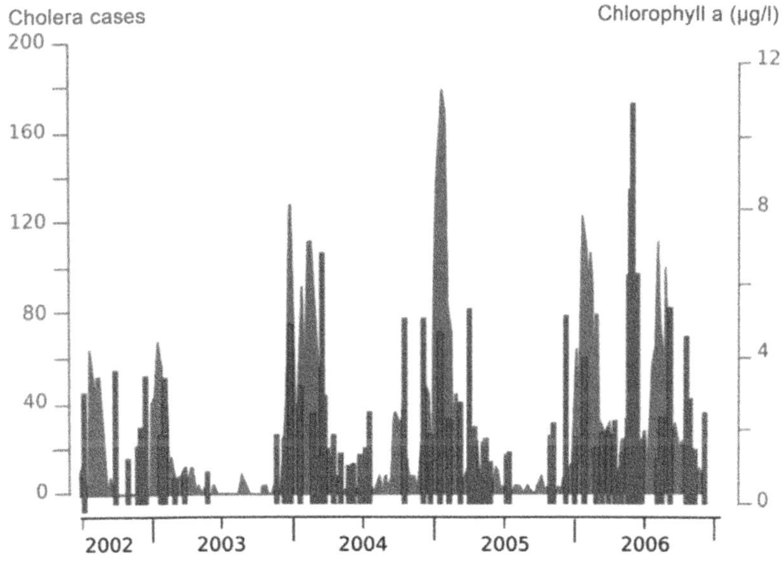

Figure 7.8. *Relationship between the number of cholera cases and fluctuations in phytoplankton abundance (chlorophyll-a concentrations) in Lake Tanganyika, January 2002–December 2006 in Uvira, from Bompangue* et al. *(2011). For a color version of this figure, see www.iste.co.uk/giraudoux/socioecosystems.zip*

The idea was, after recalculating some of the series so that they corresponded to the same temporal and spatial resolution, to correlate these detrended series, directly and with a time lag (in case the effect of one or more environmental variables would occur with a time lag). The exercise was statistically quite complex and will not be detailed here. The results confirmed that cholera increased significantly during warm El Niño events but decreased or remained stable between these events. Most outbreaks occurred in a few hot spots in lake areas, where weekly cholera incidence varied with season, rainfall, plankton fluctuations and fishing activities (Figure 7.8). During lulls, metapopulation endemic foci were found around the lakes (Bompangue *et al.* 2011), confirming the results of earlier studies (Bompangue *et al.* 2008).

To carry out these analyses, it was necessary to build a new, unconventional and unplanned collaboration with the Tervuren Museum in Belgium, which had a long history of scientific investment in the Great Lakes region.

Indeed, it was impossible for us, with only a theory, to set up a multidisciplinary project of long-term data collection on Lake Tanganyika, on the basis of hypotheses that would not necessarily have been supported by the usual research funders. Going beyond our medical field and seeking collaborations to move forward became more than an obvious requirement. We now had to determine the role of these lacustrine zones in the functioning of cholera epidemics.

7.3.3. *The role of lake areas in the persistence of cholera epidemics*

Mapping of cholera outbreaks showed a higher frequency in the Great Lakes area (Figure 7.9). However, a number of the determinants of cholera epidemics remained unknown.

The first step consisted of relating the number of cholera cases per health zone to a number of environmental and societal variables that could be measured in each zone, based on several studies in the Democratic Republic of the Congo where appropriate data were available. A principal component analysis was used to select the least correlated of the potential explanatory variables.

The following criteria were used: the area, population, location (lakeside or not) and presence/absence of a city with more than 100,000 inhabitants, a commercial port and major roads or railways. The relationships between the number of cholera cases and their socio-ecological characteristics were then studied comparing, based on the Akaike criterion, several Poissonian and binomial generalized linear models, including the logarithm of the population in offset, and by progressively introducing the potentially explanatory variables. The possible spatial autocorrelation of data collected in neighboring health zones was taken into account by using only permutation tests for inferential statistics, thus eliminating the bias that could have arisen from tests based on an erroneous estimate of the degrees of freedom of the data set. The number of cholera cases, in the end, all other variables being equal, was up to 7.5 times higher in the lakeside location. In addition, the presence of a main road, a port or a railway station significantly increased the number of cases (Bompangue *et al.* 2008, 2009, 2011).

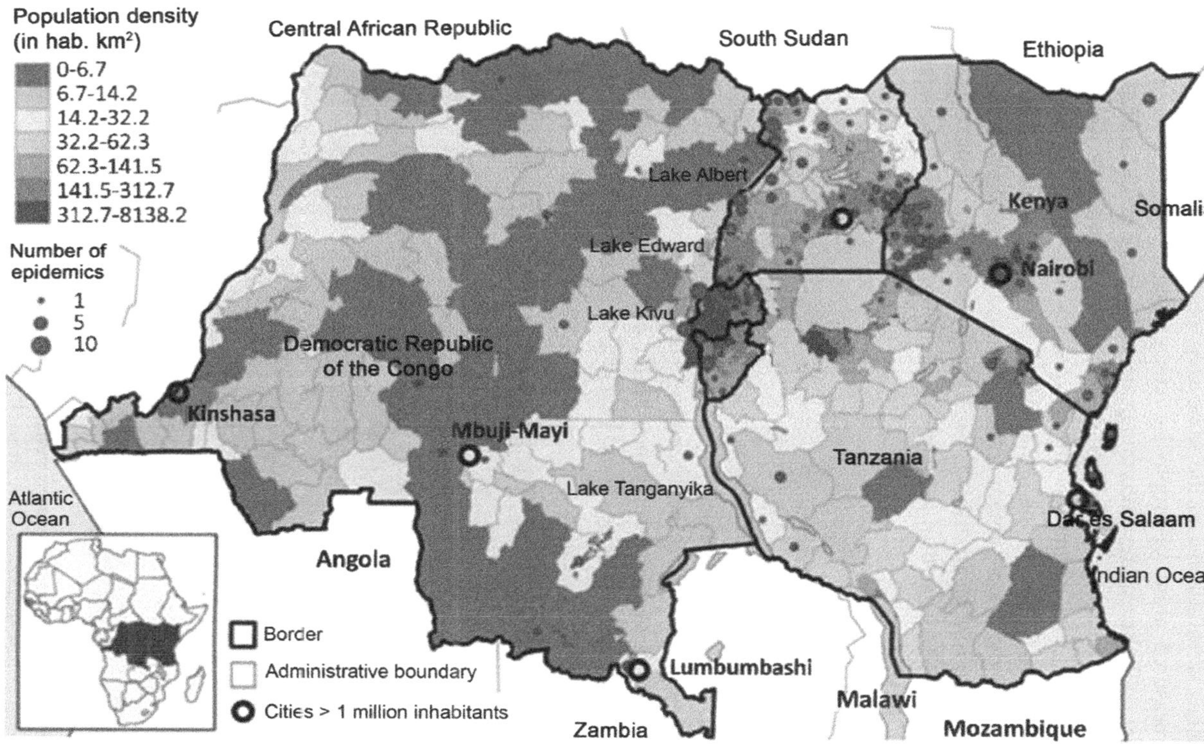

Figure 7.9. *Number of reports and location of cholera outbreaks in the African Great Lakes region, 1999–2008, from Bompangue* et al. *(2011). For a color version of this figure, see www.iste.co.uk/giraudoux/socioecosystems.zip*

It was thus statistically confirmed that lake areas were major players in the dynamics of cholera in eastern Democratic Republic of the Congo. However, not all lake localities were affected in the same way, nor at all times. Rainfall and algal blooms measured from chlorophyll-a concentrations, however significant the correlations, did not explain everything, including some of the spatial or temporal dynamics (see the exceptions in Figure 7.8). Other explorations were necessary. It was through field investigations, conversations with local staff from the cholera treatment centers, discussions with community actors and the memory of experiences in the Haut-Lomami region during investigations with MSF that I remembered the role of the fishing and trading populations in the cycle of cholera epidemics. This role was then explored and confirmed through field interviews and qualitative surveys.

7.3.4. *Other determinants*

Apart from fishing and trading activities, outbreaks in health zones sometimes far from the lake areas coincided with mining activities, particularly in the Katanga, North Kivu and South Kivu health zones. This was the case, for example, of the outbreaks that occurred in the Nyunzu health zone (Figure 7.10). It was therefore clear that other epidemic causes could be added to those that had been established as predominant.

Figure 7.10. *Artisanal miners suffering from cholera in the mining village of Kisengo in the Nyunzu health zone, Katanga in 2007 (photo D. Bompangue)*

An attempt to explore these additional causes was conducted in North and South Kivu from 2000 to 2007. During this period, both provinces experienced a major volcanic eruption, Nyiragongo, and a number of complex emergencies resulting in the displacement of populations owing to abuse from armed gangs operating in the region.

Four epidemic reactivations were then observed in 12 weeks following the war events, but simulations indicated that the number of reactivations was no greater than that obtained by a random distribution of their occurrence. The 2002 Nyiragongo volcanic eruption, on the other hand, was followed by a marked decrease in cholera incidence (Figure 7.11) (Bompangue *et al.* 2008).

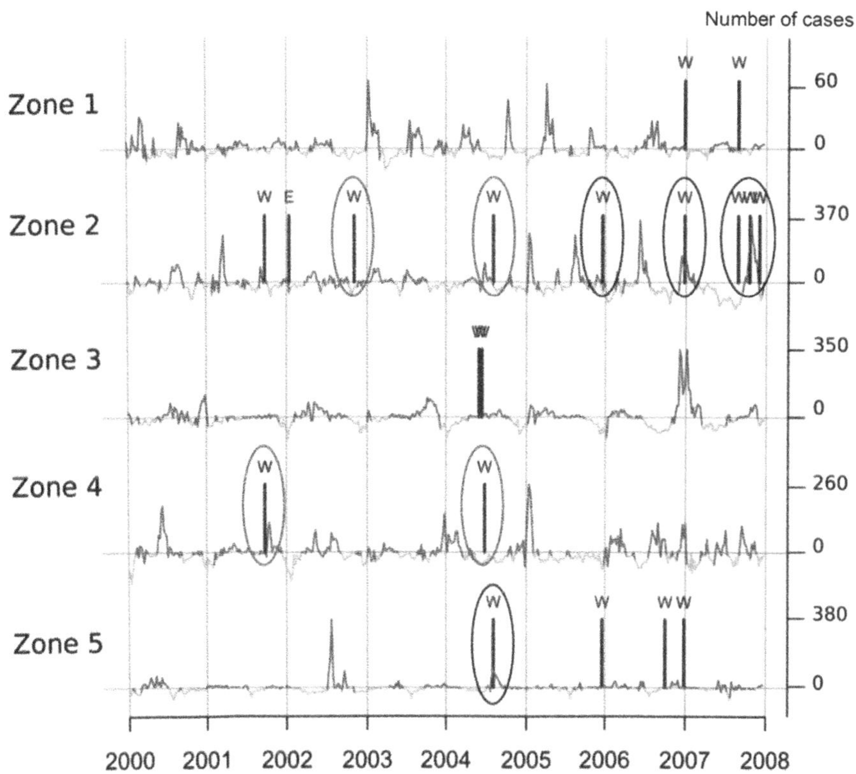

Figure 7.11. *Cholera outbreaks and main events in the five health districts, from Bompangue* et al. *(2009). For a color version of this figure, see www.iste.co.uk/giraudoux/socioecosystems.zip*

COMMENT ON FIGURE 7.11.– *War events (W) and the main natural disaster, in this case the eruption of Nyiragongo (E) observed in North and South Kivu from 2000 to 2007.*

Zone 1: Mutwanga; zone 2: Goma, Rutshuru, Kirotshe; zone 3: Bukavu, Katana; zone 4: Uvira, Nundu, Fizi; zone 5: Pinga, Walikale. The vertical axis represents the residuals of the decomposed time series (this removes the undesirable effects of seasonal and interannual trends; red: periods above the mean; green: those below the mean). War events occurring during an epidemic already in progress are circled in black. War events followed by a cholera epidemic within 12 weeks are circled in red.

These counterintuitive results clashed with the conventional wisdom that major cholera epidemics were the result of major disasters such as population displacement in war-torn areas, even though such displacement could facilitate the spread of existing epidemics, or that a geological disaster would inevitably lead to an epidemic. However, they had to be accepted as facts.

Another multidisciplinary study, involving researchers in epidemiology, hydroecology, volcanology, history and biostatistics, was to reveal the much more subtle role of volcanic activity in the region. They were able to show that the activity of the geothermal springs feeding Lake Kivu, mainly from the Nyiragongo volcano, controlled the physico-chemical characteristics of the lake and could then favor the persistence of the bacillus. The lake then became a reservoir of the pathogen, which could contaminate local residents through the consumption of water and fish. During the period from 2007 to 2012, a long-term unambiguous relationship was demonstrated between volcanic activity (not just eruptions) and cholera cases in the Katana Health Zone, as *Vibrio* contamination of lake water and fish was correlated with the physico-chemical characteristics of the lake (Batumbo *et al.* 2020).

7.3.5. *The other face of cholera*

What were sometimes only hunches, or even impressions from the field and the examination of data, could be corrected or confirmed and became scientific evidence. We were changing the face of cholera. I still remember one of Professor Jean-Jacques Muyembe's phrases at the Ibis Hotel in Besançon on October 29, 2009, when I had just defended my thesis (Bompangue 2009): "This is not cholera anymore". He knew what he was talking about. Indeed, Professor Muyembe, who has accumulated scientific distinctions at the national and international levels, can be considered the "father" of modern infectiology in the Democratic Republic of the Congo. He identified the first outbreak of Ebola in Yambuku in 1976, in Equateur, in the former Zaire, now the Democratic Republic of the Congo. He investigated the very first cases of cholera in the west of the country, during the time of Zaire, in 1974, in the port city of Matadi in Bas-Congo (called Kongo-Central today). But in the golden years of the Great Leopard's reign (that is what Marshal Mobutu called himself), when he succeeded in focusing the world's attention on Kinshasa by organizing the boxing match of the century between Muhammad Ali and George Foreman, at the height of Zaire's glory, it was not appropriate to talk about cholera. So this young microbiologist, freshly graduated from the Catholic University of Louvain in Belgium, was forbidden by the Minister of Health at the time, Ngandu Kabeya, to talk about cholera in Zaire. The disease of shame. There was no cholera in Zaire. However, the young Jean-Jacques Muyembe had indeed recognized these cases of cholera, he told me 31 years later when he was directing my thesis. He was a living witness to the history of the seventh cholera pandemic in the Democratic Republic of the Congo who spoke and knew what it was all about. Indeed, it was no longer the same cholera; new concepts had emerged and would completely revolutionize the paradigm of the fight against cholera. From then on, the disease that was said to be unpredictable and linked to poverty had become predictable and the only real poverty to which it could be linked was that of knowledge.

We had challenged many preconceptions. We were able to demonstrate that it was possible to conduct long time series studies in sub-Saharan Africa, even though there were still complaints about the lack of data. Data are only collected if they are sought. The search for data to conduct long time series studies in Africa should be compared to an archeological excavation: it takes time, you have to be patient, persistent, persevering, and you cannot conclude too quickly that there is no data. To those who say that epidemic surveillance data are of poor quality and that nothing can be done with them, again, we have shown that there is no such thing as bad data, only bad data users. It is up to the user to know what he can do with it, to know its limits, what question he can answer. To those who thought that cholera epidemics were unpredictable, that they were linked to poverty, we were able to provide an answer by highlighting the periodicity of epidemics and identifying the factors that determined this periodicity, even if this still needed to be investigated further. Many points still needed to be clarified, but we were not going to wait for further investigations to use the information available at the operational level.

7.4. Toward the elimination of cholera and the discovery of governance issues

It seemed obvious, in light of the information generated by the results of the initial work, that it was possible to envisage no more and no less than the elimination of cholera in the Democratic Republic of the Congo, as well as in similar areas. "We're in 2007 and you're talking about the elimination of cholera? What heresy!" Despite some hesitation, it had not been very difficult to convince Renaud that an "application" part related to the demonstration on one of the sanctuary sites identified in the work should be included in the thesis document. The experience of the work that was in progress was thus reported in the second part of the thesis. This pilot project to apply our initial results was conducted in Kalemie, on the Congolese shore of Lake Tanganyika, with the support of the Veolia Foundation and the NGO Solidarités International. This was an opportunity for me to become aware of another difficulty: that of getting people who think they are rational to accept the obvious. Among those who were quickly convinced by our research work and who began to draw inspiration from it for their own actions, I can mention a few friends who were still in NGOs and who are now, for the most part, in agencies such as UNICEF. Above all, there was Dr. Thierry Vandevelde of the Veolia Environnement Foundation. Thierry was quick to apply our recommendations, not hesitating to go as far as Kalemie, sometimes against the recommendations of certain agencies for whom the entire Democratic Republic of the Congo was a dangerous "red zone". The support of the Veolia Foundation for the implementation of the pilot project in Kalemie was fundamental in confirming our belief that cholera could be eliminated as a public health problem, even in complex contexts with limited resources, but I had no idea of the battle that lay ahead: how could we gain acceptance, outside our immediate circle, of the idea of a public health strategy to be implemented on a large scale to eliminate cholera as a public health problem in poor countries, and, moreover, in the midst of complex crises?

Like the previous episodes, it was a tough fight. However, I would say that tenacity, persistence and the belief that our fight was just had once again helped us to overcome major obstacles, including getting the Ministry of Health to agree to a multisectoral Strategic Plan for the Elimination of Cholera in the Democratic Republic of the Congo (PMSEC, adopted in December 2007 for the period 2008–2012), based on targeting the seven main cholera sanctuary regions in the east of the country (Figure 7.1) (Muyembe *et al.* 2013).

Once the strategy for the elimination of cholera in the Democratic Republic of the Congo was adopted, it was necessary to tackle the international level. But how to go about it? It was from Haiti, the country that some time ago was still called the "pearl of the Antilles", that the mobilization for a new impetus in the fight against cholera came. In 2010, Haiti experienced an earthquake of a magnitude greater than 7, followed 9 months later by a massive cholera epidemic, the origin of which was proven to be the importation of *Vibrio* by a UN Nepalese contingent. These tragic events accelerated the process of recognizing the cholera elimination strategy as a public health problem in poverty settings. In its document "Appeal for the Elimination of Cholera in Haiti", the United Nations implicitly recognized that cholera could be eliminated. When this appeal was launched at the height of the cholera crisis in Haiti, I was in Port-au-Prince providing technical support to the activities of the NGO Alima, in response to the cholera epidemic. Some major donors, such as the Bill and Melinda Gates Foundation (BMG), were becoming increasingly aware of the need to put some money into the fight against cholera. It was in this context that the Africhol project for cholera surveillance in Africa emerged in 2011, the first Africa-wide project funded by the BMG Foundation. It took cholera's arrival in Haiti, a short flight from the luxurious beaches of Miami, to remind us that cholera did not end with the end of the epidemics in Paris, London or New York in the 19th century. One could say that cholera was back on the international agenda. However, it is thanks to the old cholera epidemics that triggered the first international sanitary conference, held in Paris in 1851, that we can count on the International Sanitary Regulations today.

Our stay in Haiti led to an unlikely meeting with a German epidemiologist colleague, Silvan Vesenbeck, who was a post-doctoral fellow in the Global Health Department at Harvard University's School of Public Health. We first met in the conference room of the *Canne à sucre* in Port-au-Prince, during a series of conferences on cholera in the midst of the epidemic in Haiti. Things went very quickly until we were invited to a scientific visit to the Global Health Department in Boston. It was from there that we would publish a letter in the *Lancet*, arguing for the possibility of eliminating cholera (Bompangue *et al.* 2012a), and that we would also publish the map of a cholera epidemic wave originating in the endemic areas of the Great Lakes and threatening Kinshasa several thousand kilometers to the west (Bompangue *et al.* 2012b), according to the model elucidated by the previous studies (Figure 7.12).

As a result of these findings and the international recognition gained from this work, we were asked to consult with the Global Task Force on Cholera Control to conduct a larger study on the identification of potential cholera hotspots in Africa. The findings of this study were used as the basis for preparing a comprehensive strategy for cholera control in Africa, primarily through the adoption of the global strategy for cholera elimination entitled *Ending Cholera, A Global Roadmap to 2030* (Global Task Force on Cholera Control 2017). To see that it was now possible to talk about cholera elimination as a strategy at the level of an organization like WHO was a cause for immense satisfaction. But this joy was to be short-lived. We soon realized that there was a gap between generating knowledge, adopting strategies based on that knowledge and ensuring that those strategies were properly implemented. Indeed, between 2008, when the first PMSEC was launched in the Democratic Republic of the Congo, and 2017, when I managed to get Minister Oly Ilunga to accept the idea of creating a National Program for the Elimination of Cholera and the Control of Other Diarrhoeal Diseases (PNECHOL-MD), almost 10 years have passed. Another limitation to achieving the goal of cholera elimination was the difficulty of changing behaviors in the decision-making processes of key operational actors and funding partners. Generating knowledge was thus a necessary step but it was not enough.

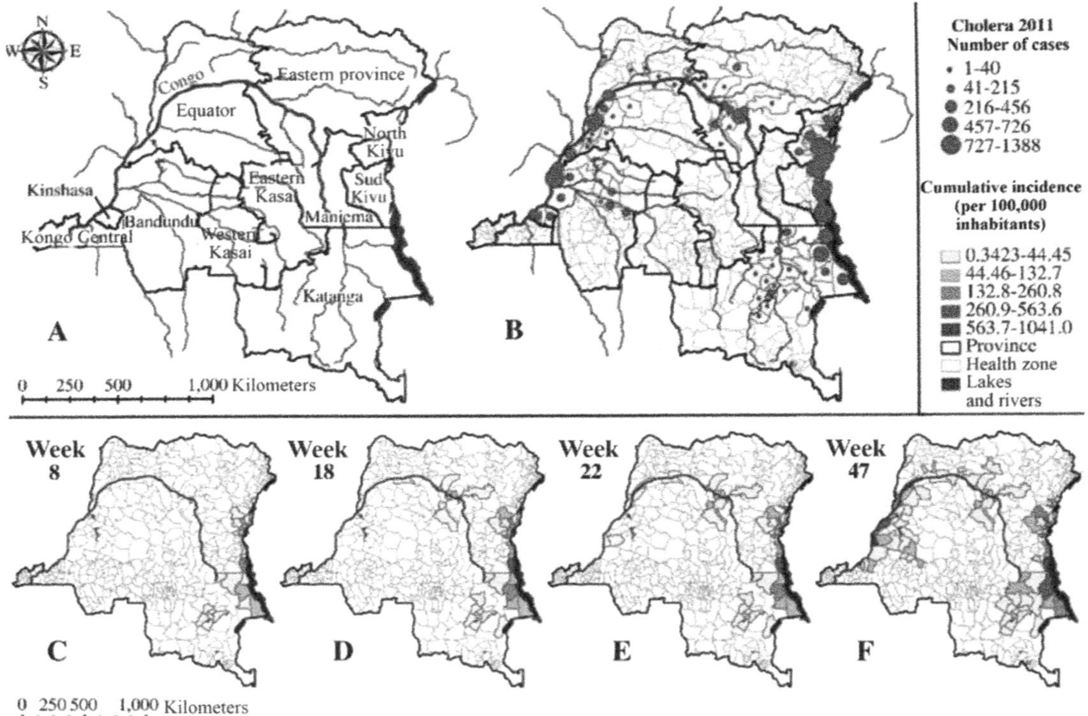

Figure 7.12. *Spatio-temporal spread of the 2011 epidemic. Map of the Democratic Republic of the Congo with its 11 provinces and large river system (A). Total number of cholera cases for 493 of the health zones as of week 47 (B). Cumulative incidence (per 100,000 population) at weeks 8, 18, 22 and 47 (C-F), from Bompangue* et al. *(2012). For a color version of this figure, see www.iste.co.uk/giraudoux/socioecosystems.zip*

I quickly realized that without rigor, discipline and ethics in operations, we would never achieve cholera elimination. As long as the people on the job continued to work in their various comfort zones, trapped in their bureaucratic routines, as long as NGOs and other agencies continued to be more concerned with their own survival than with the real suffering and problems of the people, as long as funding agencies continued to prefer to fund ready-made recipes from a short list of organizations to be funded, there was no hope of eliminating cholera or reducing the incidence of other health problems, despite our scientific knowledge and our health policies adopted at the highest level. On the other hand, we had to find ways of pointing out the kind of dictatorship of the "strongest" thinking, which is a persistence of colonization at its vilest, too often internalized by the administrations and peoples of the countries concerned, which wants all local productions to be subject to doubt, while those coming from the North are truths to be applied without verification. This is the problem we encountered in getting certain international agencies to accept PMSEC and the strategy of emergency response to major cholera outbreaks through targeted approaches, a strategy known as "gridding". This was particularly the case during the targeted cholera outbreak control program in Kinshasa in December 2017. This program, designed on the basis of infectious disease ecology concepts and implemented through PNECHOL-MD, in lieu of a vaccination campaign, resolved one of the largest cholera outbreaks in the city of Kinshasa in 9 weeks, compared to 25 weeks, the average time for cholera cases to regress during previous epidemics (Figure 7.13) (Bompangue *et al.* 2020). Despite this success, instead of applauding these approaches developed by local actors, some international agencies advised by international scientists have since tried to spend thousands of dollars trying to propose "revisions" and "adaptations" of PMSEC and the grid strategy, developing theories such as: "It's good, but there are things to adjust". Funding is never neutral. There are many local initiatives that just need to be identified, supported and promoted to advance several public health causes in disadvantaged countries. Even if, for the time being, this approach has not yet become a real strategy in development aid policies, there is reason to hope that it will one day. Indeed, there are good people everywhere, in the North and in the South. Good people like those who are still involved today in the fight to eliminate cholera in the South and to integrate ecological approaches into the understanding of disease phenomena. They invest themselves with their hearts, they break codes and cross barriers without worrying about their personal interests. Men and women, from the North as well as the South, who dream of one thing only: to see a new form of collaboration between peoples redefined, free of the negative stigma of a burdensome past from which we want to draw the necessary strength to build a shared and bright future.

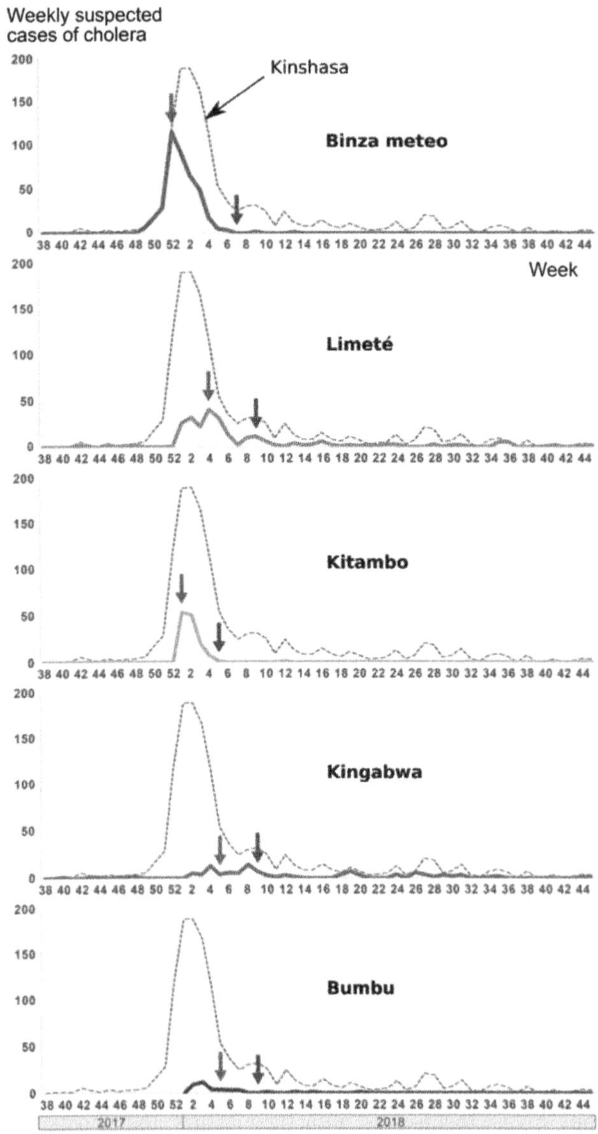

Figure 7.13. *Cholera epidemic curve by target health zone and response time. Weekly numbers of cholera cases are shown on the y-axis and weeks/years of epidemic are shown on the x-axis. The start and end points of response times in each health zone are indicated by green and red arrows, respectively, according to Bompangue* et al. *(2020). For a color version of this figure, see www.iste.co.uk/giraudoux/socioecosystems.zip*

This is where another front in research and action opened up for me: that of epidemic governance. Here again, I am starting from scratch, simply from the obvious: this is the next link to be explored, a link without which the elimination of cholera will remain a dream. We are already working on it, not only with regard to cholera but also to other infectious diseases for which the model described here could be transposable in every respect.

KEY POINTS.–

Cholera outbreaks in the Democratic Republic of the Congo are predictable and originate in the lake areas of the east of the country.

The study of correctly spatialized eco-epidemiological time series including the socio-anthropological components of the system allows us to understand seasonality and transmission dynamics at several scales.

Such studies force us to move away from the paradigms of a "colonial" vision of top-down applied knowledge, which does not take into account local realities, based on data collected *in situ*. This "colonial" vision is shared as much by actors in the South as by those in the North.

Indisciplinary approaches (in the sense that the contributions of each discipline are considered and rooted in the observed facts) are essential to overcome this situation and to detect the many commonalities and idiosyncrasies that prevail in the transmission of cholera.

Following the example of the collaborations initiated in the context of the work on cholera epidemics, it is possible to hope for a new era in international cooperation in the fight against infectious diseases, which are still major public health problems in countries with limited resources.

7.5. References

Batumbo, D., Giraudoux, P., Deniel, C., Garnier, E., Mauny, F., Kasereka, C.M., Kizungu, R., Muyembe, J.J., Bompangue, D., Bornette, G. (2020). Volcanic activity controls cholera outbreaks in the East African Rift. *PLOS Neglected Tropical Diseases*, 14(8), e0008406.

Bompangue, D. (2009). Dynamique des épidémies de choléra dans la région des Grands Lacs africains : cas de la République démocratique du Congo. PhD Thesis [Online]. Available at: http://www.theses.fr/2009BESA0020.

Bompangue, D., Giraudoux, P., Handschumacher, P., Piarroux, M., Sudre, B., Ekwanzala, M., Kebela, I., Piarroux, R. (2008). Lakes as source of cholera outbreaks, Democratic Republic of Congo. *Emerging Infectious Diseases*, 14(5), 798–800.

Bompangue, D., Giraudoux, P., Piarroux, M., Mutombo, G., Shamavu, R., Sudre, B., Mutombo, A., Mondonge, V., Piarroux, R. (2009). Cholera epidemics, war and disasters around Goma and Lake Kivu: An eight-year survey. *PLOS Neglected Tropical Diseases*, 3(5) [Online]. Available at: https://journals.plos.org/plosntds/article?id=10.1371/journal.pntd.0000436 [Accessed January 7, 2022].

Bompangue, D., Giraudoux, P., Plisnier, P.D., Tinda, A.M., Piarroux, M., Sudre, B., Horion, S., Tamfum, J.J.M., Ilunga, B.K., Piarroux, R. (2011). Dynamics of cholera outbreaks in Great Lakes Region of Africa, 1978–2008. *Emerging Infectious Diseases*, 17(11), 2026–2034.

Bompangue, D., Duc, G.L., Vesenbeckh, S., Blackwell, N. (2012a). Elimination of cholera transmission. *The Lancet*, 379(9822), e46–e47.

Bompangue, D., Vesenbeckh, S.M., Giraudoux, P., Castro, M., Muyembe, J.-J., Ilunga, B.K., Murray, M. (2012b). Cholera ante portas – The re-emergence of cholera in Kinshasa after a ten-year hiatus. *PLOS Currents Disasters*, 5 March [Online]. Available at: index.html%3Fp=621.html [Accessed December 5, 2020].

Bompangue, D., Moore, S., Taty, N., Impouma, B., Sudre, B., Manda, R., Balde, T., Mboussou, F., Vandevelde, T. (2020). Description of the targeted water supply and hygiene response strategy implemented during the cholera outbreak of 2017–2018 in Kinshasa, DRC. *BMC Infectious Diseases*, 20(1), 226.

Cleveland, R.B., Cleveland, W.S., McRae, J.E., Terpenning, I. (1990). STL: A seasonal-trend decomposition procedure based on Loess. *Journal of Official Statistics*, 6(1), 3–73.

Colwell, R.R. (1996). Global climate and infectious disease: The cholera paradigm. *Science*, 274(5295), 2025–2031.

Global Task Force on Cholera Control (2017). Ending cholera. A global roadmap to 2030. Report, WHO [Online]. Available at: http://www.who.int/cholera/publications/global-roadmap/en/ [Accessed December 24, 2020].

Muyembe, J.J., Bompangue, D., Mutombo, G., Akilimali, L., Mutombo, A., Miwanda, B., de Dieu Mpuruta, J., Deka, K.K., Bitakyerwa, F., Saidi, J.M., Mutadi, A.L., Kakongo, R.S., Birembano, F., Mengel, M., Gessner, B.D., Ilunga, B.K. (2013). Elimination of cholera in the Democratic Republic of the Congo: The new national policy. *The Journal of Infectious Diseases*, 208(suppl. 1), S86–S91.

Conclusion

Cooperation at Work: Sociology of a Scientific Standpoint

Dominique JACQUES-JOUVENOT[1] and Simon CALLA[2]

[1] *LASA, Université Bourgogne Franche-Comté, Besançon, France*

[2] *Agroecology, Université Bourgogne Franche-Comté, Besançon/Dijon, France*

The conclusion of this book is based on an analysis of the collective practices of researchers at work. It aims to report on these practices by drawing on the accounts of the authors interviewed about their research careers and the importance they attach to cooperation at work in the very special environment of the academic world. The environment is understood in the sense of "the set of multiple and reciprocal links that people, objects and institutions in the scientific world maintain" (Waquet 2019). Beyond a reflexive return on a research experience related to the environmental issue within a socioecosystem, it is the authors' commitment to ways of doing things together that we need to understand and explain. Indeed, these physicians, ecologists and sociologists are involved in a collaborative dynamic. Even if experiences of cooperation between disciplines have been increasing in French universities since the 1980s, including those that establish partnerships with field actors, these practices are still marginal today. Sociologists know, like M. Mauss, that crossing the borders of one's discipline, sharing areas of research, means venturing into that unknown which "lies at the frontiers of science, where professors 'eat each other', as Goethe says… It is generally in these little shared areas that urgent problems lie" (Mauss 1950).

Without contradicting Mauss, the writing of this book bears witness to a situation of collective research that has lasted for more than 30 years without having given rise to any act of cannibalism between medical doctors, ecologists and socio-anthropologists. The knowledge produced in this collaboration is described throughout the chapters of this book, so it is not a question here of returning to this aspect of the work but rather of hypothesizing that the researchers involved in multidisciplinary projects are the heirs of a scientific standpoint acquired prior to their entry into the profession (during their school and university training). The authors of this conclusion therefore focus on the socializing experience of each of them to highlight the importance of certain elements of their trajectory to their involvement in a cooperative experience, thus showing that this early contact with other disciplines leads them to become border-crossers. Based on the anthropological theory of M. Mauss, who considers exchanges to be gift and counter-gift logics (Mauss 1950), we will see that these heritages have led to intergenerational modes of transmission, such as the creation of a university diploma (DEA ESS in environment, health, society, then a master's degree) offered to students from all disciplines who are interested in environmental issues.

By the form given to the conclusion of this book, the authors attempt to participate in an ecological history of a small world of scholars, that is, to take into account "the environment in which men live and think, and the ties that bind them to this environment and to each other" (Waquet 2019). But before going any further, perhaps it is necessary to remind the reader of the beginnings of this or these histories.

Once upon a time… perhaps even before… The long story that begins here is the creation of the SERF (health and rural environment) laboratory in 1992 at the University of Franche-Comté (UFC) by Dominique Angèle Vuitton (DAV), who judiciously points out the lack of interest in health in rural environment by national research institutions: "The SERF labeling corresponded to an absence of work on the side of medicine in rural areas knowing that, moreover, she says, we had the *Mutualité sociale agricole*[1] (MSA) for support." This convergence of interests between an institution such as the MSA, which is a privileged place for observing health in rural areas, and scientists gives the research that follows a finalized character (Jollivet 1992). Indeed, in this type of research, "the researcher is not outside but inside the world he studies and he engages in situations of interaction with non-researchers" (Petit *et al.* 2008). The network formed by DAV around alveolar echinococcosis (AE), which was the central issue in her work at the time, was rapidly enriched by researchers from other disciplines (ecology, socio-anthropology, etc.) and partners rooted in the field: MSA, the *Institut national de la recherche agronomique* (INRA), the departments of the Besançon University

1 The MSA provides social security coverage (health, old age, etc.) for the entire agricultural population and its beneficiaries: farmers, employees of farms, companies, cooperatives and professional agricultural organizations, and employers of labor.

Hospital (hepatology and digestive surgery, in particular) and the veterinary laboratories of the Jura, Doubs and, more recently, Haute-Saône regions.

All these actors were anxious to contribute, from their place in the socioecosystem, to the reflection on what they considered to be a public health issue in the Franche-Comté region at the beginning of the 1990s: a zoonosis called "alveolar echinococcosis". This disease is particularly virulent in the region and has consequences on the health of rural populations (see Chapters 4 and 5). Over the next 30 years, other researchers from related or more distant disciplines, coming from countries as different as China or the Democratic Republic of the Congo, became involved in research related to the environmental issue, bringing with them one or two thesis students, other colleagues from other laboratories or other countries, while other researchers, for various reasons, left the group.

"Multidisciplinarity is a kind of mosaic that moves according to the subjects" (DAV). Indeed, the objects evolve according to the researchers and according to what each of them takes on and/or gives to the group in which they are involved: disease, foxes, the press, the transmission of knowledge, scientific knowledge, etc. But all the researchers share a commitment to this adventure of co-production of knowledge to solve environmental problems within a socioecosystem.

C.1. Methodological aspects

The sociologists in the current research group have set up the apparatus for writing this conclusion. Their role as socio-anthropologists is a "hybrid" one in that they are, on the one hand, the authors of two articles that make up this book (see Chapters 2 and 5) and, on the other hand, the researchers who set up the apparatus that made it possible to write this conclusion. Finally, it should be pointed out that, in this chapter, the use of the word "we" is reserved for the words of the authors of the conclusion and the initials of the authors of this work will be used when referring to their speeches.

We do not therefore conclude the book from the point of view of its content, but we would like to give an account of a practical and collective experience of research, of "acting together". The sharing of our reflections took place during two workshops organized at the end of 2020 with the co-authors of the book (in order of appearance within the collective): DAV, PG, DJJ, DB, EA, SC and AM. More specifically, we conducted focus groups with the authors, asking them about their socialization and disciplinary training, their collaborative work practices and the ways in which these practices are transmitted to the next generation. We, the authors of this conclusion, did the same exercise during these focus groups.

It is worth adding to these interviews informal discussions on these collaborative practices; exchanges which, moreover, gave rise to the production of an agreement: to free ourselves from the notion of multidisciplinarity in order to benefit the actors in this collective experience. This "gesture" that we are making resonates with the words of M. Mauss, taken up by M. Fournier (Fournier 1994) concerning the differences between sociology, ethnology and anthropology:

> It is clear that this word [ethnology] does not designate a science separate from anthropology… It is also clear that it does not designate a science separate from sociology, for ethnology also consists of the study of human groupings as groupings. Basically, what is involved in all this is rather a division of workers – and even more a historical and pedantic division of chairs and scientific institutions – than a real classification of the sciences in relation to their objects. (Mauss 1931)

As R. Frodeman points out, this is true for all disciplines. Disciplinary division is more an institutional than a scientific matter and, like Turner, is the result of "what we decide to put together as a discipline" (Turner 2000).

> Indeed, "is it not the only thing that unifies the four aspects of anthropology – physical anthropology, cultural anthropology, archaeology and linguistics? Do we classify water geochemistry and vertebrate paleontology in geology? Or do we decide that the history of philosophy belongs to philosophy rather than history?" (Frodeman 2019)

The practices of the actors, captured through their narratives, are of greater interest to us here than the institutional question. The narrative of the researchers' trajectories is an intrigue that brings together events in a "coherent and acceptable" way, in order to construct a story (Ricoeur 1985). Based on these narratives, we take a socio-anthropological look at the biographical elements that led the researchers to engage in this type of collective action. The purpose of our work is therefore to understand the singular histories and modes of relationship that facilitate the establishment of cooperation between these actors from different disciplinary backgrounds in the service of the production of knowledge for action (Darré 1999).

C.2. Legacy

The importance of education and socialization has been the subject of numerous works since the foundation of sociology and throughout its history. In his famous book *Education and Sociology* (Durkheim 1922), Emile Durkheim shows the power of education in the production of social beings through a process of internalization of social and cultural norms that shape individual trajectories. The people interviewed here are all teacher-researchers trained in a specific discipline at the university and have the particularity of having had a common research experience with colleagues from a disciplinary background other than their own. As a result, they stand somewhat on the periphery of the circle in which the majority of their colleagues find themselves, that is, at the boundaries of the various disciplines involved. In the accounts of all the teachers, it is clear that they had been exposed to other disciplines from the time they began their careers, or even during their secondary education. For some, the practice of several disciplines is inherent to the discipline itself. This is the case for medicine in particular and, to a lesser extent, for sociology as well.

C.2.1. *Initiation to collaboration: a practice acquired in the researchers' learning process*

At the time of the survey, the network was composed of two physicians (DAV and DB), two ecologists (PG and EA), two sociologists (DJJ and SC) and a philosopher (AM) who joined the working group to write the preface of the book. Many researchers have shared this multidisciplinary experience, starting with an initial research project on AE, a research project that spread from the moment the ecologists joined this team, and then some researchers left to eventually lead another one or to resume a more individual research activity. We are talking here about an experience of collaboration at work and not only about research on AE.

Within this group, we find several generations of researchers. While the first ones initiated this experience in the 1980s (DAV, PG), others participated in it from the 1990s (DJJ), then in the early 2000s (DB), while the youngest ones joined it from 2008 onward (EA, SC). Apart from the generational differences between them, these researchers have an attraction for collaboration with others, which makes sense at different key moments in their life course.

Indeed, some of them opened up to other disciplines during their university training. DAV thus thinks that her training in multidisciplinarity was part of her internship and, more specifically, in the context of her medical thesis says that:

> In Strasbourg, as surgical interns, we were obliged to have a research program because our department head had an Inserm unit, so I found myself working on an immunology program; a program on intestinal transplantation on which I did my medical thesis, which was already a research thesis... I worked a lot with an Inserm researcher, who was

> one of the few female surgeons at the time, but who was in charge of research at Inserm in this unit… I immersed myself in immunology... so I thought that crossing liver and immunology was interesting. And as I was a surgeon, using experimental models where I could tinker a bit... And that's how, being interested in this crossing of two medical disciplines – well, immunology is a biological discipline, not necessarily just a medical one – the idea came to do a PhD thesis.

Obtaining this degree, as well as the orientation in a multidisciplinary approach, is also what brings EA and SC together. The former did her doctorate in collaboration with doctors and ethologists before being recruited because of the transversality of the skills acquired during this doctoral research. While the latter mentions the multidisciplinary nature of the master's degree at the University of Franche-Comté, his interest in other disciplines essentially dates back to the time of his thesis on river pollution: "It was at this point that I saw the need to move towards other disciplines and, at the same time, to work with people who speak a different language."

Others see the link with multidisciplinarity as early as their high school years. This is the case for PG, who wants to become a teacher of natural sciences: "To pass the advanced teaching qualification in natural sciences, you do geology, you do biochemistry, you do physiology, ecology, botany, you do zoology, in short, you cover practically all the disciplines of living organisms." This is also the case for DJJ, a socio-anthropologist, who has a secondary education in an agricultural high school, a place known for its multidisciplinary teaching practices, which were considered to be very advanced for the time (1975–1980): "Knowledge of the agricultural and rural world required knowledge of genetics as well as economics and zootechnics. In addition, the lessons were often given by a joint intervention of several teachers from different disciplines, including teachers of French, etc."

Going back even further in time, others situate their openness to multidisciplinarity as early as family socialization. This is the case of DB:

> I'm a counter-example to my own family in the Congo. My father is a polyglot; I was immersed in philosophy books from a very early age and I read Karl Marx before the *baccalauréat*. I'm in a science class, but the fact that I was immersed in a different philosophy led me to take societies into account, so [to have a] very critical analysis of what's going on around us. People's lives, lifestyles, that's what interests me… Knowing what's going on around patients.

C.2.2. *The ancestors, actors enter the scene in the making of the researcher*

> There is no beginning… We are all addition. (Gary 1982)

Each interviewee cited at least one person who was a reference for them (thesis director, professor, collaborator, etc.). Whether they were a doctor, an ecologist or a sociologist, their socialization story includes, at different times, one or more mentors. This mentor always belongs to the previous generation and to the academic community. The researchers recognize them as significant figures in terms of their appetite for complex problem-solving. This can be seen as an acknowledgment of the debt owed to the previous generation. DAV says about his science thesis:

> To be able to set up an experimental model and really work on alveolar echinococcosis, I went to meet René Houin, who was the professor of parasitology at Créteil. This guy had worked in Madagascar. He was one of the first, I think, to really analyze the landscape from an airplane in order to identify I don't know which parasite, but there were several. He wanted to try to make relationships between the field data and… But above all, he thought, "I need a French research topic now that I'm back in Paris." His wife had a second home in Gray-la-Ville, so he said to himself: "We have an indigenous parasite, *Echinococcus multilocularis*, it won't be far from Gray and there is certainly something to do, to work on the functioning of the cycle." He was a parasitologist, so he was concerned with the functioning of the cycle. And that's where he got in touch with Pierre Delattre – here's where we find a common person [between PG and DAV] – so one thing led to another and I had to meet up one day at Le Souillot, which was their lab farm with Pierre Delattre.

PG defines himself as a naturalist and student since his early years at university. His meeting with P. Delattre, director of research at INRA who supervised his thesis, "bridged the gap" between ecology and medical sciences. This was an opportunity for PG to put into practice the statistical and computer skills he had acquired earlier (naturalist associations, preparation for the CAPES *agrégation* and DEA in ecology at the Sorbonne and at the Muséum national d'histoire naturelle in Paris) in the research team. Those meetings that could be considered opportune are not really. It is necessary that certain people allow, sometimes without their knowledge, the meeting and thus the construction of a network more and more widened. In this respect, DJJ attaches great importance to DAV in his integration into her research project on AE. “In this multidisciplinary history, I consider myself above all an heiress. If it hadn’t been for Dominique Vuitton in the first place, I wouldn’t have joined this team at all, I wouldn’t have submitted the research project submitted later with her and Patrick, etc.” Moreover, DJJ recognizes H. Mendras, the founder of rural sociology, as the one “who practiced multidisciplinarity, without, however, making any great speeches about the relevance of working with other disciplines. The rural studies he initiated in France were already multidisciplinary: ethnology, sociology, economics, human geography”.

The following generations, DB, EA and SC, also mention important figures in their training and emphasize the fact that each of them has had contact with other disciplines. PG is referred to by DB as her mentor in Besançon. EA, likewise, says:

> I observed Patrick a lot when we were in China… I observed in order to reproduce later… I wouldn’t have joined this group if [it hadn’t been a good fit] with Patrick in terms of the atmosphere, but also in terms of the vision of how to work together – even if you have to admit that he has quite a temper. Me, I inherited a situation and, in a way, I’m glad I inherited this network that was set up, because I don’t think I would have had the courage to start this from scratch.

As for SC, he mentions J.-M. Bessette, who was in charge of a master’s degree in Besançon, whose objective was to approach crime through its different dimensions: “We had lessons in sociology and anthropology, but also in law, psychiatry, history, forensic medicine, etc. This multidisciplinary training aspect gave me a taste, an appetite, and a curiosity for other disciplines.” But above all, he sees himself as an heir to DJJ, with whom he shares an interest in the agricultural world, and to C. Guinchard – his thesis supervisor – who taught him “the taste for sideways movement and the importance of going to see elsewhere to better understand his subject”.

C.3. Doing it together

> To cooperate is to manage to produce collective competence. (Alter 2011)

C.3.1. *Practical relationship with science: a problem to be solved*

The researchers who participated in the writing of this book share a desire to anchor their research in reality or, more precisely, to seek to solve a problem. They say it in their own way, but in their discourse, there is always a careful balance between the production of scientific knowledge and practical knowledge. By this we mean the establishment of a certain practical relationship with science: a scientific problem coupled with a social, medical, political or ecological problem, etc., hence the strength of collective action and the necessary presence of partners, such as MSA, the services of the Besançon Regional University Hospital, the veterinary laboratories of the Jura and Doubs departments and, of course, farmers. Indeed, these partnerships are essential and work all the better if their representatives feel that scientific research affects them, which was the case, for example, of J.-J. Laplante, medical doctor and director of health at the Doubs MSA, who, for DJJ, "plays a role of facilitator as well as collaborator". The partnerships in which she was already involved with the MSA brought her closer to DAV and led her to strengthen her interest in the medical problems of agricultural populations. When DAV asked her to work on AE in order to understand the reluctance of rural populations to participate in prevention campaigns, it quickly became for her a problem to be solved.

Different types of knowledge are then successively called upon in response to changing problematic situations, but no one seeks multidisciplinarity. Each expects the other to use his or her disciplinary skills to gain a better understanding of the problem, with the aim of producing practical results. Let us once again consider, as P. Tripier did, that "the researcher's work is not different from that of other professionals" and that "solving a problem is not only a matter of posing it and finding the appropriate inference engine, it is also a matter of being able to act to solve it" (Jacques-Jouvenot 1997).

The complexity of the object in relation to the need to solve a problem leads some researchers to make use of others with skills that are not yet present in the collective (Callon 1986). From this point of view, PG explains that "when you have a problem to solve, it is good to have all the tricks to solve it… And inevitably, you don't have all the tools in hand and it's others who have them; and if you have to go and get this guy who, at a given moment, has a speciality, in such and such a place and [is] useful to the project, I don't mind. I didn't look for multidisciplinarity". As for SC, he explains:

> What led me to multidisciplinarity was above all the need to understand my object. An object that might first appear to be an ecological problem – fish dying in rivers – and that I could not really understand without taking an interest in what other disciplines could say about it or what other scientists were saying about the situation.

> The people I was talking to were using technical vocabulary about the river and I had to have the right translation to be able to understand what they were saying and to interpret it in my discipline.

In the case of medical doctors, the clinical constraint orients their scientific problematic and conditions the call to other disciplines. From this point of view, DAV says:

> As soon as the object of the research, what we are focused on, for me, requires the approach of other professionals or other disciplinary specialties, it has always been a concern because I feel that I cannot think otherwise. So afterwards, I would say that calling on multidisciplinarity, i.e., other disciplines, goes without saying, somewhere. The moment I want to understand the transmission of a disease, if it affects animals, I'm not an animal specialist, so I'll call on someone who *is*.

It is the intersection of hepatological and immunological issues that led her to work on the AE cycle.

Confronted with the cholera epidemics on the African continent and more particularly in the Democratic Republic of the Congo, DB also puts forward the perspective of solving a problem. For him, multidisciplinarity is not an issue:

> From the moment people are faced with cholera, there will always be a search for a solution. For me, multidisciplinarity and all that, that was not my objective. I didn't have any idea of saying, "I'm going to work on multidisciplinarity." I'm here to solve problems. Afterwards, I look for tools to solve problems and I discover that I find myself in what is called multidisciplinarity.

Finally, pluri-, multi-, inter- and trans-disciplinarity are tools at the service of research. This is once again what we understand from EA's account:

> In fact, I'm not looking for interdisciplinarity. It's not that I don't care, but it's not my objective at all. My primary objective is really to enjoy myself, to find pleasure in the discovery of new things, in satisfying a certain curiosity and, as a result, that means answering questions that I am unable to answer.

C.3.2. *Crossing disciplinary and territorial boundaries and accepting their discomfort*

Claiming to be a border-crosser reflects the importance that everyone attaches to transmission. *Mittere*, in Latin, can mean "to let go" and takes on the popular meaning of "to put in a specific place", like *mettre* in French (Jacques-Jouvenot and Vieille-Marchiset 2012). Crossing borders is indeed an action that consists of passing objects and ideas between two people, between two generations, from one hand to another in a given time. This transmission takes place vertically between two generations and/or horizontally between colleagues.

In the context of this research, the frontiers to be crossed are first of all in the discipline: to get out of one's home, out of one's daily workspace and to meet other ways of doing and thinking. The space is getting bigger and researchers will have to find collaborators and open up to other territories if they can. This opening up to the world, made possible by one or more project leaders, leads some researchers who join a working group formed over the years to go elsewhere, sometimes far away. EA says:

> The extreme gradient is discovering other cultures. Because when you go to China, I find it quite exciting, quite fascinating to see that they don't have the same way of working, that they don't have the same way of communicating. And finally, for me, it is the extreme gradient of the discovery of the other. We are not discovering another discipline, but another culture and I find that quite fascinating.

In this case, the wandering into unfamiliar worlds and other researchers is rewarding and enjoyable, but this discovery of others can also be uncomfortable. The "doing together" is not always easy and, as DB says in this regard, "to accept some of the difficulties [of scientific cooperation], there is a discomfort to be accepted."

The sources of this discomfort can be diverse. Some of them are partly related to the differences in research practices between the actors. This reality is evident in the comments of some authors who talk about their collaboration with colleagues in China. In addition to the difficulty of working with researchers from different disciplinary backgrounds, there are cultural difficulties linked to the language barrier that makes scientific dialog complicated. PG reminds us of the importance of "not projecting our organizational fantasies onto China" or, to put it another way, of not being ethnocentric, which would lead us to think: "They should work like that, they don't work like that, so they're idiots, or else I'll teach them to work like I do", so that we can work together.

Continuing PG's comments, EA explains, for example, that the discomfort to be overcome in the context of cooperation with the Chinese is that of the confrontation of different relationships to time:

> I remember a meeting with the management and with X, where he told us: "The deadline for submitting the project is such-and-such a day, so you have to submit the project such-and-such a day." But the Chinese don't give a damn about our deadline. Things are done when they are ready to be done.

On the one hand, there is a "rigid" relationship to time which results from a "kind of injunction to efficiency", which covers both planned preparation and execution in a space-time which is also planned and constrained by personal and family norms, and, on the other hand, a more "flexible" relationship, less constrained and leaving more room for reflection to mature. This is reminiscent of the classic writings of E.T. Hall (1992) on this issue.

Through these accounts of experience, the respondents insist on the need to show humility and adaptability in order to overcome the discomfort of working together with colleagues socialized in other cultures. The same is true for the latter, who accept to work in a hurry, without counting time or energy, thus adapting to our own representations of space and time. Perhaps it is worth remembering that such dispositions are not innate but acquired. They are, in fact, the fruit of a socialization that may have taken place prior to the work situation (PG evokes in this respect the importance of his 2-year national service on the African continent) or carried out "on the job".

This last point, the socialization "on the job", allows us to point out the indispensable work of "translation" (Callon 1986), or even of "diplomacy" (Morizot 2016), which allows people to understand each other and to know each other better because "when we are in China, we work with people who have a different culture and, if we do it badly, we can hurt the feelings of the people in front of us, we can ruin the relations with our interlocutors," says EA. This work and this acculturation often require the intervention of a "mediator", transforming a state of affairs at the same time as the state of the actors (Latour 2006). For EA, this is PG, whose practices she says she has had the opportunity to observe:

> What I did during the first few years was to observe Patrick a lot when we were in meetings where he spoke with Chinese people, whether they were university presidents or heads of natural reserves. Because, in the end, this way of working with another culture can be learned and I think it's an incredible opportunity to be able to play the role of

> observer and eventually reproduce it later on, but it's something I wouldn't have learned in the same way if I had had to create a network from scratch.

In the same vein, PG says, "I can say exactly the same thing about Phil Craig – and DAV, certainly. That is, he brought us into the Chinese network and if he had not figured out the social codes that made it possible to operate in China, we would have failed". This is also the opinion of DAV, who believes that she has "learned a lot" from the man who initiated scientific cooperation with China on the problem of AE.

If a parallel is drawn between working with researchers from other disciplines and working with people from other cultures, it is because, as DAV states, "there is a whole range of behaviors, and this range of behaviors is also found when we work with other disciplines. There's a certain humility that you have to have at the beginning to understand the codes a little bit if you want it to work... It's very complicated."

DAV also states that "you feel much more comfortable in your own family where you know all the codes and that's also why the discipline is a refuge for many and some want to come back because, in the end, they are not very comfortable at the margins". However, this is not the case for everyone. Indeed, AM explains that he has "a propensity to do interdisciplinary research, to like the border areas in [his] own discipline because [he] sometimes finds the air in its centers stifling". The discipline can then appear to be a straitjacket, which SC explains having been confronted with when putting together his application for qualification as a lecturer: "I am going to apply for section 19 [sociology, demography] and section 20 [ethnology, prehistory, biological anthropology]. It is clear that I have to really fit into boxes and that I do not necessarily offer the speciality of Besançon, which is socio-anthropology".

Nevertheless, if the crossing of borders can be uncomfortable, even difficult, it is nonetheless "liberating" for those who dare to do so. By refusing to oppose "people who are looking for air and people who are looking for comfort", AM's account shows that such a move must also be seen in the context of the construction of a professional career, if not of a place within the scientific field (Bourdieu 1976):

> I went to do my training in Greek and not in philosophy. It was my way of escaping the suffocating centrality of philosophy for me. But then, I went to find the Greek professor at the Sorbonne, who may be hyper-focused and suffocating for the people who are in his discipline,

> but who was liberating for me. So, there you have it, playing with borders to build your career and your research objects.

Even more, the possibility of taking a step aside from the dominant practices and discourses of the disciplinary division – which can be considered restrictive and confining – would make it possible to "come back stronger" within one's discipline, as AM underlines:

> For me, the ideal solution would be that... Exactly like what Didier described... We put ourselves in border-crossing slots with others, just to go and look for original questions to be solved and so that we can come back to capitalize on them when we return to our own lane.

In addition to being "useful" for the production of knowledge and the resolution of practical problems, the crossing of borders and cooperation with researchers from other disciplines would produce benefits not only for the researcher himself, but also for science.

C.4. Passing on: giving and taking to give back

C.4.1. *Giving and taking (Alter 2011)*

Working together, cooperating, is not a contract; it is the result of a particular dynamic of the social link. To enter into this logic is also to accept to enter into a logic of giving made of these three obligations characterized by M. Mauss: to give, to receive and to return (Mauss 1950). This makes each of us indebted and/or creditors to one another (the public interviewed, researchers, institutions) and also to the generations that precede us or that are to come. What do the actors tell us about respecting these three obligations?

On the one hand, all of them confirm the freedom of the actors to enter or not into this cycle of giving, into this interaction that is played out in different temporalities. This is what PG underlines:

> The door has always been open to anyone who wanted to put a foot in it. So, there are those who came in and then there are those who didn't come in. Those who came in and continue to be there and have their own projects, it's hard to say it's because of me. Simply, they may have entered because they were ready to enter and function in that mode.

On the other hand, three of them are particularly attentive to a respectful functioning of the equivalence of the gift and counter-gift by the person who joins a work collective. Indeed, anyone who wants to escape this attitude and joins a group with the sole aim of taking, without ever giving or giving back, will never be a good colleague and will not survive in a project like this one. From this point of view, EA "also thinks that there is a history of giving and taking from the group. I think you have to strike a balance between putting your skills at the service of the group and getting personal benefit from it. It can be a springboard to develop your own research themes, or just enjoy working with others". Similarly, PG says that when he was leading a research unit, he kept saying that:

> When someone is in the unit, it's not about what they take into the unit but what they bring to it. So maybe I came across as a bastard, but that's still fundamental. If you see your move as a place where you take, it can't play, you have to bring something to the table and God knows if you'll be able to benefit from the group. God knows if I benefited from all the interactions that took place. You're not totally altruistic but you have to give something first.

If giving back is one of the three Maussian obligations, taking is, on the contrary, forbidden if the exchange produced in the interaction between two actors threatens the equilibrium between giving and giving back.

C.4.2. *Receive or take to return*

Exchanges between disciplines create, above all, social links. The action of taking is not described as an obligation by Mauss, and although it is similar to receiving, it does not include the giver. However, as in the obligation to receive, taking "obliges" the taker to return. Moreover, DJJ expresses his regret at not having been able to give back to the collective before the end of the 2010s:

> For me, the interest in joining a team like that was also to attract students, eventually, to work on a thesis on the relationship with animals, for example, based on the knowledge acquired on a zoonosis such as alveolar echinococcosis, or even more simply on health. In short, we could have imagined other sociological issues based on this research.

Moreover, the two cases that follow seem to us to illustrate this situation well.

While in the story we are telling here she is more of a donor, when she looks back at the beginning of her training, DAV is more of a recipient, the one who is being taught something she is passionate about in a field other than her own: "At the time, we were in science school for six months and we were taught ecology by a guy who was no joke. It was Professor Réal. I was quite fascinated by what he was saying, by the notion of the ecosystem, etc." It can be seen here that this interest in ecology triggered by teaching from another discipline, ecology, and the benefit she derives from it, creates a debt to that discipline. This debt incurred at the beginning of her training will be canceled, largely settled during her career thanks to the door opened to ecologists at a later time. Choosing to attend ecology courses is to accept to enter the cycle of gifting. And from this gift received, she will probably give back to future generations of ecologists – if we are faithful to history! – more than she has received and, by this gesture, puts the ecologists in debt to her.

Accountability returns with AM in a slightly different form. He explained that sometimes multidisciplinary involvement is more of a necessity because "one's own discipline is not comfortable". In this case, one prefers to give to another than to a colleague in one's discipline for fear of losing one's identity. In fact, by taking a step aside from the practices of some of his colleagues, he took advantage of the figure of a world-renowned forerunner in ancient history (P. Lévêque) and adopted a patrimonial standpoint to take up an unoccupied place, to appropriate a heritage that was not *a priori* dedicated to him and to make it bear fruit in the service of his own discipline. This movement on the part of the researcher can be understood in its dimension of giving back. Indeed, it is conducive to the creation of a wider network in the service of the colleagues of the group he joins:

> Five or six years ago, I created the Clisthenes Workshop with colleagues in history and literature, which is a workshop where we based ourselves on the fact that 50 years ago, Pierre Lévêque had fun tackling the great question of the Clisthenian reform that founded ancient democracy by mixing disciplines. So, we said to ourselves: "Well, we're going to take refuge behind this authority to create an international workshop in which people can only come and present a paper if we link two questions, two objects that are usually observed separately by the classics, historians and philosophers".

Being part of local history protects the researcher from the prestige of his ancestors and, at the same time, enables him to strengthen his professional identity. For sometimes, when multidisciplinary involvement meets a need, namely, to get some fresh air, to get out of the discomfort of the discipline, giving to another rather than to a colleague from the same discipline strengthens one's own identity. This does not prevent one from returning to one's discipline, enriched by the collective experience, and from creating one's own network in turn.

C.4.3. *Passing on a practice or knowledge: an art of doing*

Finally, what about the question of the transmission of collective work practices? Some researchers seem to consider each collective work experience as unique. And each one seems to have its own temporality. Several factors therefore come into play in the duration of the actors' participation in a common experience. "When you work in a multidisciplinary system, it seems that once a problem is fairly well resolved, even if imperfectly, life, funding or many other things will mean that a discipline that was very involved at time N will no longer be involved at time N+10 or even N+3," explains DAV. It is easy to understand that one can change objects, collaborators or partners in a researcher's life. So, what about the sustainability of these practices and their transmission?

The actors interviewed find it difficult to settle into the position of a multidisciplinary knowledge holder, let alone a donor. "It would be quite pretentious to say that I taught them to function like that," says PG. Like him, the other researchers find it hard to point to one or more heirs, even if reality contradicts them, since 30 years later, some of them are still around and several generations have taken part in these collective adventures. Is this a denial of transmission or researcher's modesty? The question lies elsewhere. Indeed, collaboration is not considered knowledge, but rather a way of acting together. Perhaps we should introduce this nuance between practices and knowledge and, following the example of F. Waquet, speak instead of an "art of doing" to define this practice (Waquet 2015).

This art of doing is linked to certain qualities of the researcher. These are individual skills that are most often referred to as "personal taste": trust, dialog, the ability to translate, the attention of some to the ways of saying and doing of others. These are the skills that allow one to be part of the cycle of giving within a network, but they seem to escape transmission in part. Moreover, PG believes that, within the laboratory where he was director from 2000 to 2011, no one would have benefited from or inherited the skills linked to one or more multidisciplinary projects. Even more, he and his biologist colleague affiliated to the same research unit now note a decline in the dynamics of cooperation between disciplines:

> The institutional legacy is zero. If you look at it from the point of view of [the research unit], you can see that the quadrennial discussions, for the moment, everyone wants to be in a lane to be identified in their

> discipline or in a discipline close to them – well, in a group of disciplines close to them. So, I think it's very difficult…

On the other hand, he acknowledges that he has heirs without a cost to him. In fact, he believes that these young heirs have not incurred any debt to him. "I'm too lazy for that," he replies to the question of making a successor and about his connection with DB and EA. Through this description of a social relationship that erases the cost for him and cancels the debt for the others, PG rebalances the exchange and claims a kind of equity between the actors:

> I didn't have to ask myself that question. I've never made a choice about anyone, that's not possible. That would mean you could impose your own views. I didn't even do it for my kids. It's very tricky to want to "play Pygmalion" in that way. No, I was more like "this is what I like to do" and then I associated with people who enjoyed doing it too. That's how it seems to me, anyway.

PG thus places the responsibility for the legacy on those who present him as their mentor and who have seized his outstretched hand, he who considers that every researcher is free to enter the cycle of giving or not.

There are sometimes situations that make transmission impossible because of the way the institution works, even if some researchers assume their place as givers. DJJ explains that she regrets the fact that "she did not have students on hand at the right time who were working with *her* on issues that were close enough to transmit, at least within the lab, the contacts that would have allowed her to get closer to the team that had been formed". On the other hand, DB – who thinks of himself as an heir to PG and, more broadly, to the ESS master's degree in Besançon, which he studied – was the linchpin of a multidisciplinary master's degree[2] at the University of Kinshasa, in the course of which more emphasis was placed on the transmission of the mode of production of collective knowledge than on the transmission of knowledge itself. About the master's degree in Kinshasa, he says: "We did not sell multidisciplinarity *per se*. What we sold was the possibility of solving problems that seemed insoluble, and we told them that this training would make it possible to do for other diseases what we had done for cholera. So, from that point on, from the first year, it was easy."

2 Master's degree in ecology of infectious diseases, natural hazards and risk management (ECOM-ALGER), created in 2014 in Kinshasa, whose holders claim that this training is directly derived from the ESS master's degree.

The training of students within the framework of the creation of these two master's (Besançon and Kinshasa) is a way for PG and DB to give back to the next generations the skills acquired in the collective research work. DB's career path and the way in which he transfers the skills he acquired in Besançon to the training of young people in Kinshasa are proof of this. In this way, he pays his debt and completes the cycle of giving by returning to Kinshasa what he received in Besançon: "All this started from what was done in Besançon... Maybe people don't realize it, but in Kinshasa, we are extremely grateful for that. I think that if I hadn't met Patrick, if I hadn't come across Chrono-environnement and the ESS master's degree, there wouldn't have been all this." But what has been done in Besançon? The ESS master's degree, which can of course be considered the collective product of all the disciplines involved in this research for the benefit of future generations of researchers, had as its priority objective

> [to] teach students belonging to a disciplinary field to work with researchers from different specialties, to define scientifically and socially relevant problems, and to participate in the development of an integrated multidisciplinary approach. In other words, it is a question of combining knowledge and know-how (currently fragmented between disciplines) with the actual research needs of a society questioning the environment and its degradation, based on the student's field of study, which is strengthened in the first year of the master's degree.

C.5. Conclusion

The choice we made in this concluding chapter was to focus on the actors of collective action rather than the disciplinary knowledge they convey. For knowledge does not circulate on its own; social relations must be established between researchers for knowledge to be transmitted (Jacques-Jouvenot and Vieille-Marchiset 2012). These interactions engage the actors beyond the transmission of knowledge and beyond time and space, but in a symbolic relationship specific to the logic of the gift, within which the three obligations can be read: to give, to receive and to return. The interactions between the different actors – researcher/researcher, researcher/institution, researcher/other professional – are the spaces for the co-construction of knowledge.

These accounts show that the collective experience does not exist without the willingness of an individual to look elsewhere, beyond his or her own discipline, to fill in the gaps in the discipline and to solve the scientific problem at hand. By calling on colleagues from neighboring disciplines, DAV is looking for disciplinary knowledge that will fit in with the medical knowledge produced and to come. In this way, it opens the cycle of giving and the researchers sought (ecologists, socio-anthropologists, linguists), by responding to this call and accepting this solicitation, agree to enter into a new social bond, to make an alliance. To make an alliance because to refuse the gift would be tantamount to entering into conflict. If the border is understood as a separation that calls for sharing and allows hospitality toward others (Debray 2010), then everything invites us to cross disciplinary boundaries, to receive the call of a colleague to create this community of shared knowledge.

For this, the goodwill of researchers alone is not enough. They must practice a certain scientific indiscipline. This common posture shared by the various authors of this book consists of "extracting oneself from one's disciplinary mainstream, operating in spite of it, in order to cultivate interfaces with the disciplines that are useful for resolving the questions" (see Foreword of this book). We see here the first movement of the researcher who takes the risk of entering the cycle of giving (leaving the comfort of solitary research). It will be followed by a second movement which consists of doing things together. The co-production of common knowledge presupposes a dialog between disciplinary knowledge. Beyond these, researchers must also show curiosity and have the pleasure of scientific conversation, "because we learn better with our ears than with our eyes" (Waquet 2019). This conversation is made possible by the translation efforts (Callon 1986) of both sides, that is by taking into account the language of others and their categories of thought in order to reconstruct islands of partial but shared intelligibility (Passeron 2006). In this way, a "cognitive complicity" (Alter 2011) is constructed that, in a final move, makes it possible to act collectively in the service of the common good.

C.6. References

Alter, N. (2011). *Donner et prendre : la coopération en entreprise*. La Découverte, Paris.

Bourdieu, P. (1976). Le Champ scientifique. *Actes de la recherche en sciences sociales*, 2(2), 88–104.

Callon, M. (1986). La domestication des coquilles Saint-Jacques et des marins pêcheurs dans la baie de Saint-Brieuc. *L'Année sociologique*, 1(36), 169–208.

Darré, J.-P. (1999). *La production de connaissance pour l'action : arguments contre le racisme de l'intelligence*. Éditions de la MSH/INRA, Paris.

Debray, R. (2010). *Éloge des frontières*. Gallimard, Paris.

Durkheim, É. (1922). *Éducation et sociologie*. Alcan, Paris.

Fournier, M. (1994). *Marcel Mauss*. Fayard, Paris.

Frodeman, R. (2019). *Pour un savoir soutenable : une théorie de l'interdisciplinarité.* QUAE, Versailles.

Gary, R. (1982). *Pseudo.* Mercure de France, Paris.

Hall, E.T. (1992). *La Danse de la vie : temps culturel, temps vécu.* Le Seuil, Paris.

Jacques-Jouvenot, D. (1997). *Choix du successeur et transmission patrimoniale.* L'Harmattan, Paris.

Jacques-Jouvenot, D. and Vieille-Marchiset, G. (2012). *Socio-anthropologie de la transmission.* L'Harmattan, Paris.

Jollivet, M. (1992). *Sciences de la nature, sciences de la société.* CNRS, Paris.

Latour, B. (2006). *Nous n'avons jamais été modernes.* La Découverte, Paris.

Mauss, M. (1931). Remarques à la suite de l'article de Paul Descamps : Ethnographie et ethnologie. *Revue de synthèse historique*, April–June.

Mauss, M. (1950). *Sociologie et anthropologie.* Presses universitaires de France, Paris.

Morizot, B. (2016). *Les diplomates : cohabiter avec les loups sur une autre carte du vivant.* Wildproject, Marseille.

Passeron, J.-C. (2006). *Le Raisonnement sociologique : un espace non poppérien de l'argumentation.* Albin Michel, Paris.

Petit, S., Fleury, P., Michel, V., Mougenot, C. (2008). Raconter la recherche-intervention. Retour sur trois opérations de gestion de la biodiversité. *Natures sciences sociétés*, 16(4), 326–336.

Ricoeur, P. (1985). *Temps et récit. Tome 3 : Le temps raconté.* Le Seuil, Paris.

Turner, S. (2000). What are disciplines? And how is interdisciplinarity different? In *Practising Interdisciplinarity*, Weingart, P. and Stehr, N. (eds). University of Toronto Press, Toronto.

Waquet, F. (2015). *L'Ordre matériel du savoir : comment les savants travaillent, XVIe-XXIe siècles.* CNRS, Paris.

Waquet, F. (2019). *Une histoire émotionnelle du savoir : XVIIe-XXIe siècle.* CNRS, Paris.

List of Authors

Eve AFONSO
Chrono-environnement UFC-CNRS
Université Bourgogne Franche-Comté
Besançon
France

Didier BOMPANGUE
Service d'Écologie et de Contrôle des Maladies Infectieuses
Université de Kinshasa
Democratic Republic of the Congo
and
Chrono-environnement UFC-CNRS
Université Bourgogne Franche-Comté
Besançon
France

Simon CALLA
Agroecology
Université Bourgogne Franche-Comté
Besançon/Dijon
France

Cécile CALLOU
MNHN
Paris
France

Céline CLAUZEL
LADYSS UPD-CNRS
Université Paris Diderot
France

Philip Simon CRAIG
School of Environment and Life Sciences
University of Salford
Greater Manchester
UK

Amandine DIVARET-CHAUVEAU
Chrono-environnement UFC-CNRS
Université Bourgogne Franche-Comté
Besançon
France

Patrick GIRAUDOUX
Chrono-environnement UFC-CNRS
Université Bourgogne Franche-Comté
Besançon
France
and
Yunnan University of Finance and Economics
Kunming
China

Dominique JACQUES-JOUVENOT
LASA
Université Bourgogne Franche-Comté
Besançon
France

Jean-Jacques LAPLANTE
ASEPT Franche-Comté/Bourgogne
Besançon
France

Li LI
Yunnan University of Finance
and Economics
Kunming
China

Arnaud MACÉ
Logiques de l'agir
Université Bourgogne Franche-Comté
Besançon
France

Dominique Angèle VUITTON
Université Bourgogne Franche-Comté
Besançon
France

Index

F, G

H

I, K

L

M

N, P

R

S, T

V, Y, Z

Printed and bound by CPI Group (UK) Ltd, Croydon, CR0 4YY

13/09/2026

14960994-0005